Analytical Techniques
in Materials Conservation

T0235057

Analytical Techniques in Materials Conservation

Barbara H. Stuart

*Department of Chemistry,
Materials and Forensic Sciences
University of Technology, Sydney,
Australia*

John Wiley & Sons, Ltd

Contents

Abbreviations

AAS	atomic absorption spectroscopy
ABS	acrylonitrile-butadiene-styrene
AES	atomic emission spectroscopy/Auger electron spectroscopy
AFM	atomic force microscopy
AMS	accelerator mass spectrometer
ANN	artificial neural network
ATR	attenuated total reflectance
CAT	computerised axial tomography
CE	capillary electrophoresis
CEMS	conversion electron Mössbauer spectroscopy
CGE	capillary gel electrophoresis
CI	chemical ionisation
CP	cross polarisation
CZE	capillary zone electrophoresis
DAC	diamond anvil cell
DAD	diode array detection
DETA	dielectric thermal analysis
DMA	dynamic mechanical analysis
DMTA	dynamic mechanical thermal analysis
DSC	differential scanning calorimetry
DTA	differential thermal analysis
DTG	derivative thermogram
DTGS	deuterium triglycine sulfate
DTMS	direct temperature mass spectrometry
EDS	energy dispersive spectroscopy
EDTA	ethylenediaminetetraacetic acid
EDXRF	energy dispersive X-ray fluorescence
EELS	electron energy loss spectroscopy
EGA	evolved gas analysis
EI	electron impact
EMPA	electron microprobe analysis
ESEM	environmental scanning electron microscopy
ESI	electrospray ionisation

ESR	electron spin resonance
FAB	fast atom bombardment
FD	field desorption
FI	field ionisation
FID	free induction decay
FORS	fibre optics reflectance spectroscopy
FPA	focal plane array
FTIR	Fourier transform infrared
GC	gas chromatography
GC-IR	gas chromatography – infrared
GC-MS	gas chromatography – mass spectrometry
GFAAS	graphite furnace atomic absorption spectroscopy
GPC	gel permeation chromatography
HDPE	high density polyethylene
HMDS	hexamethyldisilazane
HPLC	high performance liquid chromatography
IBA	ion beam analysis
IC	ion chromatography
ICP	inductively coupled plasma
IRMS	isotope ratio mass spectrometry
LA	laser ablation
LD	laser desorption
LDA	linear discriminant analysis
LDPE	low density polyethylene
LIBS	laser induced breakdown spectroscopy
MALDI	matrix assisted laser desorption/ionisation
MAS	magic angle spinning
MCT	mercury cadmium telluride
MRI	magnetic resonance imaging
MS	mass spectrometry
MTBS	N-methyl-N-tert-butyldimethylsilyl
NAA	neutron activation analysis
NMR	nuclear magnetic resonance
NRA	nuclear reaction analysis
OES	optical emission spectroscopy
OM	optical microscopy
OSL	optically stimulated luminescence
PAN	polyacrylonitrile
PAS	photoacoustic spectroscopy
PC	polycarbonate
PCA	principal component analysis

PDMS	poly(dimethyl siloxane)
PE	polyethylene
PEO	poly(ethylene oxide)
PET	poly(ethylene terephthalate)
PIGE	proton induced gamma emission
PIXE	proton induced X-ray emission
PL	photoluminescence
PLM	polarised light microscopy
PMMA	poly(methyl methacrylate)
PP	polypropylene
PS	polystyrene
PSL	photostimulated luminescence
PTFE	polytetrafluoroethylene
PU	polyurethane
PVA	poly(vinyl acetate)
PVAl	poly(vinyl alcohol)
PVC	poly(vinyl chloride)
Py-MS	pyrolysis mass spectrometry
RBS	Rutherford backscattering spectrometry
RI	refractive index
RP	reversed phase
RRS	resonance Raman spectroscopy
SAXS	small angle X-ray scattering
SEC	size exclusion chromatography
SERS	surface enhanced Raman spectroscopy
SFC	supercritical fluid chromatography
SG	specific gravity
SIMS	secondary ion mass spectrometry
SPME	solid phase microextraction
SSIMS	static secondary ion mass spectrometry
STEM	scanning transmission electron microscopy
STM	scanning tunneling microscopy
TEM	transmission electron microscopy
TGA	thermogravimetric analysis
TMAH	tetramethylammonium hydroxide
THM	thermally assisted hydrolysis methylation
TIC	total ion current
TLC	thin layer chromatography
TMA	thermal mechanical analysis
TMS	tetramethylsilane
TOF	time-of-flight

UV	ultraviolet
WAXS	wide angle X-ray scattering
WDXRF	wavelength dispersive X-ray fluorescence
XPS	X-ray photoelectron spectroscopy
XRD	X-ray diffraction
XRF	X-ray fluorescence

Preface

The use of scientific techniques in materials conservation has notably expanded in recent decades. There is much interest in identifying the materials used in culturally important objects. Of great importance is a clear understanding of the state and mechanisms of degradation of objects susceptible to deterioration with time and exposure to environmental factors. An understanding of the state of materials at a molecular level can provide valuable information for conservators, enabling them to decide on a conservation procedure.

There is an extensive range of analytical techniques used by scientists that may be applied to heritage materials. The application of modern analytical techniques to conservation issues continues to expand with the ability to analyse smaller and smaller quantities of sample. The development of non-destructive remote sampling methods and portable instruments will further allow for precious objects to be examined safely.

The aim of this book is to provide those with an interest in material conservation with an overview of the analytical techniques that are potentially available to them. When making the initial decision to investigate an object, it can be difficult to decide on the best analytical approach. Chapters 2 to 10 provide a straightforward background to each of the common analytical techniques. An explanation of how an instrument actually works is provided without excessive technical detail that can be overwhelming for the first-time user. The nature and size of a sample required for analysis is also an important consideration in conservation and this information is provided for each technique. Additionally, for each technique, examples of the application of the method to specific types of heritage materials are provided, with the relevant literature references provided. Clearly, a review of all the studies that have been carried out using each technique is not feasible given the extensive use of particular techniques, but suitable examples are provided in each case. An introductory chapter also provides an overview of the different types of materials that may be encountered in conservation work. An understanding of the structures of such materials aids in the interpretation of the data obtained from analytical techniques.

Naturally, conservators working in a museum environment will not necessarily have access to a laboratory equipped with all the techniques described! However, many analytical techniques are becoming more compact and less expensive to purchase with time, making it possible that a museum laboratory can have access to a suitable instrument for their interests. Additionally, for more specialised techniques involving more substantial and expensive equipment, collaboration with university and research organisations may be required. However, as will be gathered from a study of the literature provided in this book, there has been a great history of collaboration between such institutions.

I hope that this book provides a useful source of information for those with an interest in materials conservation. While it should prove an aid to those already working in this field, the book is also aimed at those entering the field who would like to know more about the analysis of heritage materials. I have not assumed an in-depth knowledge of chemistry, materials science or analytical chemistry in preparing this book, so those without a strong background in these fields will be able to cope with the scientific concepts.

I would like to thank all the conservators, curators and researchers with whom I have communicated during the preparation of this book. There is much fascinating conservation work being carried out around the world. I hope this book will help make a contribution.

Barbara H. Stuart
University of Technology, Sydney,
Australia

Trade Mark Acknowledgements

Bakelite – Bakelite AG; Dacron – DuPont; Kevlar – DuPont; Lycra – DuPont; Nomex – DuPont; Perspex – Lucite International; Parylene – Union Carbide; Teflon – DuPont; Terylene – ICI.

1

Conservation Materials

1.1 INTRODUCTION

There is no doubting the cultural importance of museum artefacts. A
vast array of objects that reflect heritage are collected, displayed and
stored by museums. Objects including paintings, ceramics, textiles, statues, glass, furniture, books, plastics and metals artefacts all find a home
in museums. Many types of materials, organic and inorganic, natural
and synthetic, have been used to make such artefacts. As many materials are susceptible to deterioration, an important issue for museum
conservators and curators is to identify and understand the degradation
processes occurring on objects. Such information can be vital in preventing further deterioration and conservation processes, such as cleaning or
consolidation, can be applied. Also of interest in museums is the origin
of an object and identifying the chemical properties of a material can
provide vital information regarding provenance.

There are many analytical techniques used in laboratories that provide
information about materials. Information regarding the elemental composition, molecular structure and physical properties can be obtained
and used to characterise a material. There are special requirements
to be considered when choosing an analytical technique suitable for
studying a museum artefact. The goal is to minimise damage to the
object of interest, so non-destructive techniques are very much favoured.
Where only a destructive technique is suitable to obtain the required
information, it should require only a very small quantity of material.
Analytical techniques for studying culturally important objects have
been developed using a wide range of established experimental methods

Analytical Techniques in Materials Conservation Barbara H. Stuart
© 2007 John Wiley & Sons, Ltd

and each of these are described in this book. Each chapter provides background on how each technique works, the sampling requirements and what information is provided by the experiment. Examples of the application of each technique to the different heritage materials encountered are provided to illustrate the information obtained by each technique.

To determine how to protect the items found in museums, an understanding of the chemical and physical properties of the material from which the item has been produced is vital. The origin of an object can be determined through an examination of its chemical structure. This enables the material to be dated and can aid in identifying forgeries. The properties of a material are also crucial to the task of detecting and preventing deterioration of precious objects. Thus, in the first instance, the material that is being dealt with must be characterised before embarking on any treatment. This chapter provides an overview of the materials that are encountered in museum materials.

Humans have always made use of materials naturally occurring around them to produce items. Naturally occurring materials may contain proteins, lipids, carbohydrates, resins, colourants and/or minerals and are all found in heritage materials. In more recent times, with the development of synthetic chemistry, man-made materials, such as polymers and colourants, have been used to produce objects. The component materials of paintings, books and manuscripts, textiles, glass, ceramic, stone and metal objects are also described.

1.2 PROTEINS

Proteins appear in many natural materials of animal origin in the museum environment, such as paintings, manuscripts, textiles, adhesives and many artefacts. Proteins are large molecules with a characteristic amide group (–NH–CO–) and consist of varying sequences of amino acids [Creighton, 1993]. The structures of amino acids are shown in Figure 1.1 and these acids can be isolated by hydrolysis of the protein. There are many proteins in existence, but only those that may be encountered in works of cultural heritage are described here.

Collagen is a fibrous protein obtained from the connective tissues, such as skin, bone, hide or muscle, of animals [Creighton, 1993; Mills and White, 1994]. The amino acid composition of collagen is listed in Table 1.1 [Bowes *et al.*, 1955; Mills and White, 1994]. Collagen contains repeating sequences of glycine–proline–hydroxylproline and

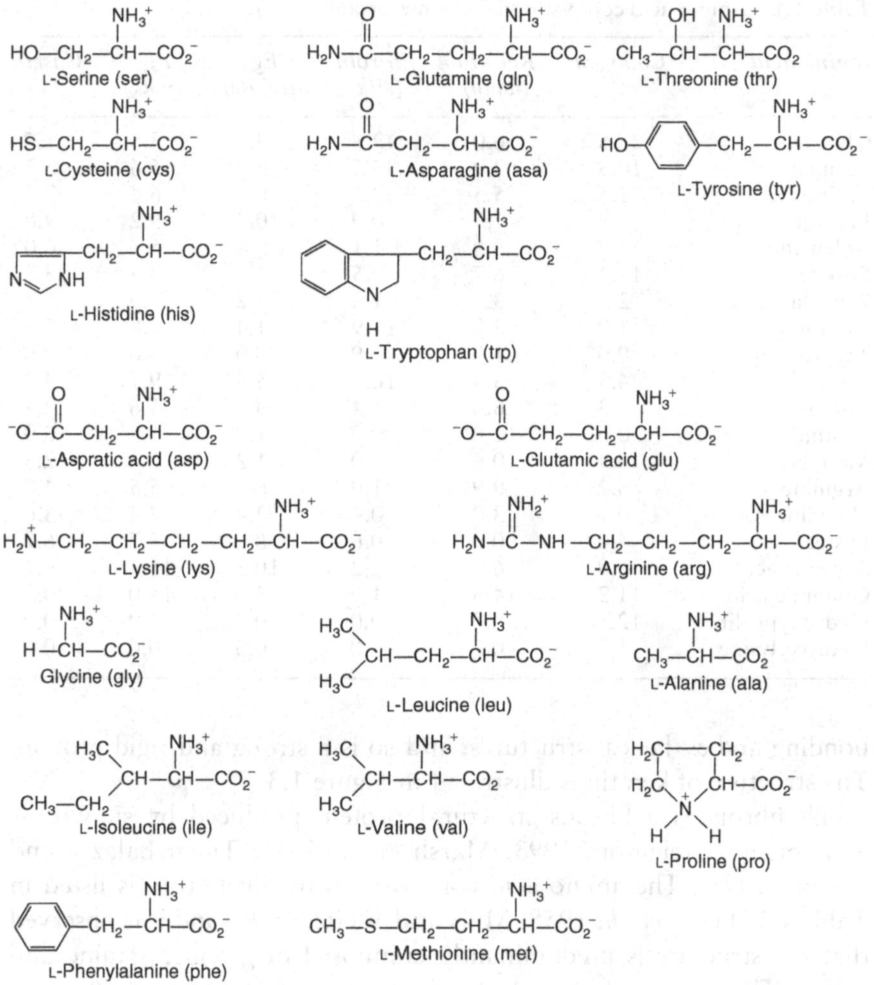

$$\underset{\text{L-Serine (ser)}}{HO-CH_2-\overset{\overset{\displaystyle NH_3^+}{|}}{CH}-CO_2^-}$$

$$\underset{\text{L-Glutamine (gln)}}{H_2N-\overset{\overset{\displaystyle O}{\|}}{C}-CH_2-CH_2-\overset{\overset{\displaystyle NH_3^+}{|}}{CH}-CO_2^-}$$

$$\underset{\text{L-Threonine (thr)}}{CH_3-\overset{\overset{\displaystyle OH}{|}}{CH}-\overset{\overset{\displaystyle NH_3^+}{|}}{CH}-CO_2^-}$$

$$\underset{\text{L-Cysteine (cys)}}{HS-CH_2-\overset{\overset{\displaystyle NH_3^+}{|}}{CH}-CO_2^-}$$

$$\underset{\text{L-Asparagine (asa)}}{H_2N-\overset{\overset{\displaystyle O}{\|}}{C}-CH_2-\overset{\overset{\displaystyle NH_3^+}{|}}{CH}-CO_2^-}$$

$$\underset{\text{L-Tyrosine (tyr)}}{HO-\text{〈benzene〉}-CH_2-\overset{\overset{\displaystyle NH_3^+}{|}}{CH}-CO_2^-}$$

L-Histidine (his)

L-Tryptophan (trp)

$$\underset{\text{L-Aspratic acid (asp)}}{^-O-\overset{\overset{\displaystyle O}{\|}}{C}-CH_2-\overset{\overset{\displaystyle NH_3^+}{|}}{CH}-CO_2^-}$$

$$\underset{\text{L-Glutamic acid (glu)}}{^-O-\overset{\overset{\displaystyle O}{\|}}{C}-CH_2-CH_2-\overset{\overset{\displaystyle NH_3^+}{|}}{CH}-CO_2^-}$$

$$\underset{\text{L-Lysine (lys)}}{\overset{+}{H_2}N-CH_2-CH_2-CH_2-CH_2-\overset{\overset{\displaystyle NH_3^+}{|}}{CH}-CO_2^-}$$

$$\underset{\text{L-Arginine (arg)}}{H_2N-\overset{\overset{\displaystyle NH_2^+}{\|}}{C}-NH-CH_2-CH_2-CH_2-\overset{\overset{\displaystyle NH_3^+}{|}}{CH}-CO_2^-}$$

$$\underset{\text{Glycine (gly)}}{H-\overset{\overset{\displaystyle NH_3^+}{|}}{CH}-CO_2^-}$$

$$\underset{\text{L-Leucine (leu)}}{\overset{\displaystyle H_3C}{\underset{\displaystyle H_3C}{>}}CH-CH_2-\overset{\overset{\displaystyle NH_3^+}{|}}{CH}-CO_2^-}$$

$$\underset{\text{L-Alanine (ala)}}{CH_3-\overset{\overset{\displaystyle NH_3^+}{|}}{CH}-CO_2^-}$$

$$\underset{\text{L-Isoleucine (ile)}}{\overset{\displaystyle H_3C}{\underset{\displaystyle CH_3-CH_2}{>}}CH-\overset{\overset{\displaystyle NH_3^+}{|}}{CH}-CO_2^-}$$

$$\underset{\text{L-Valine (val)}}{\overset{\displaystyle H_3C}{\underset{\displaystyle H_3C}{>}}CH-\overset{\overset{\displaystyle NH_3^+}{|}}{CH}-CO_2^-}$$

L-Proline (pro)

$$\underset{\text{L-Phenylalanine (phe)}}{\text{〈benzene〉}-CH_2-\overset{\overset{\displaystyle NH_3^+}{|}}{CH}-CO_2^-}$$

$$\underset{\text{L-Methionine (met)}}{CH_3-S-CH_2-CH_2-\overset{\overset{\displaystyle NH_3^+}{|}}{CH}-CO_2^-}$$

Figure 1.1   Structures of amino acids

forms a structure of three separate molecules hydrogen-bonded and coiled in a α–helical conformation (Figure 1.2) [Creighton, 1993; Fraser *et al.*, 1979]. The protein gelatin is formed from the breakdown of the intermolecular bonding of collagen.

Keratin is also a fibrous protein and is contained in horn, hoof, nail, feathers, hair and wool [Mills and White, 1994; Timar-Balazsy and Eastop, 1998; Ward and Lundgren, 1954]. The amino acid composition of keratin from wool is listed in Table 1.1, but the composition varies somewhat depending on the source. Keratin has a high degree of S–S cross-linking between the cysteine residues and extensive hydrogen

Table 1.1 Amino acid compositions of some proteins (%)

Amino acid	Collagen	Keratin (wool)	Fibroin (silk)	Egg (white)	Egg (yolk)	Casein
Glycine	26.6	6.0	42.8	3.6	3.5	1.7
Alanine	10.3	3.9	33.5	6.3	5.6	2.7
Valine	2.5	5.5	3.3	8.3	6.4	7.2
Leucine	3.7	7.9	0.9	10.3	9.2	9.0
Isoleucine	1.9	3.8	1.1	6.2	5.1	6.0
Proline	14.4	6.7	0.5	4.5	4.5	13.2
Phenylalanine	2.3	3.7	1.3	5.2	3.9	5.1
Tyrosine	1.0	5.2	11.9	1.4	2.8	5.5
Tryptophan	0.0	1.9	0.9	0.0	0.0	0.0
Serine	4.3	8.4	16.3	5.8	9.1	4.0
Threonine	2.3	6.6	1.4	3.7	5.6	2.7
Cystine	0.0	12.8	0.0	1.9	1.9	0.0
Methionine	0.9	0.6	0.0	1.2	2.3	2.3
Arginine	8.2	9.9	1.0	6.8	5.5	4.0
Histidine	0.7	3.0	0.4	2.4	2.4	3.6
Lysine	4.0	0.9	0.6	8.0	5.7	6.7
Aspartic acid	6.9	6.9	2.2	10.5	11.5	6.1
Glutamic acid	11.2	14.5	1.9	13.9	15.0	20.2
Hydroxyproline	12.8	0.0	0.0	0.0	0.0	0.0
Hydroxylysine	1.2	0.2	0.0	0.0	0.0	0.0

bonding and α–helical structures, and so is a strong and rigid protein. The structure of keratin is illustrated in Figure 1.3.

Silk fibroin is a fibrous structural protein produced by silkworms and spiders [Creighton, 1993; Marsh *et al.*, 1955; Timar-Balazsy and Eastop, 1998]. The amino acid composition of silk fibroin is listed in Table 1.1 [Lucas *et al.*, 1958; Mills and White, 1994] and it is observed that the structure is predominantly composed of glycine, alanine and serine. The nature of the amino acid sequence results in a β–sheet pleated structure for fibroin as illustrated by Figure 1.4. This type of secondary structure results in a strong silk fibre.

Albumin refers to the proteins contained in eggs [Creighton, 1993; Mills and White, 1994]. Ovalbumin is the main protein of egg white, making up 50 % of the protein content, and conalbumin and lysozyme constitute 15 % and 3 % of the proteins in egg white, respectively. These proteins form globular conformations via intramolecular hydrogen bonding. The albumins are easily denatured using heat or certain chemicals. There are small quantitative differences in the amino acid composition of the proteins in egg white and egg yolk (Table 1.1) [Keck and Peters, 1969; Mills and White, 1994] and these may be used to distinguish the two components.

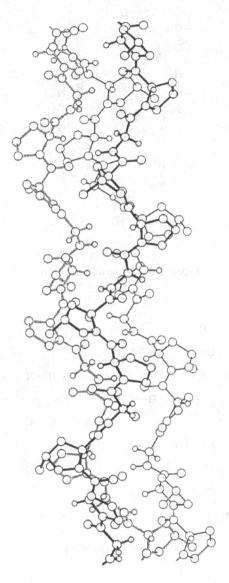

Figure 1.2 Structure of collagen. Reprinted from Journal of Molecular Biology, Vol 129, Fraser *et al.*, 463–481, 1979 with permission from Elsevier

Casein is a major protein in milk [Mills and White, 1994; Newman, 1998]. Casein contains about 1 % phosphorus, mainly with phosphoric acid esterifying the hydroxyl groups of serine and the amino acid composition of casein is also listed in Table 1.1 [Keck and Peters, 1969; Mills and White, 1994].

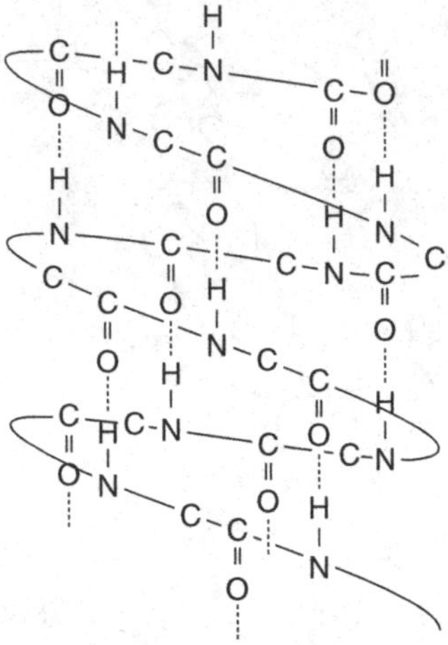

Figure 1.3   Structure of keratin

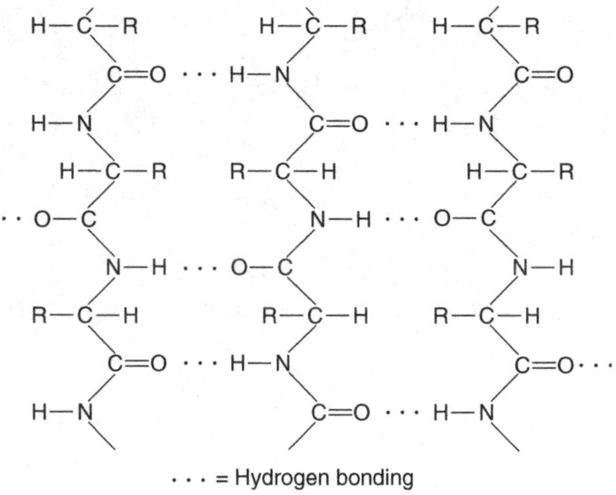

··· = Hydrogen bonding

Figure 1.4   The β–sheet structure of fibroin

1.3   LIPIDS

Lipids are a group of natural organic compounds which possess structures that make them insoluble in water but very soluble in organic solvents. The types of lipids that may be encountered in materials conservation are oils, fats and waxes [Gunstone, 2004; Mills and White, 1994]. Fats and oils are both composed of triglycerides, but are classified based on their physical state at normal temperatures. Oils are usually liquids, while fats are solid at normal temperatures. The general structure of a triglyceride is shown in Figure 1.5. Hydrolysis of a triglyceride produces glycerol and three molecules of fatty acid. Usually triglycerides have a structure that results in different fatty acids; the structures of the major fatty acids found in fats and oils are shown in Table 1.2.

Fatty acids can contain double bonds in their hydrocarbon chain and are referred to as unsaturated fatty acids. The presence of two or more double bonds makes the molecule susceptible to oxidation, but this property can be exploited when an oil is being used as a

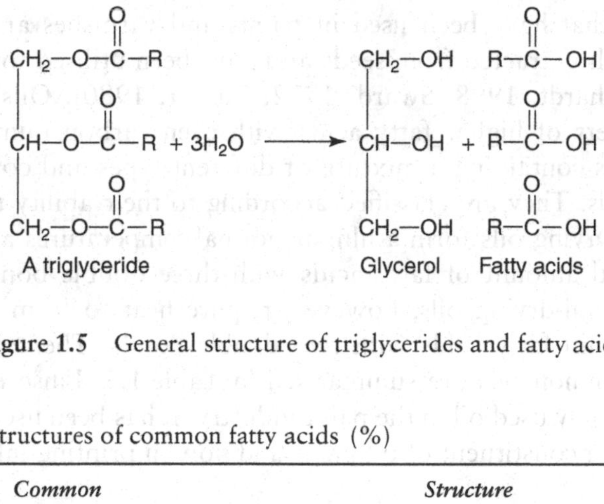

$$
\begin{array}{ccc}
\text{CH}_2\text{—O—C—R} & & \text{CH}_2\text{—OH} \quad \text{R—C—OH} \\
| & & | \\
\text{CH—O—C—R} + 3\text{H}_2\text{O} \longrightarrow & & \text{CH—OH} + \text{R—C—OH} \\
| & & | \\
\text{CH}_2\text{—O—C—R} & & \text{CH}_2\text{—OH} \quad \text{R—C—OH} \\
\text{A triglyceride} & & \text{Glycerol} \quad \text{Fatty acids}
\end{array}
$$

Figure 1.5   General structure of triglycerides and fatty acids

Table 1.2   Structures of common fatty acids (%)

Carbon atoms	Common name	Structure
Saturated		
12	Lauric	$CH_3(CH_2)_{10}COOH$
14	Myristic	$CH_3(CH_2)_{12}COOH$
16	Palmitic	$CH_3(CH_2)_{14}COOH$
18	Stearic	$CH_3(CH_2)_{16}COOH$
Unsaturated		
16	Palmitoleic	$CH_3(CH_2)_5CH=CH(CH_2)_7COOH$
18	Oleic	$CH_3(CH_2)_7CH=CH(CH_2)_7COOH$
18	Linoleic	$CH_3(CH_2)_4CH=CHCH_2CH=CH(CH_2)_7COOH$
18	Linolenic	$CH_3CH_2CH=CHCH_2CH=CHCH_2CH=CH(CH_2)_7COOH$
18	Eleostearic	$CH_3(CH_2)_3CH=CHCH=CHCH=CH(CH_2)_7COOH$
18	Licanic	$CH_3(CH_2)_3CH=CHCH=CHCH=CH(CH_2)_4CO(CH_2)_2COOH$

drying oil [Erhardt, 1998; Mills and White, 1994; Sward, 1972]. The oil transforms from a liquid to a solid as a result of free radical chain reactions. The triglyceride molecules cross-link due to oxidative and thermal polymerisation reactions. The drying process is affected by the presence of other substances such as pigments and chemical driers, including metallic salts. The oxidative process is usually modified in such cases [Erhardt, 1998].

Oil films can yellow with age and oils with a high linolenic acid content tend to be susceptible to this effect. The yellowing effect is due to the formation of oxidation products, the nature of which are complex depending on other components present in an application such as a painting [Mills and White, 1994]. The ester and the double bonds in the dried oil remain reactive in a dried oil film and so are susceptible to further oxidation. Continued oxidation may also lead to a weathering of the oil film.

The oils that have been used in paints and varnishes are generally vegetable oils extracted from seeds and have been utilised for hundreds of years [Erhardt, 1998; Sward, 1972; Turner, 1980]. Oils consist of glycerol esters of higher fatty acids with even carbon numbers, with different oils containing a mixture of different types and compositions of fatty acids. They are classified according to their ability to dry to a solid film. Drying oils form a film at normal temperatures and contain a substantial amount of fatty acids with three double bonds in their structure. Semi-drying oils, however, require heat to form a film and mainly contain fatty acids with two double bonds. The compositions of some common oils are summarised in Table 1.3. Linseed oil is the most commonly used oil in the paint industry. It has been used for many centuries as a constituent of oil paint and now in printing inks. Linseed

Table 1.3 Composition of some oils

Oil	Fatty acids, (%)				
	Saturated acids	Oleic	Linoleic	Linolenic	Eleostearic
Linseed	10	22	17	51	–
Tung	5	9	–	15	71
Poppyseed	10	15	73	2	–
Safflower	10	14	76	–	–
Soya	13	28	54	5	–
Walnut	9	17	61	13	–
Perilla	8	14	14	64	–
Oiticica	12	7	–	–	–

contains large percentages of linolenic and linoleic triglycerides and the high degree of unsaturation results in a relatively short drying time. In the past, the oil was obtained from seed by pressing, but now solvent extraction is used. As well as being used as a drying oil, linseed oil is also used as an ingredient in resins. Tung oil is also known as wood, Chinese wood or mu oil and it has a relatively high viscosity and dries quickly. It consists mainly of eleostearic acid and is used as a varnish and in alkyd paints. Walnut oil is one of the earliest oils used in painting and was a common medium in the early days of oil painting, although it is little used today. It dries more slowly than linseed oil and has a tendency to turn yellow. Poppyseed oil is used as a medium for artists' colours and it is semi-drying and resistant to yellowing. Poppyseed oil is less viscous than linseed and walnut oils and does not easily turn rancid. Safflower oil has a very low linolenic acid content, which means that it has the advantageous property of low yellowing. It is a semi-drying oil used for paint and varnishes. Soybean oil is a semi-drying oil that is used to prepare alkyd paints.

Waxes are naturally occurring esters of long chain carboxylic acids with long chain alcohols [Mills and White, 1994; Newman, 1998]. They have a characteristic smooth 'waxy' feel. Wax is used as an ingredient of polishes for paintings and wooden surfaces and in coatings for paper. Natural waxes of animal and vegetable origin such as beeswax, paraffin, carnauba and candelilla have all been used. Paraffin is obtained from crude petroleum and consists of long chain saturated hydrocarbons. Beeswax is a complex mixture of hydrocarbons, esters and free fatty acids that is obtained from the hives of honey bees [Horie, 1987; Mills and White, 1994]. It has been historically used in paintings: beeswax was used by the Egyptians to protect the surface of paintings in tombs. Beeswax is usually purified by melting and filtering and consists mainly of myricyl palmitate ester ($C_{15}H_{31}CO_2H$). Beeswax has been widely used for relining paintings, usually mixed with natural resins and fillers. It has also been used as a consolidant and in polish formulations. Beeswax has the advantage that it does not significantly deteriorate with time, but it can yellow due to oxidation.

1.4 CARBOHYDRATES

A range of heritage objects are made up of carbohydrates: paintings, adhesives, paper and wooden objects. The carbohydrates of interest are monosaccharides (simple sugars) and polysaccharides [Davis and Fairbanks, 2002; Mills and White, 1994; Newman, 1998; Timar-Balazsy

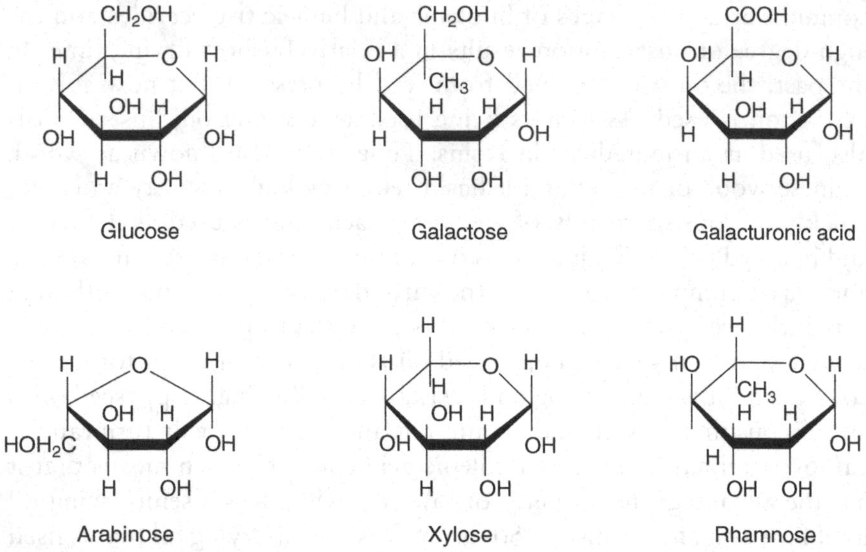

Glucose

Galactose

Galacturonic acid

Arabinose

Xylose

Rhamnose

Figure 1.6 Structures of some monosaccharides

and Eastop, 1998]. There is a large number of sugars in existence and the structures of some natural monosaccharides are illustrated in Figure 1.6. The D- and L- nomenclature is used to represent the different enantiomers and the + and − in the name of an enantiomer indicates a positive or negative sign of rotation. Monosaccharides can form bonds with one another to produce larger molecules known as oligosaccharides. Such structures can vary from two to about 10 sugar units. For example, a common sugar sucrose is made up of two hexose units.

Large molecular weight sugars are known as polysaccharides and some important polysaccharides are cellulose, starch and gums. Cellulose is a very common polysaccharide found in plant fibres [Mills and White, 1994]. It is a high molecular weight polymer of D-glucose and the structure is illustrated in Figure 1.7. The cellulose molecules form hydrogen bonds between the hydroxyl groups of the chains, allowing fibrils to be formed. Cellulose fibres are chemically stable and are

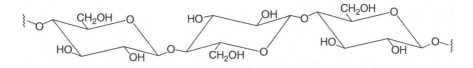

Figure 1.7 Structure of cellulose

strong, but can deteriorate due to oxidative or hydrolytic reactions. The hydroxyl groups can be oxidised to carboxylic acid groups, especially when exposed to light. Cellulose can also be hydrolysed with acids at the links to break the structure down into smaller units.

Another commonly encountered polysaccharide is starch, which is found in vegetable matter [Mills and White, 1994]. It contains glucose, as well as amylose and amylopectin units. The properties of starch vary depending upon the source due to the varying proportions of amylose and amylopectin. Starch swells on exposure to water and if heated, a thick viscous paste is formed.

Plant gums are exuded by certain plants when their barks are broken. Gums consist of complex branched polysaccharides [Mills and White, 1994; Newman, 1998]. Despite their complex structures, it is possible to identify the sugars of gums by determining the monosaccharide composition after cleavage of the glycosidic bonds. The possible sugars in plant gums are: L-arabinose, D-galactose, D-mannose, L-rhamnose, D-xylose, L-fucose, D-glucose, D-galacturonic acid and D-glucuronic acid. Gum arabic (also known as gum acacia) is the most commonly produced and is a high molecular weight polysaccharide extracted from the Acacia species [Horie, 1987]. The gum is a complex mixture of arabinogalactan oligosaccharides, polysaccharides and glycoproteins. It is transparent, brittle, soluble in water at room temperature and can form viscous solutions. Gum arabic may be cross-linked and precipitated by metal ions such as aluminium, iron, lead, mercury salts and gelatin. It is susceptible to biodeterioration. Gum arabic has long been used in inks and water based paints, as well as an adhesive for paper and textiles. Some other gums that can be encountered in objects of cultural significance are gum tragacanth, gum karaya and cherry gum.

1.5   NATURAL RESINS

Many natural resins are extracted from trees and plants and are based on terpenoid structures [Horie, 1987; Mills and White, 1994; Mills and White, 1977; Newman, 1998]. Such resins have been used extensively because of their attractive properties including adhesion, water insolubility and glassiness. Although resins are complex mixtures of terpenoid components, classification is aided by the fact that diterpenoids and triterpenoids are not found together in the same resin. Oil of turpentine is a common film forming resin derived from pine trees. The resin is composed of monoterpenoids including pinene, but the exact composition depends on the source.

Some common triterpenoid resins include dammar and mastic. Dammar is composed of a mixture of molecules including dammaradienol (Figure 1.8); it is hard and brittle and has been used for picture varnishes. Mastic resin is also a mixture of triterpenoids; Figure 1.8 illustrates a typical component, masticadienonic acid. Mastic is yellow to green in colour and was used as a picture varnish for many years for its good working and film properties. However, the resin is brittle and was superceded by dammar.

A number of diterpenoid resins have been used in heritage applications: rosin, sandarac and copals may be encountered. Rosin is a brittle, glassy resin and abietic acid is a major component (Figure 1.8). Rosin has been used for painting materials since the 9th century. Sandarac is a yellow gum resin and is also known as gum juniper. Sandarac has been used in varnish preparations and is largely composed of polymerised communic

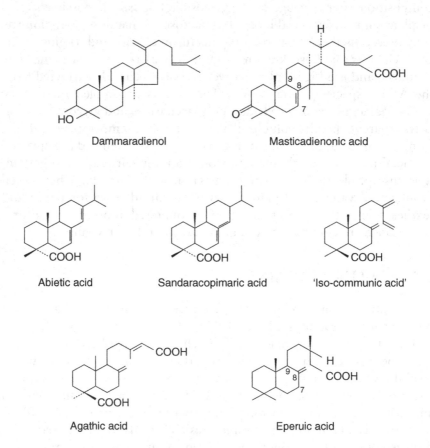

Figure 1.8   Structures of some terpenoids found in natural resins

acid with sandaracopimaric acid also present (Figure 1.8). Copal is a general term used to describe some terpenoid resins. Like sandarac, polymerised communic acid makes an important contribution to the structures of copals. Copal is sometimes referred to as immature amber and most copals originate from African, Asian or Central American sources. Agathic acid (Figure 1.8) is found in a number of copal resins and eperuic acid (Figure 1.8) in a number of African copals. Copals have been extensively used for artefacts and as a varnish for oil paintings when dissolved in solvents.

Amber is a fossilised resin and is mainly found in the Baltic region of Northern Europe. Amber is an attractive material as it is translucent and the colour can vary from pale yellow to deep brown. It is easily worked and polished and has been widely used for jewellery. Another well-known feature of amber is the presence of prehistoric insect bodies – it provides a good preservation medium. Amber can also be dispersed in oil to form a varnish. The chemical composition of amber is a complex mixture of abietic acid components and diterpenoids, with succinic acid believed to be present.

Shellac is a naturally derived polymeric resin obtained from the secretions of insects [Katz, 1994; Horie, 1987; Quye and Williamson, 1999]. The insect secretes a protective covering against predators onto twigs and a brownish resin in extracted by melting. The resin may be cooled in various shapes. Approximately 30–40 % of shellac is aleuritic acid, which is combined with sesquiterpene acids. The Egyptians used shellac thousands of years ago to coat their mummies. Shellac has been used for centuries as a lacquer for a protective and decorative finish on both metals and wood. When it was discovered that shellac could be compounded with fillers such as wood powder to produce a tough mouldable material, more applications were established in the 19th century. A particularly useful property was that shellac was capable of showing fine details and this led to the first pressing of sound records.

1.6 NATURAL MATERIALS

Animal glues are obtained from the skin, bone and other tissues of various species and have been widely used as adhesives and paint binders [Horie, 1987; Newman, 1998]. Gelatin is extracted by boiling and the extract is cooked to form a gelatin material. The gelatin can then be reliquefied with heat, which provides quick setting properties. Connective tissue is composed of various proteins, mainly collagen, and many other compounds. While collagen molecules are held together with many hydrogen bonds, these molecules are partly hydrolysed on heating in water to produce a soluble product. Parchment glues are derived from the hides of cattle and sheep and have high strength, but may contain contaminants from skin preservatives and tanning agents. Bone glues are derived from the bones of farm animals and fish glues are made from skins or swim bladders. Animal glues were the most effective adhesives until the development of synthetic adhesives. Glue and starch pastes were introduced during the 18th century for the relining of canvas paintings and were used on textiles. Glue is also used as an adhesive for conserving furniture.

The skins and hides of animals have long served as clothing or shelter for man [Lambert, 1997]. Skin is composed of collagen and hairs of keratin. Such materials require a preservation process for use and tanning is a common chemical procedure for preserving skin. A simple tanning process involves applying oils or vegetable extracts to make the skin pliable and impermeable to water. Skins were also used for written records and a smooth surface was often created by the chemical removal of hair. After soaking in water, the skins could be treated with an alkaline agent, such as an aqueous lime solution.

Horn, hoof and tortoiseshell are chemically similar and are materials based on the protein keratin [Katz, 1994; Quye and Williamson, 1994]. The thermoplastic nature of these keratotic materials has been exploited since prehistoric times. Horn was once very plentiful and the process of moulding was simple to carry out. This material has been used to produce drinking containers, shoe horns, combs, boxes, jewellery and buttons. In the 19th century a fashion for hair combs provided a demand for horn, which was eventually superceded in this application by celluloid in the 1920s. Hoof can be moulded more easily than horn and was used to produce buttons. Tortoiseshell produces an attractive mottled appearance when heated and pressed and was widely used to produce items such as decorative boxes. The techniques developed to produce decorative pieces from horn, hoof and tortoiseshell were precursors to those used in the modern plastics industry.

Ivory is the creamy white substance that forms the tusks or teeth of mammals and has been used to produce many decorative and practical objects [Cronyn, 1990; Lambert, 1997]. Although the main source of ivory is elephant tusks from Africa and India, tusks or teeth of mammoth, whale, walrus, pig, warthog and hippopotamus have all been used and described as ivory over the years. Ivory is composed of osteons, which contain lamellae of collagen and other proteins embedded in an inorganic matrix of hydroxyapatite [$Ca_{10}(PO_4)_6(CO_3).H_2O$]. The collagen provides elasticity and a high tensile strength, while hydroxyapatite provides hardness and rigidity. The properties of ivory made it a popular material for carved objects including jewellery, statues, piano keys, cutlery handles and billiard balls.

Natural rubber was first brought to Europe during the 18th century from South America, but it was not until the mid-19th century that the material was chemically modified and became suitable for manufacture [Fenichell, 1997; Katz, 1994; Mossman, 1997; Quye and Williamson, 1999]. Rubber is extracted from the latex of tropical trees and consists mainly of poly(cis-isoprene), with small amounts of the other components including proteins and lipids. Rubber in solution was used in the early 19th century as a waterproof coating for cloth in Macintosh raincoats. An important breakthrough came when Thomas Hancock discovered the vulcanisation process, the process by which rubber is lightly cross-linked by heating it with sulfur to reduce plasticity and to develop elasticity. Natural rubber has been used in a wide range of applications including shoes and tyres. When rubber is cross-linked with a greater amount of sulfur, an extremely hard dark material known as vulcanite or ebonite is produced. This process was developed in the 1840s and vulcanite was widely used to produce a range of decorative household items and its good electrical insulation properties led to use in the emerging electrical industry of the late 19th century. Gutta percha is related to natural rubber in chemical structure, but shows quite different physical properties. Gutta percha is also a tree exudate, but is mainly composed of poly(trans-isoprene). The different isomer results in a hard brittle material. Gutta percha was used from the mid-19th century until the 1930s to produce many moulded objects including containers, toys and tubing [Fenichell, 1997; Katz, 1994; Mossman, 1997; Quye and Williamson, 1999]. As gutta percha oxidises and becomes brittle with time, mouldings of this material are now scarce.

Casein is a major protein found in milk and is precipitated from skimmed milk treated with acid [Katz, 1994; Mills and White, 1994; Newman, 1998; Quye and Williamson, 1999]. The curds are reacted with formaldehyde to produce a hard material. Casein was first patented in 1899 and was produced in Europe as galalith and erinoid. The ability of casein to be surface dyed meant that decorative products such as buttons (the major product), fountain pens, jewellery, candlesticks, spoons and gaming chips could be produced in small batches. Casein has also been used as a consolidant or adhesive [Horie, 1987]. Calcium caseinite is the most useful form for adhesives and has traditionally been used for wood and plaster restoration.

Wood from plant sources has been used for the production of many objects, especially furniture, for many years. Wood is made of cellulose (40–45 %), hemicellulose (20–30 %) and lignin (20–30 %) [Hoadley, 1998; Lambert, 1997]. Like cellulose, hemicellulose is a polysaccharide, but usually has a low molecular weight. Lignin is a complex polymer molecule made up of cross-linked phenolic annamyl alcohols. The polysaccharide components of wood tend to decay faster than lignin. Wood is protected using coatings such as oils and resins.

There are several bituminous materials encountered in artefacts that are natural carbon-based products [Lambert, 1997; Mills and White, 1994]. Bitumen is a tarry petroleum product that has been used as an adhesive and for moulded artefacts. It is considered one of the earliest moulding materials and shows plastic behaviour [Mossman, 1997; Quye and Williamson, 1999]. As a property of bitumen is electrical resistance, it found application in the electrical industry at the end of the 19th century. Asphalt is similar but usually has calcite ($CaCO_3$), silica (SiO_2) or gypsum ($CaSO_4$) added. Jet, a form of brown coal, is a fossilised wood derived from an ancient species of tree [Lambert, 1997; Muller, 2003]. It contains approximately 12 % mineral oil and traces of aluminium, silica and sulfur. Jet became very popular in Britain in Victorian times, particularly for mourning jewellery. Pyrolysis of carbon-based materials such as wood, coal and peat produces aromatic compounds such as phenol, heterocyclics and polynuclear aromatic hydrocarbons, which are distilled as coal tar. The higher molecular weight residual molecules form pitch. Tar and pitch have been used for adhesives, sealants and coatings.

Bois durci is a moulding material based on cellulose and was patented in Paris in 1855 [Katz, 1994; Mossman, 1997; Quye and Williamson, 1999]. It was made from wood flour blended with albumen from egg or blood and when moulded produced a brown or black thermoset. The surface of bois durci gave it a metallic-like finish. Bois durci mouldings were used for plaques, inkwells, desksets and picture frames.

The papier mâché process was patented in the 18th century [Katz, 1994; Mossman, 1997]. In the process, pulp made from finely ground wood flour or paper is mixed with animal glue or gum arabic, then pressed into a mould and dried in an oven. The moulding was then sanded, polished and decorated. Papier mâché was popular for the production of decorative items in the 19th century, such as trays, spectacle cases and snuff boxes. Pulp ware came along in the latter part of the 19th century [Katz, 1994]. The pulp was made from purified ground wood and linseed oil and phenolic resin and melamine used to coat the surface. As pulp ware was impact and water resistant, it found use for inexpensive kitchenware.

1.7　SYNTHETIC POLYMERS

Synthetic polymers are a 20th century phenomena and have been used for an enormous range of commodities. Despite their reputation as cheap and disposable, there are many polymeric products that are of interest to conservators. Polymers have been used to produce culturally important materials held in museums, such as sculptures, paintings, toys, clothing, jewellery, furniture, household goods and cars [Fenichell, 1997; Katz, 1994; Mossman, 1997; Quye and Williamson, 1999].

Synthetic polymers are synthesised from their constituent monomers via a polymerisation process and most are based on carbon compounds. Polymers have been commonly referred to as 'plastics'. Plastic refers to one class of polymers known as thermoplastics, which are polymers that melt when heated and resolidify when cooled. Thermoplastics tend to be made up of linear or lightly branched molecules. Thermosets are polymers which do not melt when heated, but decompose irreversibly at high temperatures. Thermosets are cross-linked; the restrictive structure preventing melting. Some cross-linked polymers may show rubber-like characteristics and these are known as elastomers. Elastomers can be stretched extensively but rapidly recover their original dimensions.

When examining a polymeric material it is important to be aware that it may be composed of more than one component. Copolymers are comprised of chains containing two or more different types of monomers. Polymer blends are mixtures of polymeric materials and consist of at least two polymers or copolymers. Composites are materials composed of a mixture of two or more phases and polymers can constitute the fibre or matrix component of a composite.

The identification and characterisation of polymers is made more complex by the fact that they often contain additives to modify their properties. Fillers are added to polymers to improve their mechanical properties, such as strength and toughness, and usually consist of materials such as calcium carbonate, glass, clay or fibres. Colourants such as dyes or pigments are used to impart a specific colour. Polymers may have their flexibility improved by the addition of plasticisers, which tend to be low molecular weight liquids such as phthalate esters. Stabilisers are used to counteract degradation of polymers by exposure to light and oxygen and include lead oxide, amines and carbon black. The flammability may also be minimised by the addition of flame retardants such as antimony trioxide.

The means by which polymers are processed can also affect the properties of the material produced [Osswald, 1998]. A common method for processing thermoplastics is injection moulding, where the polymer, in the form of a powder or granules, is heated by a rotation screw until soft then forced through a nozzle into a cooler mould. Extrusion is also common for thermoplastics, where the polymer is melted and formed into a continuous flow of viscous fluid then forced through a die. Thermoset polymers can be formed by transfer moulding, where the powder is heated and compressed in a chamber and enters a mould cavity in a flowing state. Compression moulding is also used to produce thermosets and involves placing partially polymerised thermoset powder into a two-part mould. The mould is closed, heat and pressure are applied, and the polymer adopts the mould shape as it cross-links and hardens.

Cellulose nitrate (or nitrocellulose) (Figure 1.9) was an early synthetic plastic developed during the 1840s [Fenichell, 1997; Katz, 1994; Mossman, 1997; Quye and Williamson, 1999]. Although initially recognised as an explosive, it was soon realised that cellulose nitrate was also a hard elastic material which could be moulded into different shapes. In the 1860s, Alexander Parkes plasticised cellulose nitrate with oils and camphor. However, his material, known as Parkesine and used for mouldings, such as combs and knife handles, soon resulted in warping and cracking. In the United States, the Hyatt brothers plasticised cellulose nitrate with camphor and patented celluloid in 1869. Celluloid was a commercial success, leading to the development of photography and the cinema industry. Celluloid was also used to produce toys, combs and other moulded household goods. Cellulose nitrate has been superseded in this type of application because of its flammability and degradability, but now finds use in the field of coatings as a lacquer. By the late 19th century other modifications of cellulose had been developed. When cellulose is dissolved via a particular chemical reaction and then reprecipitated as pure cellulose, the product is known as regenerated cellulose. When regenerated cellulose is prepared as a fibre, it is known as viscose or viscose rayon and has been widely used for textile fibres. When prepared as a film, regenerated cellulose is known as cellophane, a well known packaging and wrapping material. The development of the commercially important cellulose acetate early in the 20th century was a result of the esterification of cellulose [Fenichell, 1997; Katz, 1994; Mossman, 1997; Quye and Williamson, 1999]. When complete acetylation is carried out, cellulose triacetate is formed (Figure 1.9).

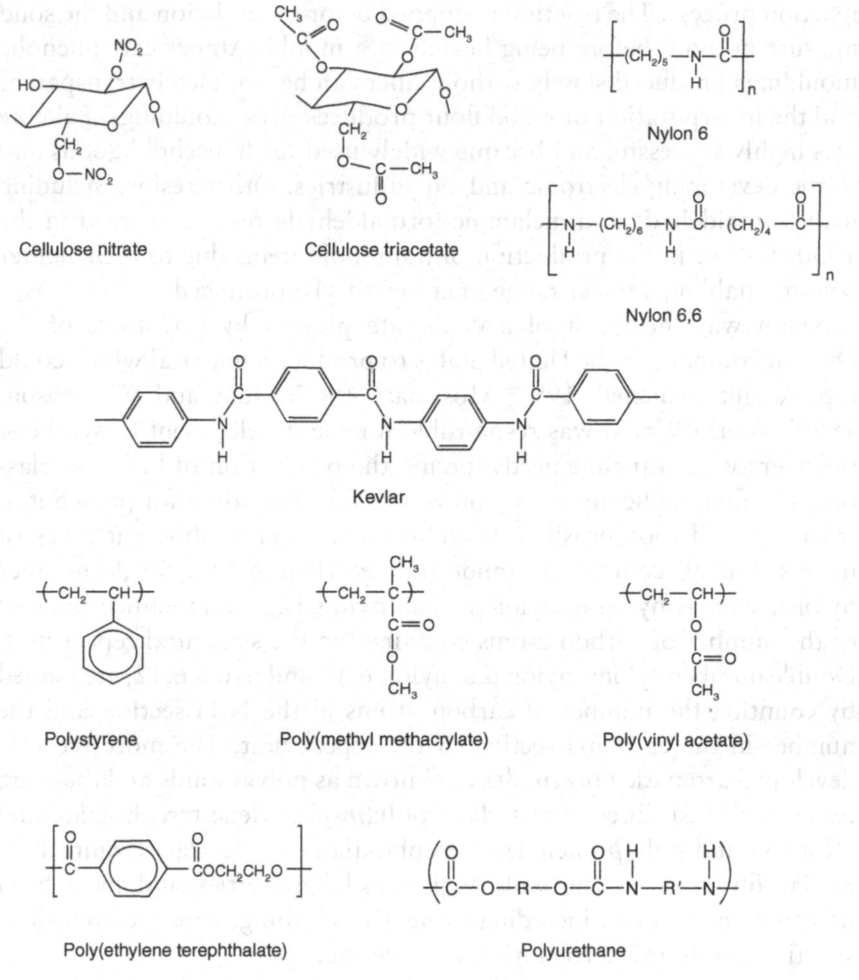

Figure 1.9 Structures of some common polymers

However, the acetylation reaction may also be reversed to a point where cellulose diacetate is formed, which is more suitable for use as a fibre. Cellulose acetates are employed in a wide range of forms including fibres, moulded products, films and packaging.

Phenol-formaldehyde or phenolic resins are thermosetting polymers made by reacting phenol and formaldehyde and adding filler. Phenolic resins are better known by the tradename Bakelite [Fenichell, 1997; Katz, 1994; Mossman, 1997; Quye and Williamson, 1999]. At the beginning of the 20th century, Leo Baekeland in the United States was the first to successfully commercialise the polymerisation of phenolic resin by incorporating wood flour into the phenol and formaldehyde reaction process. The reaction is stopped before completion and the solid mixture ground, before being heated in a mould. Amber cast phenolic mouldings produced slowly without filler can be completely transparent and the incorporation of wood flour produces dark mouldings. Bakelite was highly successful and became widely used for household goods and in the developing electronic and car industries. Other resins, including urea-formaldehyde and melamine-formaldehyde resins, emerged in the 1930s for use in the production of household items due to their lighter colour, enabling a wider range of colours to be produced.

Nylon was the result of a deliberate process by Carothers of the DuPont company in the United States to produce a material which could replace silk [Fenichell, 1997; Mossman, 1997; Quye and Williamson, 1999]. World War II was responsible for the development of synthetic polymers with war-time needs forcing the production of low-cost plastics. The first application of nylon was as fibres in fabric for parachutes, stockings and toothbrushes. There are a number of different types of nylons, but all contain an amide linkage (Figure 1.9). Single number nylons, such as nylon 6, nylon 11 and nylon 12, are so named because of the number of carbon atoms contained in the structural repeat unit. Double number nylons, nylon 6,6, nylon 6,10 and nylon 6,12, are named by counting the number of carbon atoms in the N-H section and the number in the carbonyl section of the repeat unit. The more recently developed aromatic polyamides are known as polyaramids and there are two established fibres in this class, poly(m-phenylene terephthalamide) (Nomex) and poly(p-phenylene terephthalamide) (Kevlar) (Figure 1.9). Kevlar fibres show unusually high tensile properties and have been used for applications including protective clothing, ropes, composites, sporting goods and aeronautical engineering.

Commercially important vinyl polymers were developed from the 1930s onwards. Chemists at ICI in Britain, while experimenting with ethylene at different temperatures and pressures, established polyethylene (PE) [–(–CH_2–CH_2–)–] [Fenichell, 1997; Mossman, 1997; Quye and Williamson, 1999]. PE, often referred to as polythene, is the major general purpose thermoplastic and is widely used for packaging, containers, tubing and household goods. There are two main types of mass produced PE: low density polyethylene (LDPE) has a branched chain structure and tends to be used for bags and packaging; high density polyethylene (HDPE) has a mostly linear structure and finds uses in bottles and containers. Polypropylene (PP) shows a similar structure to PE, but with a substituted methyl group [–(–CH_2–$CH(CH_3)$–)–] [Quye and Williamson, 1999]. This polymer is used for a wide range of applications such as chairs, bottles, carpets, casings and packaging.

Poly(vinyl chloride) (PVC), first commercialised in the 1940s, has a structure containing a chlorine atom on alternate main chain carbons [–(–CH$_2$–CHCl–)–] [Fenichell, 1997; Mossman, 1997; Quye and Williamson, 1999]. The lack of flexibility in PVC molecules means that this polymer is commonly processed with plasticisers. PVC has been widely used: toys were made from PVC from the 1940s and shoes and clothing were made of PVC from the 1960s.

Polystyrene (PS) is a clear, rigid and brittle material unless it is modified with rubber (Figure 1.9) [Fenichell, 1997; Mossman, 1997; Quye and Williamson, 1999]. PS is used widely in packaging and appliance housings. In the 1940s, acrylonitrile–butadiene–styrene (ABS) blends were developed to improve the impact resistance of PS and used for decorative items and in cars. ABS blends are composed of two copolymers, with a matrix consisting of a styrene–acrylonitrile copolymer and a rubbery phase consisting of a styrene–butadiene copolymer.

Polyacrylates (or acrylics) are derived from acrylic acid [CH$_2$=CHCOOH] or methacrylic acid [CH$_2$=C(CH$_3$)COOH] and are valuable for use as varnishes and transparent plastics [Fenichell, 1997; Mossman, 1997; Mills and White, 1994; Quye and Williamson, 1999]. Poly(methyl methacrylate) (PMMA) is a well known type of acrylic and was developed in the 1930s as an alternative to glass (Figure 1.9). Well known by the trade name Perspex, this polymer is a transparent, hard and rigid material, making this polymer particularly useful in glazing. A number of acrylic copolymers, such as methyl acrylate – ethyl methacrylate copolymers, have been employed in conservation as varnishes due to their stability and clarity.

Polytetrafluoroethylene (PTFE) [–(–CF$_2$–CF$_2$–)–] is perhaps better known by its trade name of Teflon [Fenichell, 1997; Quye and Williamson, 1999]. There is a common misconception that 'the only good thing that came out of the space race was the non-stick frying pan', referring to the emergence of PTFE as a common surface coating in the 1960s. In fact, PTFE was originally discovered in the 1930s at DuPont, when tetrafluorine gas was accidentally polymerised; the polymer was later commercialised in the 1950s. PTFE is a thermoplastic which shows remarkable chemical resistance, electrical insulating properties and a low friction coefficient. PTFE is used for non-stick coatings, electrical components, bearings and tape.

Poly(vinyl acetate) (PVA) is commonly used in the form of an emulsion (Figure 1.9) [Mills and White, 1994]. PVA is tough and stable at room temperature, but becomes sticky and flows at slightly elevated temperatures. PVA is a quick drying polymer and so can be used in the production of water-based paints and is also used commonly as an adhesive. Alcohols are added to PVA to produce poly(vinyl alcohol) (PVAl) [–(CH_2–CH(OH)–)–], which is a water soluble polymer used for fibres, adhesives and as thickening agents.

Polyethylene oxide (PEO) [–(–CH_2–CH_2–O–)–] is a water soluble polymer. PEO has been used in conservation by impregnating the polymer into water affected materials such as wood and leather [Mills and White, 1994]. Another polymer that has been used to impregnate fragile objects, such as paper and books, is poly-p-xylylene (Parylene). The p-xylylene monomer is diffused into a material of interest placed in a vacuum and the polymerisation occurs at room temperature. The strength of the object is improved with no change in appearance [Mills and White, 1994].

Poly(ethylene terephthalate) (PET) (Figure 1.9) is a polyester used to form fibres known by the trade names Dacron and Terylene [Fenichell, 1997; Mills and White, 1994; Quye and Williamson, 1999]. PET is widely used in film form for bottles and cinema film base. Another type of polyester, alkyd resins, is widely used in paint and varnishes [Mills and White, 1994]. Alkyds are composed of a polyfunctional alcohol (e.g. glycerol), a polyfunctional acid (e.g. phthalic anhydride) and an unsaturated monoacid (e.g. drying oil). The presence of drying oil fatty acids means that further cross-linking can occur to produce a cross-linked network. Unsaturated polyesters can be cross-linked to form thermosets and are commonly used with glass fibres to form high strength composites. Linear polyesters are cross-linked with vinyl monomers, such as styrene, in the presence of a free radical curing agent. As polyester resins are low in viscosity, they can readily be mixed with glass fibres.

Epoxy resins contain epoxide groups and a common type of epoxy prepolymer is based on glycidyl ethers [Mills and White, 1994]. The resins are cured using catalysts or cross-linking agents such as amines and anhydrides. The epoxy and hydroxyl groups are the reaction sites for cross-linking and can undergo reactions which result in no by-products. This results in low shrinkage during hardening and they are used in coatings and composites.

Polyurethanes (PUs) are versatile thermosets which are used as foams, elastomers, fibres, adhesives and coatings. There is a range of chemical compositions in PUs, but all contain the common urethane group (Figure 1.9). Urethanes are formed by a reaction between a hydroxyl containing molecule and a reactant containing an isocyanate group. Another type of elastomer, silicone, is based on a silicon, rather than a carbon backbone. The structures of silicones are based on silicon and oxygen and the most common silicone elastomer is poly(dimethylsiloxane) (PDMS), which may be cross-linked to form $Si–CH_2–CH_2–Si$ bridges. These elastomers are thermally stable and water resistant and are used for medical applications and sealants.

Despite their reputation as materials that last forever, polymers do deteriorate with time. Degradation affects the appearance and physical properties of polymers, with some common effects being discoloration and embrittlement [Allen and Edge, 1992; Lister and Renshaw, 2004; McNeill, 1992; Quye and Williamson, 1999]. The factors that can cause polymers to degrade include exposure to light, oxygen and moisture in the atmosphere and the presence of additives, such as plasticisers and fillers.

There are various types of chemical change that may occur within a polymer that lead to decay. The polymer chains can be shortened due to the breakdown of bonds within the chains, which leads to poorer properties compared to the original polymer. Cross-linking of polymer chains may also occur. Reactions that produce bonds between the polymer chains can produce a polymer that is brittle and less flexible. Reactions associated with the side groups in polymers can also lead to degradation. Such reactions may involve the release of small molecules such as water or acids. Not only do these reactions change the chemical structure, the molecules released can produce further changes or catalyse other reactions.

1.8   DYES AND PIGMENTS

Colourants are used in a broad range of museum objects including paintings, textiles, polymers, written works, inks and ceramics. Dyes and pigments are the compounds that are used to create an array of colours [Doerner, 1984; Feller, 1986; Lambert, 1997; Mills and White, 1994; Needles, 1986; Roy, 1993; West Fitzhugh, 1997; Timar-Balazsy and Eastop, 1998]. Such materials must be reasonably stable to light. Pigments are coloured compounds that come in the form of solid particles suspended in a medium that binds to a surface, such as the canvas of a painting. Dyes are dissolved in a liquid and are usually bound directly to the surface, such as in the case of textile fibres. Pigments and dyes are obtained from naturally occurring minerals, extracted from plants or insects or produced by chemical synthesis. Some organic dyes can be converted to a pigment by incorporating them into kaolinite or chalk and then powdering the mixture to produce a 'lake'. Most early pigments were inorganic, but today both organic and inorganic pigments are used. Tables 1.4 and 1.5 show a summary of the names and structures of a number of inorganic and organic colourants. The chemical structures of a number of the organic molecules found in dyes and pigments are illustrated in Figure 1.10. The range of synthetic dyes and pigments is extensive. The synthetic colourants are standardised by a CI (Colour Index) number, referenced by professional colourists' associations [Colour Index, 2002].

Table 1.4   Chemical compositions of inorganic pigments

Colour	Common name	Chemical Composition	Origin
White	Anatase	TiO_2	20th century
	Antimony white	Sb_2O_3	Synthetic
	Bone white	$Ca_3(PO_4)_2$	Antiquity
	Chalk/Calcite/Whiting	$CaCO_3$	Mineral
	Gypsum	$CaSO_4.2H_2O$	Mineral
	Kaolin	$Al_2(OH)_4SiO_5$	Mineral
	Lead white	$2PbCO_3.Pb(OH)_2$	Synthetic, Antiquity
	Lithopone	$ZnS, BaSO_4$	Synthetic, 19th century
	Permanent white /Barium white/Barite/ Barytes/Barium sulfate	$BaSO_4$	Synthetic, 19th century /Mineral
	Rutile	TiO_2	20th century
	Titanium white	TiO_2	Synthetic, 20th century
	Zinc white	ZnO	Synthetic, 19th century
Yellow	Barium yellow /Lemon yellow	$BaCrO_4$	Synthetic, 19th century
	Cadmium yellow	CdS	Synthetic, 19th century
	Chrome yellow	$2PbSO_4.PbCrO_4$ or $PbCrO_4$	Synthetic, 19th century
	Chrome yellow orange /Chrome yellow deep	$PbCrO_4.PbO$	Synthetic
	Cobalt yellow	$K_3[Co(NO_2)_6]$	Synthetic, 19th century
	Lead-tin yellow (type I)	Pb_2SnO_4	Synthetic, 14th century
	Lead-tin yellow (type II)	$Pb_2Sn_{1-x}Si_xO_3$	Synthetic
	Litharge/Massicot	PbO	14th century
	Naples yellow	$Pb(SbO_3)_2, Pb_3(SbO_4)_2$	Synthetic, 1500–1300 BC
	Orpiment /Auripigmentum	As_2S_3	Mineral
	Pararealgar	As_4S_4	13th century
	Strontium yellow	$SrCrO_4$	Synthetic, early 19th century
	Yellow ochre/Limonite	$Fe_2O_3. nH_2O$, Clay, Silica	Mineral
	Zinc yellow	$ZnCrO_4$	Synthetic, early 19th century
Red	Cadmium red	$CdSe, CdS$	Synthetic, 20th century
	Chrome red	$PbCrO_4.Pb(OH)_2$	Synthetic, 19th century
	Haematite/Mars red	Fe_2O_3	Synthetic, 19th century
	Litharge	PbO	Antiquity
	Molybdate red	$7PbCrO_4 .2PbSO_4.PbMoO_4$	
	Realgar	As_4S_4	Mineral

Table 1.4 (*continued*)

Colour	Common name	Chemical Composition	Origin
	Red lead/Minium	Pb_3O_4	Synthetic, Antiquity
	Red ochre/Red earth	$Fe_2O_3 . nH_2O$, Clay, Silica	Mineral
	Vermillion/Cinnabar	HgS	Synthetic, 13th century /Mineral
Blue	Azurite	$2CuCO_3.Cu(OH)_2$	Mineral
	Cerulean blue	$CoO.SnO_2$	Synthetic, 19th century
	Cobalt blue	$CoO.Al_2O_3$	Synthetic, 18th century
	Cobalt violet	$Co_3(PO_4)_2$	Synthetic, 19th century
	Egyptian blue	$CaO.CuO.4SiO_2$	Synthetic ca 3000 BC
	Han blue	$BaCuSi_4O_{10}$	Mineral
	Han purple	$BaCuSi_2O_6$	Mineral
	Lazurite/Ultramarine /Lapis lazuli	$Na_{8-10}Al_6Si_6O_{24}S_{2-4}$	Synthetic, 19th century /Mineral
	Manganese blue	$BaSO_4.Ba_3(MnO_4)_2$	20th century
	Posnjakite	$CuSO_4.3Cu(OH)_2.H_2O$	Mineral
	Prussian blue	$Fe_4(Fe[CN]_6)_3$	Synthetic, 18th century
	Smalt	K_2O, SiO_2, CoO	16th century
Green	Atacamite	$CuCl_2.3Cu(OH)_2$	Mineral
	Basic copper sulfate	$CuSO_4.nCu(OH)_2$	Synthetic
	Brochantite	$Cu_4(OH)_6SO_4$	Mineral
	Chromium oxide	Cr_2O_3	Synthetic, early 19th century
	Chrysocolla	$CuSiO_3 . nH_2O$	Mineral, Antiquity
	Cobalt green	$CoO.5H_2O$	Synthetic, 18th century
	Copper chloride	$CuCl_2$	Synthetic
	Emerald green	$Cu(CH_3COO).3Cu(AsO_2)_2$	Synthetic, 19th century
	Malachite	$CuCO_3.Cu(OH)_2$	Mineral, Antiquity
	Scheele's green	$Cu(AsO_2)_2$	Synthetic, 18th century
	Terre-verte /Green earth	Variations on $K[(Al^{3+}, Fe^{3+}) (Fe^{2+}, Mg^{2+})], (AlSi_3,Si_4)O_{10}(OH)_2$	Mineral
	Veronese green	$Cu_3(AsO_4)_2.4H_2O$	Synthetic, 19th century
	Verdigris	$Cu(CH_3COO)_2.nCu(OH)_2$	Synthetic, Antiquity
	Viridian/Guignet's green	$Cr_2O_3.2H_2O$	Synthetic, 19th century
Black	Antimony black	Sb_2S_3	Mineral
	Carbon black/Lamp black/Charcoal black	C	Antiquity
	Galena	PbS	Mineral
	Ivory black/Bone black	C, $Ca_3(PO_4)_2$	Antiquity
	Iron black/Black iron oxide	Fe_3O_4	Mineral/
	/Mars black		Synthetic, 19th century
	Manganese oxide /Manganese black	MnO, Mn_2O_3	Mineral
	Plattnerite	PbO_2	Mineral
Orange /Brown	Burnt sienna	$Fe_2O_3 . nH_2O$, Al_2O_3	Mineral
	Cadmium orange	Cd, S, Se	19th century
	Mars orange	Fe_2O_3	Synthetic, 19th century
	Ochre/Goethite	$Fe_2O_3.H_2O$, Clay	Mineral

Table 1.5   Chemical compositions of some organic dyes and pigments

Colour	Common name	Chemical composition	Origin
Yellow	Berberine	Berberinium hydroxide	Mahonia stems, 19th century
	Gamboge	Gambogic acid	Gum resin, 17th century
	Hansa yellow	Azo class	Synthetic, 20th century
	Indian yellow	Mg salt of euxanthic acid	Cow urine, 15th century
	Quercitin	Quercitin	Quercus oak bark
	Saffron	Crocetin	Crocus flower, Antiquity
	Turmeric	Curcumin	
	Weld	Luteolin	Plant foliage, Stone age
Red	Carmine, Cochineal	Carminic acid	Scale insect, Cochineal, Aztec
	Kermes	Kermesic acid	Scale insect, Kermes, Antiquity
	Madder	Anthraquinones including Alizarin and purpurin	Plant root 3000 BC, Synthetic alizarin 19th century
	Permanent red	Azo class	Synthetic, 19th century
	Toluidine red	Azo class	Synthetic
	Basic red	Azo class	Synthetic
	Acid red	Azo class	Synthetic
Blue	Indigo/Woad	Indigotin	Plant leaf, Antiquity
	Copper phthalocyanine	Phthalocyanine class	Synthetic, 20th century
	Direct blue	Azo class	Synthetic
Purple	Tyrian purple	6,6'-dibromoindigotin	Mollusc 1400 BC, Synthetic, 20th century
Brown	Sepia	Melanin	Cuttlefish ink, 19th century
	Van Dyck brown	Humic acids, Allomelanins	Lignite, 16th century

Various pigments and dyes have been introduced at different times in history, often enabling the age of an object to be estimated [Doerner, 1984; Feller, 1986; Roy, 1993; West Fitzhugh, 1997]. In the ancient world, pigments derived from clay or burnt stick, such as red earth, yellow earth, chalk and carbon black, were used. The technology of colour developed in ancient Egypt and Egyptian blue was the first synthetic colour. Other pigments used in ancient Egypt included malachite, azurite, cinnabar and orpiment. The Greeks and Romans developed new colours such as white lead and vermillion. Dyes such as indigo, madder and Tyrian purple were also used. Through the medieval and renaissance periods new pigments were introduced as painting techniques developed. The number of pigments available to artists had

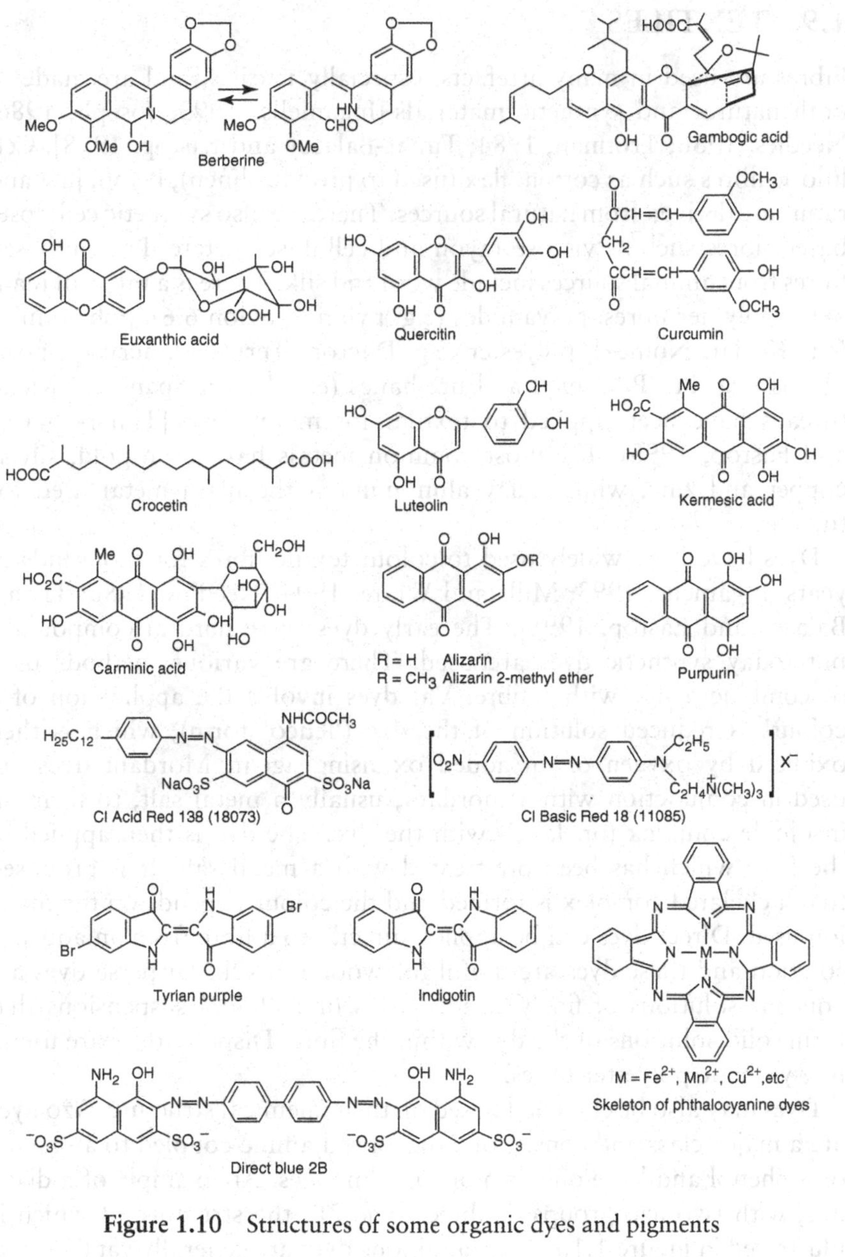

Figure 1.10 Structures of some organic dyes and pigments

considerably expanded by the time of the industrial revolution. Most of the metal-based colours were developed by the 19th century. The 20th century saw the development of pigments such as titanium white, Hansa yellows and cadmium red. Hundreds of synthetic colours are now available.

1.9 TEXTILES

Fibres are used in many artefacts, especially textiles, and are made of both natural and synthetic materials [Ingamells, 1993; Joseph, 1986; Needles, 1986; Trotman, 1984; Timar-Balazsy and Eastop, 1998]. Cellulosic fibres such as cotton, flax (used to produce linen), hemp, jute and ramie are derived from natural sources. There are also synthetic cellulose-based fibres such as viscose rayon and cellulose acetate. Protein-based fibres from animal sources include wool and silk. There is a range of man-made polymer fibres: polyamides (e.g. nylon 6, nylon 6,6), polyaramids (e.g. Kevlar, Nomex), polyester (e.g. Dacron, Terylene), acrylics, polyolefins (e.g. PE, PP), vinyl and urethanes (e.g. Lycra, Spandex). Metal threads have been applied to textiles for many years [Timar-Balazsy and Eastop, 1998]. The most common metals have been gold, silver, copper and zinc, while today aluminium is the main metal used for threads.

Dyes have been widely used to colour textile fibres for thousands of years [Ingamells, 1993; Mills and White, 1994; Needles, 1986; Timar-Balazsy and Eastop, 1998]. The early dyes were natural compounds, but today synthetic dyes are used. There are various methods used to combine a dye with a fibre. Vat dyes involve the application of a colourless reduced solution of the dye ('leuco' form), which is then oxidised by oxygen or an added oxidising agent. Mordant dyes are used in conjunction with a mordant, usually a metal salt, to form an insoluble complex (or 'lake') with the dye. The dye is then applied to the fibre which has been pre-treated with a metal salt. It is proposed that a chelated complex is formed and the colour depends on the metal ion used. Direct dyes can be applied directly to a fibre from an aqueous solution and these dyes are useful for wool and silk. Disperse dyes are aqueous solutions of finely divided dyes or colloidal suspensions that form solid solutions of the dye within the fibre. Disperse dyes are useful for synthetic polyester fibres.

Dyes may also be classified based on their chemical structure. Azo dyes are a major class and consist of a diazotised amine coupled to an amine or a phenol and have one or more azo linkages. An example of a diazo dye, with two azo groups, is direct blue 2B, the structure of which is illustrated in Figure 1.10. Anthraquinone dyes are generally vat dyes and an example is alizarin. Many red dyes are quinones and are derivatives of naphthaquinone and anthraquinone (Figure 1.11). Indigoid dyes are also vat dyes and indigo is an example. Triphenylmethane dyes are derivatives of a triphenylmethyl cation and are basic dyes for wool or silk. An example is malachite green.

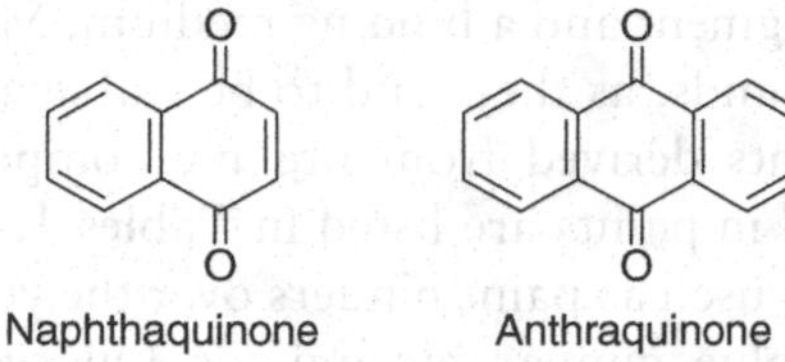

Naphthaquinone   Anthraquinone

Figure 1.11 Structures of naphthaquinone and anthraquinone

1.10 PAINTINGS

Much of the world's cultural heritage is contained in paintings. An array of materials have always been employed in paintings [Doerner, 1984; Gettens and Stout, 1966; Mayer, 1991; Wehlte, 1975]. The support, the main structural layer, of a painting has been principally canvas, wood or wall. Canvas is usually linen, although cotton, hemp and silk have all been used. A typical cross section of the paint layers on the surface of a painting is illustrated in Figure 1.12. Generally a painted surface consists of two paint layers on top of a ground layer on the support. The ground layer, also known as the primer or the preparation layer, is the layer on which a drawing is made before the paint is applied and acts as a barrier between the paint and the support. The primer is often gypsum, chalk or a white paint in animal glue and is about 0.5–2 mm in thickness. A transparent varnish layer is used to protect the paint layers, and also provides gloss and colour improvement. Varnishes are natural or synthetic resins and are applied from a solution using an organic solvent. The solvent is volatile so after evaporation a glassy and hard film is formed on the surface.

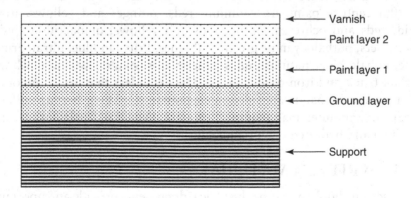

Figure 1.12 Cross-section of the surface of a painting

Paint consists of pigment and a binding medium. Most paint pigments are inorganic compounds, as they tend to be stable and light-fast when compared to pigments derived from organic compounds. Commonly used pigments found in paints are listed in Tables 1.4 and 1.5. A range of materials has been used as paint binders over the years: egg white, egg yolk, casein, animal glue, gum arabic and oils. Oils were used in binders (in oil paints) from the 15th century until the 20th century and linseed oil has been the most widely employed. Tempera paints use an oil-water emulsion as a binder. The emulsifying agent is usually gum arabic, glue or egg yolk. In watercolour paint, the pigment is finely ground and suspended in water with a binder such as gum arabic. In fresco painting, the binder is applied first as damp plaster made from lime (CaO). The pigments are mixed with lime and water and then applied to the damp surface. The lime reacts with air to form calcium carbonate ($CaCO_3$) and the pigments become encased.

The main components of 20th century paints are binders, pigments and extenders [Learner, 2004]. The main types of binders used are oils, alkyds, acrylics and PVA. Oils and acrylics are found in artists' paints, while alkyds and PVA are used in commercial paints. Linseed oil is still the most commonly used oil, but soya and safflower oils are more commonly used in light colours to avoid the yellowing effect of linseed oil. Alkyds are often described as oil-modified polyester paints and have been used as the binding media in the majority of commercial paints since the 1930s. Acrylics, in the form of an emulsion, were introduced in the 1950s. Early acrylic emulsions were based on copolymers of ethyl acrylate and methyl methacrylate, but in recent times copolymers of butyl acrylate and methyl methacrylate have been used. PVA-based paints were first used in the 1940s. Although PVA is now not used for artists' paint, it is used in commercial household paints. The important pigments used in 20th century paint are cadmium reds, oranges and yellows, iron oxide reds and ochres, naphthol reds and yellows, quinacridone reds and violets, phthalocyanine greens and blues, Prussian blue, ultramarine blue, cobalt blue, natural and synthetic iron oxides, titanium white, carbon black and iron oxide blacks. The extenders that are commonly used in modern paints are calcium carbonate ($CaCO_3$), barium sulfate ($BaSO_4$), kaolinite, magnesium carbonate ($MgCO_3$), calcium sulfate ($CaSO_4$) and hydrated magnesium silicate.

1.11 WRITTEN MATERIAL

Manuscripts, books and other historical documents provide an important record of cultural heritage. The types of materials used have evolved with time. Papyrus plants were one of the most important sources of writing material in ancient cultures and are first used over 5000 years ago in the Nile Valley Kingdoms. The plants are swamp reeds supported by a hollow fibrous stem. To create writing material two layers of stem are laid over each other at right angles, then pounded and smoothed into a single sheet.

Parchment and vellum, produced from the dermis of animal skin, were widely used after the 2nd century AD to produce ancient manuscripts or book-bindings. Vellum is used to describe fine parchment made from calf skin. Parchment is prepared by cleaning the skin and removing the hair by a mechanical treatment that involves the scraping of fat from the skin [Covington et al., 1998; Vandenabeele and Moens, 2004]. The main structural component of parchment is collagen and an understanding of the state of this protein is required when developing a conservation treatment for parchment. A choice of the optimum storage conditions for the long term storage of documents and manuscripts made of parchment is crucial. If parchment is stored in a high humidity environment, the collagen component can denature to form gelatin and the parchment will become soft and sticky. There is also the possibility of microorganisms growing in a high humidity environment and attacking the material. On the other hand, if the humidity is too low, parchment can shrink and deform and will become brittle. The preservation of animal skins following the removal of fatty material involved drying by salting with sodium chloride (NaCl) or potassium chloride (KCl) and ammonium chloride (NH_4Cl) or sulfate addition with lime to adjust the pH. This was followed by treatment with potash alum mixed with flour and egg yolk to produce a supple substrate. Poorer parchments were treated with preparations such as sodium sulfite suspended in natural oils in order to mimic the more translucent vellums.

Paper is believed to date back to China in 200 AD [Hon, 1989]. Sheets of paper were first made by drying fibres, such as flax and rice stalks, macerating in water and draining on moulds. Wood became the main source of paper fibres during the 19th century after the development of a grinder for the mechanical pulping of wood. During the manufacture of paper, sizing agents are usually added to the pulp or the paper surface after formation to make the cellulose component more hydrophobic [Carter, 1997]. This is required to prevent printing inks from running. Cellulose contains hydroxyl groups which make it a naturally hydrophilic material. Until the 1950s, a rosin precipitated by alum ($Al_2(SO_4)_3.18H_2O$) was used as a sizing agent. However, the acid hydrolysis of the complex hydrated aluminium ions produced by this salt of cellulose results in the depolymerisation of the cellulose chain and the paper becomes brittle as a consequence. Deacidification is used in an attempt to protect paper from degradation. The goal is to establish an alkaline buffer that compensates for the acidic species responsible for degradation. Organic basic solutions are usually used to minimise reactions which can affect the appearance of paper and those that may involve inks.

Writing inks of various compositions have been used for millennia [Brunelle and Reed, 1984]. Carbon inks consisting of fine grains of carbon black suspended in liquid were used as early as 2500 BC. Such inks are known as India ink or Chinese ink. A red version used cinnabar instead of carbon and was used on the Dead Sea Scrolls. Carbon inks are still used and the carbon is usually suspended in a gum or glue solution which also acts as a binder. Another ink that has been used since the middle ages is iron gall ink. The main ingredients of this ink are gallic acid, tannic acid (extracted from gall nuts), iron[II] sulfate ($FeSO_4.7H_2O$) and gum arabic. When iron gall ink is applied to paper or parchment, the ink darkens due to the oxidation of Fe^{2+} to Fe^{3+}. Iron gall ink was regarded as the most important ink up until the 20th century. However, the corrosive nature of some compositions led to iron gall inks falling out of favour. Modern inks contain an array of synthetic resins and pigments.

Medieval manuscripts have survived well with time, largely because such materials have always been regarded as valuable and so were handled with care. The materials used for the production of these manuscripts were parchment, pigment, binder and ink [Vandenbeele and Moens, 2004]. Parchment was the important writing material during medieval times, although during the early middle ages, papyrus was used as well. Some pigments used during this period were vermillion, red lead, haematite, azurite, lapis lazuli, smalt, lead–tin yellow (type I), massicot, limonite, verdigris, malachite, copper phosphate, basic copper sulfate, copper resinate, Veronese green earth, chrysocolla, carbon black and iron black [Vandenbeele and Moens, 2004]. Protein-based materials such as casein, egg white and egg yolk were used as binding media. Starch and gum arabic were also employed as binders. Animal glues were not used at the time for binders, but were used for applying gold leaf. Iron gall ink and charcoal were the most commonly used medieval inks.

1.12 GLASS

Glass is an amorphous solid that has been used widely in the production of heritage objects [Henderson, 2000; Lambert, 1997]. Man-made glass is believed to have originated in northern Mesopotamia (Iraq) prior to 2500 BC. The mass production of glass objects began in the second millennium BC in Mesopotamia and in the Mediterranean region. Glass technology was further developed during the Roman period and green glass was mass produced during this era. Glass was also produced in Asia from early times. Glass production spread to centres throughout Europe during medieval times.

Glass is made by melting a mixture of three main components [Cronyn, 1990; Davison, 2003; Henderson, 2000; Lambert, 1997; Pollard and Heron, 1996]. The basic glass forming material is silica (SiO_2), which comprises 60–70 % of the mixture. An alkaline flux, such as sodium or potassium derived from minerals or plant ash (Na_2O, K_2O), is added to lower the melting temperature of the batch. Fluxes interrupt some of the Si–O bonds and so disrupt a continuous network produced by silica. The unattached oxygen atoms become negatively charged and loosely hold the monovalent cations in the spaces of the network (Figure 1.13). As the bonding is weak, the cations can migrate out of the network in water. A stabiliser, such as lime (CaO) or magnesia (MgO), is needed to make the glass water resistant. The stabiliser is doubly charged so is held more tightly than the monovalent ions, thus the fluxes are held within the network. The most commonly produced glass is soda-lime, which contains silicon dioxide (SiO_2), sodium oxide (Na_2O) and calcium oxide (CaO). From about 1675, lead oxide was used in glass, acting both as a flux and a stabiliser. The batch is melted in a furnace at a temperature of approximately 1500 °C until it reaches a liquid state. The qualities and characteristics of glass may be changed by varying the proportions of the main components and the addition of other components. These variations affect the manner in which the hot glass behaves when it is shaped.

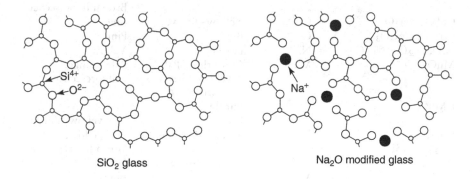

SiO$_2$ glass   Na$_2$O modified glass

Figure 1.13   Unmodified and Na$_2$O modified silica glass networks

The colour of glass is determined by the presence of metallic oxides. The silica used to produce glass generally contains natural impurities of iron that result in a green colour in the glass. Pure silica must be used in order to produce colourless glass. Alternatively, the natural green can be neutralised by the addition of a small quantity of manganese. The green tint of reduced iron is diminished by the pink manganese ions because the iron is oxidised to a yellow colour that then appears colourless in the presence of the pink manganese. If a surplus of manganese is added, a pink coloured glass results. Other colourants include copper or cobalt oxides for blue, copper or iron oxides for green, gold oxide for ruby red and uranium oxide for a radiant green. Table 1.6 summarises some of the metal ions responsible for colouring glass [Pollard and Heron, 1996]. The colour produced by an ion depends on its oxidation state and on the position it occupies in the glass structure. For instance, it is significant as to whether the metal forms a tetrahedral or an octahedral coordination. Opacifiers, such as tin oxide and calcium antimonite, can be added to create opal-like or opalescent glass. Glass objects can be decorated by firing on gilding or unfired paint applied in a lacquer, varnish or oil. Window glass is painted by enamelling, firing on a mixture of powdered glass and iron oxide, or stained yellow by firing with silver sulfide.

Table 1.6 Metal ions in glass

Raw material	Colouring ion	Colour in tetrahedral coordination	Colour in octahedral coordination
Cr_2O_3	Cr^{3+}	–	Green
$K_2Cr_2O_7$	Cr^{6+}	Yellow	–
CuO	Cu^+	–	Colourless to red, Brown fluorescence
$CuSO_4.5H_2O$	Cu^{2+}	Yellow–Brown	Blue
Co_2O_3, $CoCO_3$	Co^{2+}	Blue	Pink
NiO, Ni_2O_3, $NiCO_3$	Ni^{2+}	Purple	Yellow
MnO_2	Mn^{2+}	Colourless–Pale yellow, Green fluorescence	Pale orange, Red fluorescence
$KMnO_4$	Mn^{3+}	Purple	–
Fe_2O_3	Fe^{3+}	Brown	Pale yellow–Pink
U_3O_3, $Na_2U_2O_7.3H_2O$	U^{6+}	Yellow–Orange	Pale yellow, Green fluorescence
V_2O_5	V^{3+}	–	Green
V_2O_5	V^{4+}	–	Blue
V_2O_5	V^{5+}	Colourless–Yellow	–

The deterioration of glass is dependent upon its composition, age, firing history and storage environment [Davison, 2003]. When glass is in contact with water, the alkali metal ions are leached out and replaced by hydrogen ions. The surface of the glass will change its refractive index and appear dull. Colouring metal ions may also be leached from glass by water and this leads to discolouration. Alternatively, the ions may change colour by oxidation. For example, Mn^{2+} can be deposited by black manganese dioxide (MnO_2). In addition, colouring ions from the environment may be taken up by the glass and result in a colour change. Encrustation may also be observed due to the deposition of insoluble salts. Excess lime can leach out of the glass and deposit as a white crust on the surface.

1.13 CERAMICS

Ceramic objects have been produced for tens of thousands of years by many cultures and so form an important aspect of cultural heritage. Ceramics are produced from fired clay [Cronyn, 1990; Henderson, 2000; Lambert, 1997; Pollard and Heron, 1996]. Clay is a widely available material with plastic properties that enable it to be shaped when wet, producing a hard object when heated. The main components of clays are minerals such as kaolinite, montmorillonite and illite. Different clay types contain layers of silica, alumina (Al_2O_3) and water in different proportions. The structures of kaolinite and montmorillonite are illustrated in Figure 1.14. Kaolinite is composed of a silicate sheet ionically bonded to a sheet of $AlO(OH)_2$ and is often represented by the formula $Al_2Si_2O_5(OH)_4$. Montmorillonite ($Al_2(SiO_5)_2(OH)_2$) contains two silicate sheets around an $AlO(OH)_2$ layer. Clays also contain fillers (or temper) that can aid in the physical properties of the ceramic produced. A variety of materials including sand, limestone, mica, ash and organic matter have been used.

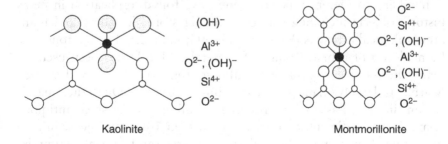

Figure 1.14 Kaolinite and montmorillonite clay structure

The formation of ceramics when clay is heated is a complex process due to the mixture of components present. When wet clay is heated, water molecules that are not bound are first lost by evaporation at 100 °C. At temperatures near 600 °C, the clay loses water bound to the clay structures. The aluminosilicate component rearranges its structure to accommodate the loss of the water molecules and the material loses its plasticity. Above about 850 °C, vitrification occurs where glass-like regions are formed in the aluminosilicate structure.

The porosity of a fired ceramic is dependent on how the clay minerals fuse together, on whether the pores have been filled with glass, and on the original size of the components. Terracotta describes the oldest form of man-made ceramic and is fired at about 900 °C. The addition of iron and heating in the range 900–1100 °C produces red earthenware, which is stronger and less porous. The addition of lime, containing calcium, and heating at 1100–1200 °C produces a cream colour in earthenware. Faience or majolica is earthenware covered with a glaze. Stoneware is produced by heating at 1200–1300 °C, which intensifies vitrification and produces a strong material which is non-porous. Heating above 1300 °C produces porcelain, which is highly vitrified and translucent.

A number of surface finishes can be applied to ceramics. Colour may be added to the surface using graphite (black), haematite (black–red) or mica powder (golden). A gloss may be added by the application of a thin coat of diluted fluid clay, known as a slip. Glazes, which are glasses, are frequently added for decoration or to improve the impermeability of the ceramic. Glaze is made of quartz and is formed by heating, often applied to an object and created in a second firing. Paint can be fixed to unglazed ceramics by applying pigments such as iron ores that contain silica or clay minerals that bind the colour to the ceramic during firing.

1.14   STONE

A number of different types of stone have found application in many historic objects including statues, building stone, jewellery and many archaeological artefacts [Henderson, 2000; Lambert, 1997]. Stones can be made up of a broad range of minerals. Marble has been used for many thousands of years as a building stone and for carved statues. Marble is largely made up of calcium carbonate. Quartzite, mainly silicon dioxide, has also been used for building stone since antiquity. Some rocks, sedimentary rocks, are formed from marine sediments over time. Sandstone is formed from quartz sand and also contains clays such as kaolinite. Over time and exposure to pressure and heat, sandstone is transformed into quartzite. Like quartzite, flint is a stone mainly composed of silica. As flint was widely used for tool making and only mined in particular locations, this material is of great interest to archaeologists. Limestone is formed from shells and coral and can be transformed into marble over time. Granite is highly heterogenous and more difficult to characterise. The presence of trace elements in stones can be used to identify the source of many stones.

These types of stone can be susceptible to deterioration due to environmental factors. Compounds in air pollution are responsible for deleterious reactions on the surface of certain stones. For instance, the sulfur from burnt coal forms sulfur oxides which deposit on marble and transform into calcium sulfate ($CaSO_4$). The calcium sulfate is more water-soluble than calcium carbonate and so detail on carved objects is lost. Attempts to counteract such degradation are made by applying materials known as consolidants. Consolidants are often made of polymers such as silicones.

A stone that has been widely studied is obsidian because of its wide use for the manufacture of tools and weapons [Henderson, 2000; Lambert, 1997]. Obsidian is a naturally occurring glassy material that is found in volcanic regions of the world. Obsidian is predominantly made of silicon dioxide, but also contains aluminium oxide (Al_2O_3), sodium oxide (Na_2O), potassium oxide (K_2O) and calcium oxide (CaO). However, the composition of obsidian will vary depending on the source and the presence of trace elements can also be used to determine the origin.

There are many gemstones used in the creation of precious objects, such as jewellery [O'Donoghue and Joyner, 2003]. A gemstone is a mineral or petrified material that is cut or faceted. An array of minerals produce such precious stones and Table 1.7 lists the structures and properties of some common gemstones. Most gemstones are mined, but there are some stones derived from other sources. Pearl is an iridescent gem produced by molluscs including oysters. Pearls are produced of layers of calcium carbonate in the form of the minerals aragonite and/or calcite. Nacre, also known as mother-of-pearl is a naturally occurring organic-inorganic composite. Mother-of-pearl contains a combination of calcium carbonate and the protein conchiolin and is found in the inner surface of mollusc shells.

1.15 METALS

Metals have been used to produce many historic artefacts such as coins and tools over many thousands of years [Cronyn, 1990; Henderson,

Table 1.7 Properties of common gemstones

Common name	Chemical structure	Properties
Agate	SiO_2	White–Grey, Blue, Orange–Red, Black
Emerald	$Be_3Al_2(SiO_3)_6$ with Cr	Green–Blue colour
Garnet	$A_3B_2(SiO_4)_3$ A = Ca^{2+}, Mg^{2+} or Fe^{2+} B = Al^{3+}, Fe^{3+} or Cr^{3+}	Red and other colours
Jade	Nephrite $Ca_2(Mg,Fe)_5Si_8O_{22}(OH)_{22}$ Jadeite $NaAlSi_2O_6$	Green and other colours
Lapis lazuli	$(Na,Ca)_8(AlSiO_4)_6(S,SO_4,Cl)_{1-2}$	Mineral lazurite with calcite, Sodalite and pyrite, Blue
Opal	$SiO_2.nH_2O$	Iridescent, Range of colours
Ruby	Al_2O_3 with Cr	Red colour
Sapphire	Al_2O_3	Blue, Pink, Yellow, Green or White
Topaz	$Al_2SiO_4(F,OH)_2$	Translucent, Yellow with other colours due to impurities
Turquoise	$CuAl_6(PO_4)_4(OH)_8.5H_2O$	Blue–green colour, Opaque

2000; Lambert, 1997]. Originally metal was used as found, but later it was recognised that metal containing ores could be mined and the metal isolated. It was also recognised that with heating, the physical properties of metals could be controlled and the metal worked into various shapes.

Metals form crystalline solids due to the regular packing of the atoms. When describing metallic structures, the repeating structures are defined using unit cells. Each unit contains a specific number of atoms in a particular pattern depending upon the element. Metals are made of crystals known as grains and the physical properties are affected by the size and shape of such grains. While pure metals are used, it is also common to find alloys, which are mixtures of metals and other elements.

The ability to deform metals has enabled them to be used in a wide range of applications. Metals can also be joined mechanically as well as by soldering, which involves the formation of an alloy between the surface of the metal and another metal known as the solder. Metal surfaces are sometimes coated with a thin layer of a different metal in order to change the appearance or to protect against corrosion.

Metals are susceptible to deterioration. The good electrical conductivity of metals also means that they can undergo oxidation-reduction reactions. Corrosion can occur resulting in tarnishing, the formation of patinas or the significant disintegration of the surface. A layer of oxide or sulfide can form on the surface and may form a protective layer. More damaging is corrosion in a wet environment. A combination of water and oxygen cause severe deterioration of particular metals. The nature of corrosion products depends on the potential reactants in the metal and the environment and crystals are usually formed on the surface.

Gold is an attractive malleable metal that is highly prized due to its comparatively low abundance. The yellow lustrous qualities make it popular for use in jewellery and coins. Gold is also used in the form of gold leaf in picture frames and in illuminated manuscripts. The metal is also found in alloys with silver and copper to improve the mechanical properties. Gold is very resistant to corrosion.

The appearance of silver and its relatively low abundance make this a valuable metal used for jewellery, coins and cutlery. Sterling silver (92.5 % silver/7.5 % copper) is commonly used to produce jewellery and cutlery. Silver plate involves the deposition of a thin layer of silver on a base metal (e.g. copper, brass). Silver objects tarnish due to the formation of a black silver sulfide (Ag_2S) layer. Buried objects or items removed from the sea may be coated with grey silver chloride (AgCl) and copper corrosion products.

Copper is a malleable and lustrous red metal that is commonly used in alloys. Copper with zinc is brass and bronze is copper with tin. Some common corrosion products are copper oxides, carbonates and sulfates, but they can form a protective layer. However, copper corrosion products can breakdown in the presence of chlorides from sea water or ground water. The formation of light blue–green growths due to the reaction with chlorides in a humid environment is referred to as bronze disease: the surface deposits crumble and a pitted surface remains.

Iron is an abundant metal that is found in many museum collections. This metal is used in a variety of forms: cast iron (2.5–5 % carbon), wrought iron (0–0.07 % carbon) and steel (0.07–0.9 % carbon). Iron may also be plated with zinc. Iron and most of its alloys are susceptible to rust formation by oxygen in the presence of moisture. Red–brown iron oxides form that do not provide a protective layer and, in fact, accelerate corrosion.

Lead is a very malleable metal and was used to produce pewter by combining with tin prior to the 19th century. The major corrosion products of lead are white–grey lead carbonates which provide a protective layer. A grey–black patina of lead sulfide (PbS) may also be observed. Acetic acid produced by wood can react with lead to form lead acetates.

REFERENCES

N.S. Allen and M. Edge, *Fundamentals of Polymer Degradation and Stabilisation*, Elsevier, London (1992).

J.H. Bowes, R.G. Elliot and J.A. Moss, The composition of collagen and acid-soluble collagen of bovine skin, *Biochemical Journal*, **61** (1955), 143–150.

R.L. Brunelle and R. Reed, *Forensic Examination of Ink and Paper*, Charles C. Thomas Publishing, Springfield (1984).

H.A. Carter, The chemistry of paper preservation Part 4. Alkaline paper, *Journal of Chemical Education*, **74** (1997), 508–511.

Colour Index, Society of Dyers and Colourists, London, and The American Association of Textile Chemists and Colourists, 4th Online ed (2002).

A.D. Covington, G. Lampard and M. Pennington, Nothing to hide: the chemistry of leathermaking, *Chemistry in Britain*, **34** (1998), 40–43.

T.E. Creighton, *Proteins: Structures and Molecular Properties*, W.H. Freeman, New York (1993).

J.M. Cronyn, *The Elements of Archaeological Conservation*, Routledge, New York (1990).

B.G. Davis and A.J. Fairbanks, *Carbohydrate Chemistry*, Oxford University Press, Oxford (2002).

S. Davison, *Conservation and Restoration of Glass*, 2nd ed, Butterworth–Heinemann, Oxford (2003).

M. Doerner, *The Materials of the Artists and Their Use in Painting*, Revised ed, Harcourt, Orlando (1984).

D. Erhardt, Paints based on drying-oil media, in *Painted Wood: History and Conservation* (eds V. Dorge and F.C. Howlett), Getty Conservation Institute, Los Angeles (1998), 17–32.

R.L. Feller (ed), *Artists' Pigments. A Handbook of their History and Characteristics*, Vol. 1, Cambridge University Press, Cambridge (1986).

S. Fenichell, *Plastic: The Making of a Synthetic Century*, Harper Collins, New York (1997).

R.D.B. Fraser, T.P. MacRae and E. Suzuki, Chain conformation in the collagen molecules, *Journal of Molecular Biology*, **129** (1979), 463–481.

R.J. Gettens and G.L. Stout, *Painting Materials: A Short Encyclopedia*, Dover, New York (1966).

F.D. Gunstone, *The Chemistry of Oils and Fats: Sources, Composition, Properties and Uses*, Blackwell, Oxford (2004).

J. Henderson, *The Science and Archaeology of Materials: An Investigation of Inorganic Materials*, Routledge, New York (2000).

R.B. Hoadley, Wood as a physical surface for paint applications, in *Painted Wood: History and Conservation* (eds V. Dorge and F.C. Howlett), Getty Conservation Institute, Los Angeles (1998), 2–16.

D.N.S. Hon, Critical evaluation of mass deacidification processes for book preservation, in *Historic Textile and Paper Materials II*, (eds S.H. Zeronian and H.L. Needles), ACS Symposium Series 410, American Chemical Society, Washington DC (1989), 13–33.

C.V. Horie, *Materials for Conservation: Organic Consolidants, Adhesives and Coatings*, Architectural Press, Oxford (1987).

W. Ingamells, *Colour for Textiles: A User's Handbook*, Society of Dyers and Colourists (1993).

M.L. Joseph, *Introductory Textile Science*, Holt, Reinhardt and Winston, New York (1986).

S. Katz, *Early Plastics*, Shire Publications, Princes Risborough (1994).

S. Keck and T. Peters, Identification of protein-containing paint media by quantitative amino acid analysis, *Studies in Conservation*, **14** (1969), 75–82.

J.B. Lambert, *Traces of the Past: Unravelling the Secrets of Archaeology Through Chemistry*, Perseus Publishing, Cambridge (1997).

T.J.S. Learner, *Analysis of Modern Paint*, The Getty Conservation Institute, Los Angeles (2004).

T. Lister and J. Renshaw, *Conservation Chemistry – An Introduction*, Royal Society of Chemistry, London (2004).

E. Lucas, J.T.B. Shaw and S.G. Smith, The silk fibroins, *Advances in Protein Chemistry*, **13** (1958), 107–242.

R.E. Marsh, R.B. Corey and L. Pauling, An investigation of the structure of silk fibroin, *Biochimica et Biophysica Acta*, **16** (1955), 1–34.

R. Mayer, *Artists' Handbook of Materials and Techniques*, 5th ed, Faber and Faber, London (1991).

I.C. McNeill, Fundamental aspects of polymer degradation, in *Polymers in Conservation* (eds N.S. Allen, M. Edge and C.V. Horie), Royal Society of Chemistry, Cambridge (1992), 14–31.

J.S. Mills and R. White, Natural resins of art and archaeology: their sources, chemistry and identification, *Studies in Conservation*, **22** (1977), 12–31.

J.S. Mills and R. White, *The Organic Chemistry of Museum Objects*, 2nd ed, Butterworth–Heinemann, Oxford (1994).

S. Mossman (ed.), *Early Plastics: Perspectives 1850–1950*, Leicester University Press, London (1997).

H. Muller, *Jet Jewellery and Ornaments*, Shire Publications, Princes Risborough (2003).

H.L. Needles, *Textile Fibres, Dyes, Finishes and Processes: A Concise Guide*, Noyes Publications, Park Ridge (1986).

R. Newman, Tempera and other non-drying-oil media, in *Painted Wood: History and Conservation* (eds V. Dorge and F.C. Howlett), Getty Conservation Institute, Los Angeles (1998), 33–63.

M. O'Donoghue and L. Joyner, *Identification of Gemstones*, Butterworth–Heinemann, Oxford (2003).

T.A. Osswald, *Polymer Processing Fundamentals*, Hanser, New York (1998).

A.M. Pollard and C. Heron, *Archaeological Chemistry*, Royal Society of Chemistry, Cambridge (1996).

A. Quye and C. Williamson (eds), *Plastics: Collecting and Conserving*, NMS Publishing, Edinburgh (1999).

A. Roy (ed), *Artists' Pigments. A Handbook of their History and Characteristics*, Vol. 2, Oxford University Press, Oxford (1993).

G.G. Sward (ed), *Paint Testing Manual: Physical and Chemical Examination of Paints, Varnishes, Lacquers and Colours*, 13th ed, American Society for Testing and Materials, Philadelphia (1972).

A. Timar-Balazsy and D. Eastop, *Chemical Principles of Textile Conservation*, Butterworth–Heinemann, Oxford (1998).

E.R. Trotman, *Dyeing and Chemical Technology of Textile Fibres*, 6th ed, John Wiley & Sons, Inc., New York (1984).

G.P.A. Turner, *Introduction to Paint Chemistry and Principles of Paint Technology*, 2nd ed, Chapman and Hall, London (1980).

P. Vandenbeele and L. Moens, Pigment identification in illuminated manuscripts, in *Non-Destructive Microanalysis of Cultural Heritage Materials* (eds S.K. Janssens and R. van Grieken), Elsevier, Amsterdam (2004), 635–662.

H.W. Ward and H.P. Lundgren, The formation, composition and properties of keratin, *Advances in Protein Chemistry*, 9 (1954), 242–297.

K. Wehlte, *The Materials and Techniques of Paintings*, Van Nostrand Reinhold, New York (1975).

E. West Fitzhugh (ed), *Artists' Pigments. A Handbook of their History and Characteristics*, 3, Oxford University Press, Oxford (1997).

2

Basic Identification Techniques

2.1 INTRODUCTION

Before embarking on more sophisticated methods of analysis, it is
often useful for the conservation scientist to carry out some basic
tests to identify the materials of interest. This approach means that
a preliminary identification may be made and then a decision about
more detailed analyses taken. An obvious first step is to use visual
examination. There are also many simple chemical tests that require
only small quantities of sample which may be used to characterise a
material of interest. Density and specific gravity are commonly used
as a means of differentiating materials. Solubility tests can be used
to characterise organic substances. In addition, heating tests including
flame tests, pyrolysis tests and melting temperature can be used to
identify materials. A choice of one or a combination of these tests can
provide a useful start to the characterisation of materials.

2.2 VISUAL EXAMINATION

An initial visual examination of a material may provide useful information before embarking on more detailed experiments. Apart from
important identifying marks on an object, an examination with a magnifying glass can provide information regarding colour, surface finish,
degradation and production method [Quye and Williamson, 1999].

Different types of lighting an object can also assist a visual examination [Ianna, 2001]. Standard lighting from the front of an object
provides information regarding colour, opacity and gloss. Light from

Analytical Techniques in Materials Conservation   Barbara H. Stuart
© 2007 John Wiley & Sons, Ltd

the side (raking light) reveals information about texture, cracking and planar distortion. The use of transmitted light from the back of an object can reveal information regarding tears, mending and watermarks.

2.3 CHEMICAL TESTS

There are numerous simple chemical tests that may be employed to identify materials. Such qualitative analysis can be carried out on various scales, but semi-micro and microanalysis are the most useful for conservation scientists. In these analyses, generally a reagent solution is mixed with the unknown sample and any reaction occurring is observed. In semi-micro analysis about 0.05 g of sample and 1 ml of solution are used, while in microanalysis the corresponding quantities are about 5 mg and 0.1 ml. The tests provide a straightforward preliminary step in the characterisation process as they require basic equipment (such as test tubes and droppers). There are detailed references describing the established tests that may be used to identify metals, organic and inorganic materials [Braun, 1996; Feigl, 1958; Feigl, 1966; Odegaard *et al.*, 2001; Svehla, 1996; Sward, 1972; Vogel, 1978]. An excellent collection of tests is outlined in Odegaard *et al.* (2000).

The measurement of pH also provides a useful test for differentiating materials. The hydrogen ion (H^+) concentration of a sample solution is measured by the pH value. Indicator papers or pH meters may be used and are simple to operate. For pH papers the colour of an acid-base indicator in the paper changes in the presence of an acid or base. The papers can be immersed in a solution or passed over a reaction tube to measure the pH of the evolved gases.

2.3.1 Paintings

A number of useful, simple chemical tests can be used to identify a range of pigments and binders found in paints [Fiegl, 1958; Odegaard *et al.*, 2001]. The following summarises some tests used for pigment and dye identification:

- Iron using potassium ferrocyanide [Odegaard *et al.*, 2001]
  A small quantity of pigment is placed on filter paper or on a spot test plate and a drop of concentrated hydrochloric acid is added. The sample is dried in an oven or under an infrared lamp and a drop of potassium ferrocyanide ($K_4Fe(CN)_6$) solution (1 g in 25 ml water) added. The reaction that takes place is:

  $$4Fe^{3+}(aq) + 3K_4Fe(CN)_6(aq) \rightarrow Fe_4(Fe(CN)_6)_3(s) + 12K^+(aq)$$

$Fe_4(Fe(CN)_6)_3$ (Prussian blue) is a bright blue complex and indicates the presence of iron. If copper ions are present in the sample, they may interfere with the interpretation of this test.

- Lead using Plumbtesmo test papers [Odegaard et al., 2001]
  A piece of Plumbtesmo test paper is wetted with a small quantity of deionised water. A few grains of pigment are placed on the paper. The presence of lead is shown by the appearance of a pink–red colour.

- Copper using Cuprotesmo test papers [Odegaard et al., 2001]
  A piece of Cuprotesmo test paper is wetted and a few grains of pigment placed on the paper. Small spots of pink colour will form around the pigment if copper is present, but magnification may be required.

- Copper using nitric acid and ammonia [Odegaard et al., 2001]
  A small quantity of pigment is placed on damp filter paper and a drop of 8M nitric acid (HNO_3) is added. The paper is held over a concentrated ammonia (NH_3) solution. The formation of a blue colour indicates the presence of copper:

$$Cu^{2+}(aq) + 6\ NH_3(g) \rightarrow [Cu(NH_3)_6]^{2+}(aq)\ (blue)$$

- Zinc using diphenylthiocarbazone [Odegaard et al., 2001]
  A drop of 1M sodium hydroxide solution is placed on a piece of filter paper and a small amount of pigment is placed on the paper. After two minutes a drop of diphenylthiocarbazone solution (0.01 g in 100 ml dichloromethane (CH_2Cl_2)) is added. If zinc is present a pink–red colour will appear after two minutes.

- Mercury using aqua regia [Odegaard et al., 2001; Schramm, 1995]
  A small amount of pigment is placed on a watch glass and several drops of aqua regia (distilled water / nitric acid / hydrochloric acid 1:1:3) are added. The sample is dried in an oven or under a lamp and a drop of distilled water is then added to the residue. A piece of aluminium foil is abraded using emery cloth, and a drop of 10 wt- % sodium hydroxide is added. The foil is dried with blotting paper. One drop of the sample solution is added to the abraded aluminium surface. If mercury is present, then the aluminium will rapidly corrode.

- Indigo using sodium hydrogen sulfite [Odegaard et al., 2001; Hofenk de Graaff, 1974]

Add 3–5 drops of a solution of 2.5 g sodium hydroxide and 2.5 g sodium hydrogen sulfite ($NaHSO_3$) in 50 ml distilled water to dye in solution in a test tube. When the dye turns yellow, five drops of ethyl acetate are added. The sample is shaken and will separate into two layers. If indigo is present, the upper layer will turn blue.

- Barium sulfate using sodium rhodizonate [Fiegl, 1958]
  A few grams of pigment are mixed with 0.5 g ammonium chloride in a micro crucible and heated over a flame until the fumes no longer appear. The residue is transferred to a test tube and several drops of 0.2 % sodium rhodizonate solution are added. If barium sulfate ($BaSO_4$) is present, a red-brown precipitate forms.

- Chromium using diphenylcarbazide [Odegaard et al., 2001]
  One drop of concentrated nitric acid is placed on the sample on a piece of filter paper. After about one minute, a drop of diphenylcarbazide solution (0.01 g in 10 ml ethanol) is added to the sample. The presence of chromium is indicated by the formation of a violet colour.

The variety of binders produced from natural materials can be identified using simple tests. Some examples are:

- Protein using calcium oxide [Fiegl 1966; Odegaard et al., 2001]
  A micro spatula quantity of calcium oxide is added to a small quantity of sample in a test tube. The tube is heated over a flame and a pH paper is held at the opening. If the pH paper indicates a higher pH (green to violet colour), a protein is present (ammonia is formed).

- Protein using copper sulfate (Buiret test) [Odegaard et al., 2001]
  One drop of 2 wt-% copper sulfate ($CuSO_4$) solution is added to a small quantity of sample on a spot test plate. After several minutes the sample will absorb solution and turn slightly blue. Excess solution is removed using blotting paper and a drop of 5 wt-% sodium hydroxide is added. Protein is present if a purple colour appears. It may take up to an hour for the colour to appear and observation using magnification may be required.

- Starch using iodine/potassium iodide (Lugol's test) [Odegaard et al., 2001]
  A small quantity of sample is ground into a fine powder using a mortar and pestle. A drop of I_2/KI solution (2.6 g potassium iodide and 0.13 g iodine in 5 ml distilled water) is added to the powder

on a spot-test plate. The presence of starch is indicated by the appearance of a dark blue colour.

- Simple carbohydrates using o-toluidine [Odegaard *et al.*, 2001; Stulik and Florsheim, 1992]
  Place 5 mg of finely ground sample in a vial containing 0.5 ml distilled water. The solution is boiled for 1–2 minutes then allowed to cool to room temperature. The solution is microcentrifuged for one minute. Three drops of supernatant is added to o-toluidine reagent and heated for 10 minutes in a water bath. The presence of simple sugars is indicated by the appearance of a blue–green colour.

- Complex carbohydrates using o-toluidine [Odegaard *et al.*, 2001; Stulik and Florsheim, 1992]
  5 mg of sample in 0.5 ml 0.5 M sulfuric acid is heated in a 100 °C oven for two hours. The sample is microcentrifuged for one minute. One or two drops of 7.5 M ammonium hydroxide are added to neutralise the acid in the supernatant. 0.5 ml o-toluidine reagent is added to the supernatant and boiled in a water bath for 10 minutes. The appearance of a blue-green colour indicates the presence of carbohydrates. Certain pigments such as red ochre, yellow ochre and manganese dioxide can give false negatives.

- Carbohydrates using triphenyltetrazolium chloride [Odegaard *et al.*, 2001]
  One drop of concentrated hydrochloric acid is added to a small quantity of sample on a spot-test plate and allowed to evaporate in an oven or under a lamp. Two drops of distilled water are added to dissolve the sample, which is then transferred to a test-tube. Two drops of 10 wt-% triphenyltetrazolium chloride and one drop of 1.2 M sodium hydroxide are added and boiled for 1–2 minutes in a water bath. The presence of carbohydrate is indicated by the formation of a red colour.

- Triglycerides using triglyceride reagent [Odegaard *et al.*, 2001; Stulik and Florsheim, 1992]
  A small quantity of sample is placed on filter paper and 1–2 drops of triglyceride reagent is added. A pink–purple colour indicates that oils or fats are present. The reagent separates the glycerine in the triglycerides and produces characteristic coloured species.

- Unsaturated oils using potassium permanganate [Odegaard *et al.*, 2001].

A few drops of sample are placed in a test tube, or, if the sample is solid, it is dissolved in ethanol. 1 wt-% potassium permanganate (KMnO$_4$) is added dropwise with shaking. If an unsaturated oil is present, a brown precipitate will form within a minute.

2.3.2 Written Material

Simple chemical tests can be employed to make a preliminary investigation of the materials used to produce written documents. There are tests that can identify the components of materials, such as paper, and also identify treatments that may have been carried out on such substrates. Some examples follow:

- Cellulose using aniline acetate [Browning, 1969; Odegaard *et al.*, 2001]
  One drop of concentrated phosphoric acid (H$_3$PO$_4$) is added to a small quantity of sample placed in a test tube. The test tube is covered with a piece of filter paper and a drop of aniline acetate solution (1:1 glacial acetic acid and distilled water added to aniline) is placed on the paper. The tube is heated until the sample is carbonised. If cellulose is present, a pink colour is observed.

- Cellulose and its derivatives using 1-naphthol (Molisch test) [Braun, 1996; Odegaard *et al.*, 2001]
  Several drops of 2 wt-% 1-naphthol in ethanol are added to a small quantity of sample dissolved in acetone on a microscope slide. 1–2 drops of concentrated sulfuric acid are placed next to the sample. At the boundary a red–brown colour appears after 10–15 minutes if cellulose is present. A green colour appears if cellulose nitrate is present and lignin and other sugars can cause a brown or black colour.

- Lignin using phloroglucinol [Browning, 1969; Odegaard *et al.*, 2001]
  One drop of 5 wt-% phloroglucinol solution (0.25 g phloroglucinol in 15 ml methanol, 15 ml distilled water and 15 ml concentrated hydrochloric acid) is added to a small quantity of sample. If lignin is present, a red–violet colour will appear after a minute.

- Starch using iodine/potassium iodide (Lugol's test) [Browning, 1969; Odegaard *et al.*, 2001]
  The test is similar to that for paint binders, but instead of grinding to a powder, a drop of reagent is placed on the paper or a few fibres from a paper sample.

- Protein using copper sulfate (Buiret test) [Browning, 1969; Odegaard *et al.*, 2001]
  The same test that was described for paint binders can be used for identifying protein in materials such as vellum.

- Aluminium (Al^{3+}) ions using aluminon [Feigl, 1958; Odegaard *et al.*, 2001]
  A drop of aluminon solution (0.1 g aurintricarboxylic acid in 100 ml distilled water) is placed on a small piece of sample or on an inconspicuous part of an object. The formation of a red-pink colour indicates the presence of Al^{3+}, and hence, the presence of alum sizing on paper.

Inks can be identified through a number of simple chemical tests [Feigl, 1958]. For instance, iron, chromium chloride and sulfate in inks can be detected. A useful test for iron in ink is:

- Iron using α,α'-dipyridyl in thioglycolic acid.
  One drop of 2 wt-% α,α'-dipyridyl in thioglycolic acid is placed on ink writing. The appearance of a pink or red colour indicates the presence of iron, the colour depending on the quantity present. The reagent reduces Fe^{3+} to Fe^{2+} during the reaction.

2.3.3 Natural Materials

There is an extensive list of simple chemical tests that may be used to indicate the presence of a wide range of natural materials used in heritage objects [Feigl, 1966; Odegaard *et al.*, 2001]. There are several useful tests that can be used to test for proteins in objects, including those already described for identifying proteins in paint binders:

- Protein using calcium oxide [Feigl, 1966; Odegaard *et al.*, 2001]
  The same method described for paint binders can be used for materials such as adhesives.

- Protein using copper sulfate (Buiret test) [Odegaard *et al.*, 2001]
  The same method as that described for paint binders can be utilised for materials such as leather, horn and tortoiseshell. It is possible that such materials may need to be soaked in solution for an hour before the sodium hydroxide is added.

- Sulfur using lead acetate [Odegaard *et al.*, 2001]

This method can be used to indicate the presence of sulfur in proteins present in natural materials such as hide glues and leather. A small piece of sample is introduced into the tapered section of a Pasteur pipette that has had its tip sealed with a flame. A 2- cm strip of lead acetate test paper wetted with distilled water is placed at the opening of the pipette. The pipette is heated and fumes are allowed to reach the test paper. The observation of a brown colour indicates the presence of sulfur.

There are a number of spot tests available for the characterisation of treatments used for leather [Feigl, 1958; Feigl, 1966; Odegaard *et al.*, 2001; Reed, 1972]. A variety of tanning agents are used for leather and can be identified using such tests:

- Phenols in vegetable-tanned leather using lead acetate [Odegaard *et al.*, 2001]
  A 0.01–0.02 g sample is placed in a test tube with 2 ml 50 vol- % aqueous acetone solution for four hours. The extract is filtered and 1 ml of 2 wt- % lead acetate is added to 1 ml of the extract. The appearance of a light brown precipitate indicates the presence of vegetables tannins (phenolic hydroxyl groups in tannins form salt complexes with lead).

- Vegetable-tanned leather using iron sulfate [Odegaard *et al.*, 2001]
  A small quantity of sample is placed on a glass slide and a drop of 2 wt- % iron sulfate ($FeSO_4$) solution is placed on the sample. The appearance of a blue–green colour indicates the presence of phenols and, thus, the presence of vegetable tannins.

Other tests including a test for aluminium (Al^{3+}) ions using aluminon (described for studying written material) and a test for chromium using diphenylcarbazide (used for studying pigments) can be used to identify tanning agents.

There are some simple tests that may be used to identify natural resins [Feigl, 1966; Odegaard *et al.*, 2001]. For example:

- Rosin using sulfuric acid (Raspail test) [Odegaard *et al.*, 2001]
  A drop of a saturated sugar solution (35–40 g sugar in 25 ml distilled water) is added to a small sample on a spot test plate. After 10 seconds the excess solution is removed with blotter paper. One drop of concentrated sulfuric acid is added and the appearance of a red colour after 1 minute indicates the presence of rosin.

The appearance of wood components can be carried out using straightforward chemical tests [Feigl, 1958; Odegaard *et al.*, 2001]. Some examples are:

- Lignin using phloroglucinol [Browning, 1969; Odegaard *et al.*, 2001]
  The same method as that used for paper is applicable.

- Cellulose using 1-napthol (Molisch test) [Odegaard *et al.*, 2001]
  The same method is used as that described for cellulosic polymers in the next section.

2.3.4 Synthetic Polymers

There is a range of specific simple chemical tests for the identification of polymers [Braun, 1996; Feigl, 1966; Odegaard *et al.*, 2001].

- Cellulose using aniline acetate [Browning, 1969; Odegaard *et al.*, 2001]
  The same method as that used for written material can be used for plastic materials. This test can be used to detect the presence of cellulose in cellulose acetate or cellulose nitrate. Cellulose nitrate forms a yellow colour in the test.

- Cellulose and its derivatives using 1-naphthol (Molisch test) [Braun, 1996; Odegaard *et al.*, 2001]
  The same method as that described for written material can be used for polymers including cellulose acetate, cellulose nitrate and regenerated cellulose.

- Cellulose nitrate using diphenylamine [Braun, 1996; Odegaard *et al.*, 2001].
  One drop of 20 mg diphenylamine in 1 ml concentrated sulfuric acid is added to a small quantity of sample on a spot-test plate. The appearance of a dark blue colour indicates that cellulose nitrate is present.

- PVAl using iodine/potassium iodide [Braun, 1996; Odegaard *et al.*, 2001]
  Two drops of acetone are added to a small quantity of sample on a spot-test plate and allowed to evaporate. Add one drop of 7 parts of a solution of 1 g of potassium iodide and 0.9 g of iodine (both dissolved in 2 g glycerol) in 40 ml distilled water to 10 parts of 1:1 glacial acetic acid and distilled water. If a colour appears, the sample is PVA or PVAl.

- Polyesters using hydroxylamine hydrochloride [Odegaard *et al.*, 2001]

  Ten drops of 10 wt-% potassium hydroxide in methanol and six drops of 10 wt-% hydroxylamine hydrochloride in methanol are added to a small quantity of sample in a test tube. The test tube is heated in a water bath at 50 °C for two minutes. After removal from the water bath, 10 drops of 4M hydrochloric acid and one drop of saturated iron [III] chloride solution (about 1.0 g iron [III] chloride in 1 ml water) are added to the sample. The appearance of a brown–violet colour indicates ester groups. Some epoxies and ester plasticisers in PVC will produce colour.

- Polyamides and polycarbonates using *p*-dimethylaminobenzaldehyde [Odegaard *et al.*, 2001]

  A small piece of sample is introduced into the tapered section of a Pasteur pipette that has the tip sealed with a flame. A 2 cm strip of filter paper is folded and inserted at the opening of the open end. Two drops of 14 wt-% *p*-dimethylaminobenzaldehyde solution and one drop of concentrated hydrochloric acid are added. The open end of the pipette is covered with wrapping film and the pipette is heated. The smoke produced causes a red colour in the filter paper if the sample is a polyamide. The appearance of blue colour indicates that the sample is polycarbonate.

- Sulfur using lead acetate paper [Odegaard *et al.*, 2001]

  The same method as described for natural materials can be used to determine if an object is made of natural rubber or ebonite.

- Epoxy resins using sulfuric acid, nitric acid and sodium hydroxide [Braun, 1996]

  About 100 mg of resin is dissolved in 10 ml concentrated sulfuric acid in a test tube and 1 ml of concentrated nitric acid is then added. After five minutes, several drops of 5 wt-% sodium hydroxide are carefully added to the top of the solution. The appearance of a red colour at the interface indicates the presence of bisphenol A due to an epoxy resin.

- Urea resins using phenylhydrazine [Feigl, 1966]

  A few mg of sample and a drop of concentrated hydrochloric acid are placed in a micro test tube and dried in an oven at 110 °C until dry. On cooling, one drop of phenylhydrazine is added and the sample is heated to 195 °C for five minutes. On cooling, three drops of 50 vol-% ammonia and 5 drops of 10 wt-% nickel sulfate are added and then shaken with 10 drops of trichloromethane ($CHCl_3$). The appearance of a red–violet colour indicates that urea is present.

- Formaldehyde resins using sulfuric acid [Braun, 1996; Feigl, 1966] 2 ml of concentrated sulfuric acid and several crystals of chromotropic acid are added to a small quantity of sample and heated for 10 minutes at 60–70 °C. The formation of a violet colour indicates the presence of formaldehyde.

2.3.5 Textiles

A range of simple chemical tests is available for the identification of the fibres and dyes used in textiles [Braun, 1996; Feigl 1966; Odegaard et al., 2001]. Tests are available for both natural and synthetic fibres. The following are some tests that can be used to characterise fibres:

- Differentiation of wool and cotton using iodine–azide [Feigl, 1966] The sample fibre is moistened with water on a watchglass and 1–2 drops of iodine–azide solution (3g sodium azide (NaN_3) in 100 ml 0.05M iodine solution) is placed on the fibre. If sulfides are present, small bubbles due to nitrogen gas will appear as the sulfides catalyse the reaction between sodium azide and iodine:

$$2NaN_3(s) + I_2(aq) \rightarrow 2NaI(s) + 2N_2(g)$$

Wool will be covered in small bubbles, while cotton shows none.

- Sulfur using calcium oxalate [Odegaard et al., 2001] This test can be used to identify sulfur in fibres such as wool or silk. A microspatula quantity of calcium oxalate (CaC_2O_4) is added to a small amount of sample in a test tube. A small piece of lead acetate paper is folded and inserted at the top of the test tube, which is heated in a flame until the sample burns. If sulfur is present, the paper will turn brown. The sulfur compounds in the fibres are decomposed to hydrogen sulfide (H_2S) on heating in the presence of calcium oxalate. The hydrogen sulfide reacts with lead acetate paper to form a brown sulfide:

$$H_2S(g) + Pb^{2+}(aq) \rightarrow PbS(s) + 2H^+(aq)$$

- Cellulose and its derivatives using 1-napthol (Molisch test) [Braun, 1996; Odegaard et al., 2001]

This test may be employed to identify cellulose in fibres such as cotton, linen, viscose rayon and cellulose acetate. The same method as that described for written material can be applied to fibres.

- Polyester groups using hydroxylamine hydrochloride [Odegaard *et al.*, 2001]
  This method can be used to identify polyester fibres. The same method as that described for polymeric materials can be employed.

Certain dyes used to colour textiles may be detected using simple chemical tests [Feigl, 1966]. Some examples are:

- Azo dyes using *p*-dimethylaminobenzaldehyde [Feigl, 1966]
  The sample is treated with a drop of 1M hydrochloric acid and a piece of zinc in a micro test tube. After five minutes the solution is placed on filter paper then dried in a 110 °C oven. A drop of *p*-dimethylaminobenzaldehyde in benzene is placed on the paper. The appearance of yellow–red colour indicates an azo dye is present.

- Anthraquinone dyes using sodium hydrogen sulfite [Feigl, 1966]
  A small quantity of sample is placed in a micro test tube and one drop of 5M sodium hydroxide and a small quantity of sodium hydrogen sulfite are added. The appearance of a red colour indicates the presence of anthraquinone.

- Distinction between acid and basic dyes [Feigl, 1966]
  Urea and stearic acid are melted in a test tube and a small quantity of sample is added. The tube is shaken and two layers are formed. If the lower urea layer is coloured, an acid dye is present. If the upper stearic acid layer is coloured, a basic dye is present.

- Indigo using sodium hydrogen sulfite [Odegaard *et al.*, 2001; Hofenk de Graaff, 1974]
  The same method as that described for pigments can be used to identify indigo in blue textiles.

2.3.6 Stone

A number of chemical tests are available for identifying the inorganic species present in stone structures, as well as for identifying decomposition products.

- Carbonate (CO_3^{2-}) ions using hydrochloric acid and barium hydroxide [Odegaard *et al.*, 2001; Sorum, 1960]

This method is useful for detecting calcium carbonate on stone surfaces. A small quantity of sample is placed in a test tube and 2–3 drops of 6M hydrochloric acid. The production of effervescence indicates the presence of CO_3^{2-}:

$$CO_3^{2-}(s) + 2H^+(aq) \rightarrow CO_2(g) + H_2O(l)$$

To confirm the production of carbon dioxide, a dropper containing several drops of 3 wt-% barium hydroxide ($Ba(OH)_2$) is held over the effervescence. If the barium hydroxide turns cloudy, the gas is carbon dioxide:

$$CO_2(g) + Ba(OH)_2(aq) \rightarrow BaCO_3(s) + H_2O(l)$$

- Calcium (Ca^{2+}) ions using nitric acid and sulfuric acid [Odegaard *et al.*, 2001; Schramm, 1995]
  5–10 drops of 0.5 M nitric acid are added to a small quantity of sample. A drop of the solution is placed on a glass slide and evaporated in an oven. A drop of 2 M sulfuric acid is added to the sample. Magnification is used to observe the formation of gypsum crystals:

$$Ca^{2+}(aq) + SO_4^{2-}(aq) \rightarrow CaSO_4.2H_2O(s)$$

- Sulfate (SO_4^{2-}) ions using sulfate test paper [Odegaard *et al.*, 2001]
  1–2 drops of distilled water are placed on a hidden region of the object and a sulfate test paper is held in contact for 2–5 seconds. The appearance of a yellow colour indicates that SO_4^{2-} is present.

- Sulfate (SO_4^{2-}) ions using barium chloride [Odegaard *et al.*, 2001; Sorum, 1960]
  Several drops of distilled water followed by several drops of 3M hydrochloric acid are placed on a small sample on a spot test plate. pH indicator paper is used to ensure an acidic solution. Two drops of 5 wt-% barium chloride ($BaCl_2$) are added. The appearance of a white precipitate after a few minutes indicates that SO_4^{2-} is present.

- Nitrate (NO_3^-) ions using iron sulfate [Odegaard *et al.*, 2001]
  A drop of concentrated sulfuric acid is added to a small quantity of sample in a test tube. A drop of iron sulfate solution (6g $FeSO_4.7H_2O$ in 25 ml distilled water) is placed on top. The appearance of a brown colour indicates that NO_3^- is present due to the formation of $Fe(NO)^{2+}$:

$$NO_3^-(aq) + Fe^{2+}(aq) + 4H^+(aq) \rightarrow Fe^{3+}(aq) + NO(g) + 2H_2O(l)$$

$$NO(g) + \text{excess } Fe^{2+}(aq) \rightarrow Fe(NO)^{2+}(aq)$$

2.3.7 Ceramics

A number of spot tests exist that can be used to examine the surfaces of ceramics, particularly deposits [Odegaard *et al.*, 2001].

- Carbonate $(CO_3{}^{2-})$ ions using hydrochloric acid and barium hydroxide [Odegaard *et al.*, 2001; Sorum, 1960]
  The same method as that described for stones can be utilised for ceramics.

- Chlorine (Cl^-) ions using silver nitrate [Odegaard *et al.*, 2001; Sorum, 1960]
  A small quantity of sample is placed in a test tube. 5–8 drops distilled water, 2 drops 8M nitric acid and one drop of 0.2M silver nitrate are added to the sample. The formation of a white precipitate indicates the presence of Cl^-:

$$Ag^+(aq) + Cl^-(aq) \rightarrow AgCl(s)$$

- Phosphate (PO_4^{3-}) ions using ammonium molybdate and ascorbic acid [Odegaard *et al.*, 2001]
  A small sample is placed on filter paper elevated on the top of a beaker. Two drops of ammonium molybdate solution (2.5 g ammonium molybdate in 50 ml distilled water with 115 ml 3M hydrochloric acid added). After 30 seconds, two drops of 0.5 wt-% ascorbic acid are added to the sample. The appearance of a blue colour indicates the presence of $PO_4{}^{3-}$.

- Lead using Plumbtesmo test papers [Odegaard *et al.*, 2001]
  A similar method to that used to identify pigments can be used, but the test can be performed directly on the object. A piece of Plumbtesmo test paper wetted with a drop of deionised water is held with tweezers on the surface of the object.

2.3.8 Glass

Many chemical tests used for identifying glasses have been superceded by other analytical techniques. As the tests often involve etching with hydrofluoric acid (HF) or fusion with sodium carbonate ($NaCO_3$) they can be quite destructive [Davison, 2003; Feigl, 1958]. However, non-destructive tests such as the test for lead using Plumbtesmo test papers described for identifying lead in ceramics, have been successfully applied to the examination of leaded glass [Odegaard *et al.*, 2001].

2.3.9 Metals

There is an extensive range of simple chemical tests available for identifying the presence of particular metals in objects [Feigl, 1958; Odegaard et al., 2001]. Some examples are provided by the following:

- Antimony using antimony test papers [Odegaard et al., 2001]
  A hidden region of the object studied is cleaned with acetone. An antimony test paper that has been dipped in 0.1M hydrochloric acid is placed on the object for 1–2 minutes. The appearance of a pink colour indicates that antimony is present. The test area should be cleaned with distilled water following the test.

- Arsenic or phosphorus in copper alloys using iron chloride [Odegaard et al., 2001]
  One drop of iron chloride solution (1g iron chloride in 1 ml hydrochloric acid added to 9 ml distilled water) is placed on a hidden region of the object studied. The appearance of a dark spot indicates that arsenic or phosphorous is present. After 30 seconds the region is rinsed with distilled water and allowed to dry.

- Copper using Cuprotesmo test papers [Odegaard et al., 2001]
  A similar test to that described for pigments is used, but the test can be carried out directly on a metallic object. A Cuprotesmo test paper is wetted with deionised water and applied to the object and the appearance of a pink–purple colour on the paper indicates that copper is present.

- Copper using nitric acid and ammonia [Odegaard et al., 2001]
  A similar method to that described for pigments can be used, but the test is carried out directly on the metallic object. A drop of 8M nitric acid is placed on a hidden region of the object. The resulting dissolution at the surface is touched with filter paper and held over concentrated ammonia solution. The appearance of a blue–green colour indicates the presence of copper.

- Copper using rubeanic acid [Odegaard et al., 2001; Schramm, 1995]
  This test can be carried out on a hidden surface of a metal object or on a small sample removed from the object. Two drops of 3M hydrochloric acid are added to a small quantity of sample on a spot test plate. Filter paper is dipped into the solution and held over concentrated ammonium hydroxide. When no more fumes are formed, a drop of rubeanic acid is placed on the paper. If a green colour appears, copper is present.

- Gold quality using nitric acid [Odegaard *et al.*, 2001]
  A small piece of filter paper is dipped into a drop of concentrated nitric acid on a spot test plate. The paper is placed on a hidden region of the object. The appearance of colour is observed after about one minute and the acid washed with distilled water. Gold alloys of greater that 18 karat (k) (75 % gold) will not be affected. Alloys of 9–14 k (37.5–58.3 % gold) turn brown and alloys of less than 9 k turn black. Copper alloys will turn the filter paper green.

- Iron using potassium ferrocyanide [Odegaard *et al.*, 2001]
  A similar method to that described for testing pigments can be used to examine metal corrosion products.

- Iron using hydrochloric acid [Odegaard *et al.*, 2001]
  This test can be used to test for iron in a metallic object or in corrosion products. For metallic objects, a drop of concentrated hydrochloric acid is placed on the surface which is then touched with a piece of filter paper. The object is cleaned with distilled water. The appearance of a yellow colour on the paper indicates that iron is present. If corrosion products are being examined, then a small quantity is placed on a spot test plate and 1–2 drops of hydrochloric acid added. A yellow colour forms.

- Lead using Plumbtesmo test papers [Odegaard *et al.*, 2001]
  Corrosion products can be examined using the same method as that described for the investigation of pigments. The tests can also be carried out directly on a metallic object as was described for ceramic objects.

- Lead using potassium dichromate [Odegaard *et al.*, 2001]
  1–2 crystals of potassium dichromate ($K_2Cr_2O_7$) are placed on a hidden region of the object that has been cleaned with acetone. A drop of glacial acetic acid is placed on the crystals and after 1 minute a drop of distilled water is added. The observation of a yellow colour indicates that lead is present:

$$Pb^{2+}(aq) + CrO_4{}^{2-}(aq) \rightarrow PbCrO_4(s)$$

  The object is cleaned with distilled water after examination.

- Nickel using nickel test papers [Odegaard *et al.*, 2001]
  A drop of 2.5M nitric acid is placed on a hidden region of the object. After one minute, a piece of nickel test paper is held at the drop with non-metallic tweezers. If the nickel ion (Ni^{2+}) is present,

a red colour will appear on the paper. The object is cleaned with distilled water after the test.

- Nickel (Ni^{2+}) using dimethylglyoxime [Odegaard *et al.*, 2001; Vogel, 1978]
  A piece of filter paper is held with non-metallic tweezers on a drop of concentrated nitric acid that has been placed on the surface of the object under investigation. The paper is held over concentrated ammonium hydroxide for one minute. A drop of glacial acetic acid then a drop of dimethylglyoxime solution (0.1 g in 10 ml ethanol) are added to the paper. If the nickel ion (Ni^{2+}) is present, a pink–red colour is observed. The test area should be cleaned with distilled water and dried after examination.

- Silver (Ag^+) ions using potassium dichromate [Odegaard *et al.*, 2001]
  A drop of 2M sulfuric acid is placed on 1–2 crystals of potassium dichromate ($K_2Cr_2O_7$) on a hidden part of the artefact under investigation. If a red precipitate is observed after one minute, the silver ion (Ag^+) is present. The test area should be rinsed with distilled water and dried.

- Silver quality using nitric acid [Odegaard *et al.*, 2001]
  A small piece of filter paper is placed on a drop of concentrated nitric acid on a spot test plate. The paper is touched on an inconspicuous region of the artefact under study. The filter paper will show a green–blue colour if the object is silver plate on nickel–silver. Nitric acid dissolves silver plating revealing the metal below which is a copper alloy. Nickel–silver contains no silver, but has enough copper to react with nitric acid. Other alloys with silver will result in a cream colour.

- Tin using sulfuric acid [Odegaard *et al.*, 2001]
  One drop of a reagent of 1 part 6M sulfuric acid and 4 parts of 7.8g sodium hydrogen sulfite in 100 ml with 4 ml concentrated sulfuric acid is placed on the surface of the metal object. If tin is present, a black colour is formed:

$$H_2SO_3(aq) \rightarrow SO_2(g) + H_2O(l)$$

$$Sn^{2+}(aq) + SO_2(g) \rightarrow SnS(s) + O_2(g)$$

The formation of a white precipitate indicates the presence of tin. The test area is cleaned with distilled water and dried as soon as the test is finished.

- Zinc using ammonium mercuric thiocyanate [Odegaard *et al.*, 2001]
  A drop of 3.6M sulfuric acid is placed on an inconspicuous region of the object. When the evolution of gas bubbles has ceased, a drop of liquid is transferred by dropper to a test plate. A drop of ammonium mercuric thiocyanate solution (0.8g mercury chloride ($HgCl_2$) and 0.9g ammonium thiocyanate (NH_4SCN) in 10 ml distilled water) is added to the sample. The formation of a white precipitate indicates the presence of zinc:

$$Zn(s) + H_2SO_4(aq) \rightarrow ZnSO_4(aq) + H_2(g)$$

$$ZnSO_4(aq) + Hg(NH_4)_2(SCN)(aq) \rightarrow ZnHg(SCN)_4(s)$$

- Zinc using diphenylthiocarbazone [Odegaard *et al.*, 2001]
  A similar method to that described for pigments, but the test can be performed directly on the metallic object. A small piece of filter paper that has been dipped in 1M sodium hydroxide is held on the surface of the object that has been cleaned with acetone. The paper is placed on a large piece of filter paper to make several wet spots. Several drops of diphenylthiocarbazone solution (0.01g in 100 ml dichloromethane (CH_2Cl_2)). An orange colour will appear on the paper if zinc is present. The test area should be cleaned with distilled water and dried.

Some spot tests for metals may require the sample be in solution and this can be achieved by the use of electrolysis with minimal damage [Odegaard *et al.*, 2001]. The metal object is connected into an electrical circuit as the positive electrode and reagent treated paper is connected as the negative electrode. The cations produced are transferred to the reagent treated paper.

There are also tests that can identify corrosive gases that can be emitted in the museum environment that may cause damage to metallic artefacts:

- Formic acid using chromotropic acid [Feigl, 1966; Zhang *et al.*, 1994]
  This test can be carried out on materials such as wood to be used for storage. A modified stoppered flask with a 2 mm deep reaction dish suspended from the stopper is used for the test. 2 g of sample is placed in the stoppered flask and 0.2 ml 1 wt- % chromotropic acid in concentrated sulfuric acid is added to the reaction dish. The flask is placed in an oven at 60 °C for 30 minutes. The appearance of a purple colour in the reagent indicates that formic acid is present.

Formic acid is readily reduced to formaldehyde, which reacts with chromotropic acid.

- Volatile organic acids using the iodide-iodate test [Feigl, 1966; Zhang et al., 1994]
  This test uses the same set-up as that described for the chromotropic test. 2 g of sample is placed in the flask and 2 drops each of 2 wt % potassium iodide (KI), 4 wt- % potassium iodate (KIO_3) and 0.1 wt-% starch solution are added to the reaction dish. The flask is placed in a 60°C oven for 30 minutes. The appearance of a blue colour in the test solution indicates the presence of volatile organic acids.

- Sulfides using iodine–azide [Daniels and Ward, 1982; Feigl, 1958]
  A drop of iodine–azide solution (3g sodium azide in 100 ml 0.05 M iodine solution) is placed on a small quantity of sample on a glass slide or watchglass. If sulfides are present, small bubbles due to nitrogen gas will appear.

2.4 DENSITY AND SPECIFIC GRAVITY

The density (ρ) of a substance is the ratio of the mass (m) to the volume (V):

$$\rho = \frac{m}{V} \tag{2.1}$$

Determining this value for an unknown material provides a useful means of identification. Density measurements may be made by exploiting the Archimedes' principle, which states that an object immersed in a fluid is buoyed by a force equal to the weight of a displaced fluid. The volume of the displaced fluid is equal to the volume of the immersed object. It follows that the apparent loss in weight of the immersed object is equal to the weight of an equal volume of the fluid. Water is commonly used as the fluid.

Specific gravity is also a quantity that may be used to characterise a material. Specific gravity is the ratio of the density of a substance to the density of water. This value can be obtained by comparing the weight of the material of interest to the weight of an equal volume of water.

Density and specific gravity measurements can be made using a modified mass balance [Odegaard et al., 2001]. The device consists of a hanging support with a hook that rests on the balance pan, a platform that rests above the balance pan and a foil basket with hanging wires that can be submerged into a beaker of liquid while on the balance pan [Odegaard et al., 2001]. Measurements of the weights of the sample in air and when immersed are made at the same temperature. The density of the sample is then measured using the equation:

$$\rho = \frac{w_1}{(w_1 - w_2)} \times \rho_{liquid} \qquad (2.2)$$

where w_1 is the weight of the sample in air, w_2 is the weight of the sample when immersed in the liquid and ρ_{liquid} is the density of the liquid in which the sample is immersed. Similarly, the specific gravity of a sample (SG) may be determined using the equation:

$$SG = \frac{w_1}{(w_1 - w_2)} \times SG_{liquid} \qquad (2.3)$$

where SG_{liquid} is the specific gravity of the liquid employed.

Water is usually the liquid chosen for making such measurements. However, organic liquids may be used as an alternative if there is concern that the sample of interest is water soluble. It is important to realise that density is dependent upon temperature. However, the density of a liquid can be measured at the specific laboratory temperature using a pycnometer. A pycnometer is a glass vessel with a stopper which has a capillary in its centre to allow the excess liquid to emerge from the vessel when the stopper is put in place. The volume of the vessel level with the stopper in place is known accurately to the level of the top of the capillary in the lid. The mass of the dry pycnometer is subtracted from the mass of the pycnometer containing liquid to obtain the mass of the liquid. The density of the liquid is the mass of the liquid divided by the volume of the pycnometer.

The densities of a range of materials are listed in Table 2.1 [Braun, 1996; Callister, 2006; Lide, 2005]. Specific gravity will generally be very close to the density as the density of water is close to one at normal temperatures. It should be noted that materials containing additives and/or treated to different processing methods will show variation in density and specific gravity. For example, polymers containing fillers will have different values compared to the pure polymer. Also, polymer foams will have significantly lower densities compared to the moulded version of the polymer.

It should also be noted that a simple semi-quantitative method for determining the density, a flotation procedure may be applied to materials such as polymers [Braun, 1996]. This approach involves floating the polymer sample in a liquid of known density. For instance, samples can be immersed in columns of methanol ($\rho = 0.79 \ g\,cm^{-3}$), water ($\rho = 1.00 \ g\,cm^{-3}$), saturated aqueous magnesium chloride ($MgCl_2$) solution ($\rho = 1.34 \ g\,cm^{-3}$) or saturated aqueous zinc chloride ($ZnCl_2$)

Table 2.1 Densities of some common materials at room temperature

Material	Density/g cm^{-3}
Polymers	
ABS	1.04–1.08
Amino resins	1.25–1.52
Cellulose acetate	1.25–1.42
Cellulose nitrate	1.34–1.70
Epoxy resins	1.10–1.40
HDPE	0.94–0.98
Kevlar	1.44
LDPE	0.89–0.93
Natural rubber	0.92–1.00
Nylon 4,6	1.18
Nylon 6	1.12–1.15
Nylon 6,6	1.13–1.16
Nylon 6,10	1.07–1.09
Nylon 11	1.03–1.10
Nylon 12	1.01–1.04
PC	1.20–1.22
Phenolic resins	1.03–1.20
Polyesters	1.10–1.40
PET	1.38–1.41
PMMA	1.16–1.20
PP	0.85–0.92
PS	1.04–1.07
PTFE	2.10–2.30
PU	1.10–1.26
PVA	1.17–1.20
PVAl	1.21–1.32
PVC	1.38–1.41
Silicone rubber	0.80
Viscose	1.31–1.52
Metals	
Ag	10.49
Al	2.71
Au	19.32
Cu	8.94
Cu alloys	7.45–8.94
Cast Fe	7.10–7.30
Fe	7.87
Ni	8.90
Pb	11.35
Pt	21.45
Sn	7.30
Steels	7.80–7.90
Ti	4.51
Zn	7.13
Zn alloys	5.00–7.17

Table 2.1   (*continued*)

Material	Density/g cm^{-3}
Natural materials	
Bone	1.94–2.10
Casein	1.04
Coral	2.60–2.70
Elephant ivory	1.70–1.98
Jet	1.33
Pearls, cultured	>2.17
Pearls, natural	<2.17
Tortoiseshell	1.29
Wood	0.31–0.75
Oils	
Linseed	0.924–0.934
Poppyseed	0.920–0.925
Safflower	0.912–0.930
Tung	0.935–0.940
Walnut	0.903–0.930
Waxes	
Beeswax	0.96–0.98
Carnauba	0.99
Paraffin	0.89
Resins	
Amber	1.05–1.09
Dammar	1.05–1.06
Mastic	1.07
Rosin	1.07–1.08
Sandarac	1.08
Shellac	1.11–1.20
Natural fibres	
Cotton	1.40–1.55
Flax	1.50
Hemp	1.50
Jute	1.50
Silk	1.33
Wool	1.31
Pigments	
Hansa yellow	1.48
van Dyck brown	1.61
Carbon black	1.77–1.79
Prussian blue	1.83
Ivory black	2.29
Ultramarine/Lapis lazuli	2.34
Gypsum	2.36
Calcite	2.70
Ochre/Goethite	2.98–2.99
Lithopone	3.10–4.30
Haematite	3.63–5.08

Table 2.1 (*continued*)

Material	Density/g cm^{-3}
Anatase	3.80–3.90
Rutile	3.80–4.20
Cobalt blue	3.83
Cobalt black	3.83
Titanium white	3.90
Cadmium yellow	4.25
Cadmium red	4.30
Barium white	4.36–4.45
Iron black	4.83
Chromium oxide	5.10
Antimony white	5.60–5.75
Zinc white	5.65–5.78
Zirconium oxide	5.69
Chrome yellow	5.83–5.96
Chrome yellow orange/Chrome yellow deep	6.69–7.04
White lead	6.70–6.80
Red lead	8.73–8.90
Litharge	9.40
Minerals and stones	
Agate	2.6
Amethyst	2.65–2.66
Aquamarine	2.68–2.80
Diamond	3.51
Emerald	2.63–2.91
Feldspar	2.55–2.76
Garnet	3.1–4.3
Jade	2.9–3.1
Obsidian	2.6
Opal	2.0–2.3
Quartz	2.65
Ruby	4.0
Sapphire	3.95–4.03
Topaz	3.4–3.6
Turquoise	2.6–2.9
Ceramics and glasses	
Borosilicate glass	2.23
Glass ceramics	2.40–2.70
Soda–lime glass	2.50

solution ($\rho = 2.01$ g cm^{-3}). The behaviour of the samples is then observed, that is, whether the sample remains on the liquid surface, floats within the liquid or sinks. The behaviour determines whether the polymer has a lower or higher density than the liquid in which it is immersed.

2.5   SOLUBILITY

Solubility tests are useful for differentiating many substances [Braun, 1996; Derrick *et al.*, 1999]. Such tests involve placing the sample, a particle seen with the naked eye or a swab, in a micro test tube with several drops (10–100 μl) of solvent. Larger quantities can be used if available. The sample may be left for several hours as some take longer to dissolve. A useful flowchart for solubility tests for a range of materials is illustrated in Figure 2.1 [Derrick *et al.*, 1999]. This demonstrates a sequence of solvents that may be used to selectively remove one or more components in a sample.

Solubility tests are useful for identifying lipids. Lipids do not dissolve in water, but do dissolve in non-polar solvents such as chloroform, ethanol or ether. For example, in the emulsion test for lipids, a small quantity of the test sample is placed in a test tube with about 4 ml ethanol and shaken. The ethanol solution is decanted into a test tube containing water, leaving any undissolved substances behind. If lipids are dissolved in the ethanol, then they will precipitate in the water and show a cloudy white emulsion.

2.6   HEAT TESTS

Heating a sample can provide a number of preliminary approaches to analysis. Flame tests provide a quick and simple method for identifying the presence of certain elements in a sample [Braun, 1996; Svehla, 1996]. A sample can be held in a flame with a spatula, a pair of tweezers or wetted onto a wire. A low flame of a Bunsen burner is used for such a test. The flame test is based on the appearance of light by an element and metallic ions produce a characteristic colour in the flame. Table 2.2 lists the colours produced by different metals in a flame tests. The use of this test can be limited by interferences or ambiguities. For example, a barium (Ba) flame can mask calcium (Ca), lithium (Li) or strontium (Sr), and potassium (K), rubidium (Rb) and caesium (Cs) produce similar colours in a flame. If such a problem arises, the use of a more sophisticated technique such as atomic absorption spectroscopy or atomic emission spectroscopy (described in Chapter 5) is recommended.

The degradation behaviour of organic materials can be studied using simple pyrolysis tests [Braun, 1996; Katz, 1994; Odegaard *et al.*, 2001; Quye and Williamson, 1999; Stuart, 2002]. Thermal degradation produces low molecular weight fragments which are often flammable or

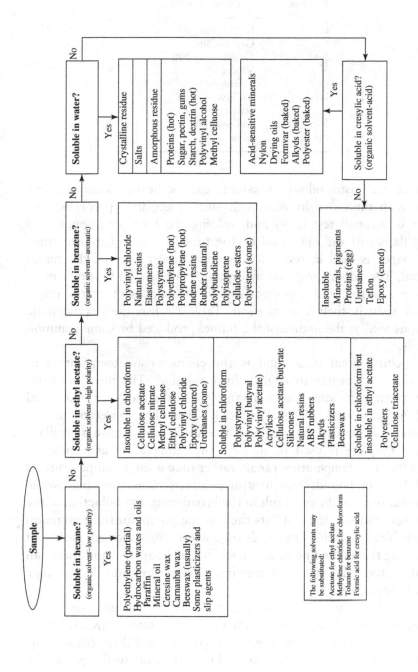

Figure 2.1 Solubility flowchart for materials. Reproduced by permission from Infrared Spectroscopy in Conservation Science, M R Derrick, D Stulik & J M Landry, 1999

Table 2.2 Flame test colours

Element	Flame colour
Ca	Red–Orange
Cu	Green–Blue
Na	Orange–Yellow
K	Violet
Li	Crimson
Ba	Yellow–Green
Sr	Scarlet
Mb	Yellow–Green
Pb	Blue

have a characteristic odour. A sample can be heated without direct contact with the flame by placing a small piece in a pyrolysis tube (similar to a small test tube) and holding the tube with tongs in a Bunsen flame. At the open end of the tube, a piece of moist litmus or pH paper is held in order to determine the nature of the fumes produced. The odour produced by organic materials on heating can be quite characteristic and a comparison of the odours of heated materials can assist in their identification. Table 2.3 lists the characteristic odours, as well as the nature of the flame, produced by some common materials.

The melting point of a solid is the characteristic temperature at which the solid begins to change into a liquid. The most common melting point apparatus houses a sample and thermometer in a metal block [Odegaard *et al.*, 2001]. The block is heated by an electric current and a small quantity of the sample of interest is contained in a capillary tube. An eyepiece is used to observe the temperature at which the solid melts and a temperature range, rather than a single temperature, is quoted. Pure compounds show melting points over a narrow range, while contamination or additives result in the broadening and reduction of the temperature range. Table 2.4 lists the melting points for some common materials [Braun, 1996; Stuart, 2002]. More sophisticated means of determining the melting point are described in Chapter 9 and these can be used to study materials with melting temperatures outside the range of simple apparatus.

Another simple pyrolysis test useful for examining precious pieces involves applying a hot pin to a hidden part of an artefact [Katz, 1994]. This may be used to differentiate thermoplastics and thermosets: if the material stays hard it is a thermoset, but if it yields to the pin it is a thermoplastic.

Table 2.3 Characteristics of pyrolysis tests on materials

Material	Flame	Odour
Amino resins	Difficult to ignite, Bright yellow	Ammonia, Amines, Formaldehyde
Casein	No flame	Burnt milk or cheese
Cellulose	Easily ignited, Sooty, Yellow–Orange	Burnt paper, Camphor, Nitric acid
Cellulose acetate	Easily ignited, Sparks, Light green	Vinegar (acetic acid)
Cellulose nitrate	Easily ignited, Bright	Nitrogen oxides
Epoxy resin	Yellow	Phenol
Horn	Smoulders	Burnt hair (keratin)
Parkesine	No flame	Camphor
Phenolic resins	Difficult to ignite, Bright, Sooty	Phenol, Formaldehyde, Burnt paper (filler)
Polyamides	Difficult to ignite, Yellow–Orange, Blue smoke	Burnt horn
PE	Yellow, Blue centre	Paraffin
PC	Difficult to ignite, Shiny, Sooty	No odour
Polyester resin	Shiny, Sooty	Sharp
PET	Yellow–Orange, Sooty	Sweet, Aromatic
PMMA	Easily ignited, Shiny, Blue centre	Sweet, Fruity
PP	Yellow, Blue centre	Paraffin
PS	Easily ignited, Shiny, Sooty	Sweet, Natural gas
PU	Yellow, Blue edge	Stinging
PVA	Easily ignited, Dark yellow, Slightly sooty	Vinegar (acetic acid)
PVAl	Shiny	Irritating
PVC	Difficult to ignite, Green edge	HCl
Rubber	Easily ignited, Dark yellow, Sooty	Burnt rubber
Rubber, Chlorinated	Difficult to ignite, Green edge	HCl
Shellac	No flame	Sealing wax
Silicone	Difficult to ignite, Yellow, Green smoke	No odour
Tortoiseshell	Burns with flame	Burnt hair (keratin)
Vulcanite	No flame	Sulfur

Table 2.4 Melting temperatures for some common materials

Material	Melting temperature(°C)
Polymers	
PDMS	−54
Natural rubber	28
PVA	35–85
LDPE	115
PMMA	120–160
Cellulose acetate	125–175
Cellulose nitrate	160–170
HDPE	130–137
PP	160–170
Nylon 12	170–180
Unplasticised PVC	175
Nylon 11	180–190
Nylon 6,10	210–220
Nylon 6	215–225
PS	240
PET	245–265
Nylon 6,6	250–260
PC	265
PVAl	265
PTFE	327
Oils	
Linseed	−24
Sunflower	−17
Poppyseed	−15
Tung	−2.5
Waxes	
Paraffin	48–74
Beeswax	62–65
Candelilla	67–69
Carnauba	83–86
Metals	
Sn	232
Pb	327
Zn	418
Al	660
Brass	932
Ag	961
Cu	1083
Cast Fe	1149–1232
Stainless steel	1427
Ni	1441
C steel	1515
Fe	1538

REFERENCES

D. Braun, *Simple Methods for Identification of Plastics*, Hanser, Munich (1996).

B.L. Browning, *Analysis of Paper*, Marcel Dekker, New York (1969).

W.D. Callister, *Materials Science and Engineering: An Introduction*, 7th ed, Wiley, New York (2006).

V. Daniels and S. Ward, A rapid test for the detection of substances which will tarnish silver, *Studies in Conservation*, **27** (1982), 58–60.

S. Davison, *Conservation and Restoration of Glass*, 2nd ed, Butterworth–Heinemann, Oxford (2003).

M.R. Derrick, D. Stulik and J.M. Landry, *Infrared Spectroscopy in Conservation Science*, Getty Conservation Institute, Los Angeles (1999).

F. Feigl, *Spot Tests in Inorganic Analysis*, 5th ed, Elsevier, Amsterdam (1958).

F. Feigl and V. Anger, *Spot Tests in Organic Analysis*, 7th ed., Elsevier, Amsterdam (1966).

J.H. Hofenk de Graaff, A simple method for the identification of indigo, *Studies in Conservation*, **19** (1974), 54–55.

C. Ianna, *Non-destructive techniques used in materials conservation*, Proceedings of the 10th Asia-Pacific Conference on Non-Destructive Testing (2001) (online publication).

S. Katz, *Early Plastics*, 2nd ed, Shire Publications, Princes Risborough (1994).

D.R. Lide (ed), *CRC Handbook of Chemistry and Physics*, 86th ed, CRC Press, Boca Raton (2005).

N. Odegaard, S. Carroll and W.S. Zimmt, *Material characterisation tests for objects of art and archaeology*, Archetype Publications, London (2000).

A. Quye and C. Williamson (eds), *Plastics: Collecting and Conserving*, NMS Publishing, Edinburgh (1999).

R. Reed, *Ancient Skins, Parchments and Leathers* , Seminar Press, London (1972).

H.P. Schramm, *Historische malmaterialien und ihre identifizierung*, Ferdinand Enke Verlag, Stuttgart (1995).

C.H. Sorum, *Introduction to Semi-micro Quantitative Analysis*, Prentice–Hall, Englewood Cliffs (1960).

D. Stulik and H. Florsheim, Binding media identification in painted ethnographic objects, *Journal of the American Institute for Conservation*, **31** (1992), 275–288.

B.H. Stuart, *Polymer Analysis*, John Wiley & Sons, Ltd, Chichester (2002).

G. Svehla, *Vogel's Qualitative Inorganic Analysis*, 7th ed, Longman, Harlow (1996).

G.G. Sward (ed), *Paint Testing Manual: Physical and Chemical Examination of Paints, Varnishes, Lacquers and Colours*, 13th ed, American Society for Testing and Materials, Philadelphia (1972).

A.I. Vogel, *Vogel's Practical Organic Chemistry, Including Qualitative Organic Analysis*, Longman, New York (1978).

J. Zhang, D. Thickett and L. Green, Two tests for the detection of volatile organic acids and formaldehyde, *Journal of the American Conservation Institute*, **33** (1994), 47–53.

3

Light Examination and Microscopy

3.1 INTRODUCTION

Visual examination of a culturally important object can provide valuable information. However, sometimes important details about an object may not be visible to the naked eye. This is where other regions of the electromagnetic spectrum can provide more information about surface properties and also about what lies beneath the surface. Infrared, ultraviolet, X-ray and γ-ray radiation can be utilised. Another means of observing properties not seen by the naked eye is the use of optical (or light) microscopy. Optical microscopy has been widely available for many years and enables an object to be magnified many times. Even higher magnifications can be attained using electron microscopy. More recently, atomic level techniques, such as atomic force microscopy and scanning tunnelling microscopy, have developed and provide a means of examining materials on the atomic scale.

3.2 INFRARED TECHNIQUES

Infrared photography has been used for many years to study art objects [Dorrell, 1989; Mairinger 2000a; Mairinger, 2004]. For infrared examinations, either film or electronic recording devices may be used. Commonly used cameras with films using emulsions sensitised to infrared radiation can be employed. Electronic recording uses video cameras equipped with infrared sensitive electronic image converters.

Analytical Techniques in Materials Conservation   Barbara H. Stuart
© 2007 John Wiley & Sons, Ltd

The technique of infrared reflectography enables the ground layer under paint layers of a painting to be examined non-destructively [Faries, 2005; Hain *et al.*, 2003; Mairinger, 2004; Mairinger, 2000a; van Asperen de Boer, 1968]. This technique involves irradiating the painting of interest with near infrared radiation (0.8–2 μm). While the pigment layers of a painting absorb much less near infrared light compared to visible light, the carbon-based underdrawing strongly absorbs the near infrared radiation. In the near infrared region the paint layers become transparent depending on their composition and thickness. An image is produced by detecting the radiation reflected from the picture using a near infrared camera. An electronic recording of the painting may be obtained using a video camera with infrared sensitive detectors. The resulting reflectogram has the appearance of a black-and-white image and may be used to gain information regarding the artist's technique. It also provides information about the type of pigments used and earlier alterations to a painting. The use of black-and-white film for infrared photography has been widely used in the past to record reflectograms. However, there are limitations with the emulsions used for such films and the improved technology of video cameras has meant that still infrared photography has been superceded in recent years.

3.2.1 Paintings

Infrared techniques have been extensively used to examine paintings [Faries, 2005; Mairinger, 2000a, 2004]. Infrared photography is helpful in detecting of alterations and later additions to paintings. Differences in the infrared behaviour of the original paint and the later added paint can be detected. Infrared photography is particularly useful for examining paintings with discoloured varnish layers. It is a more suitable approach than ultraviolet fluorescence (described in the next section) for such studies: paint layers are masked by the strong luminescence of the varnish produced in the ultraviolet region. Infrared reflectography is less suitable for studies of paint layers as many of these layers become transparent near 2 μm.

Artists since medieval times have made drawings with pencil, pen or brush to define form and composition in paintings. These drawings can provide art historians with valuable information about the origins of a painting. Although parts of the underdrawings (or ground layer) can be observed by infrared photography, drawings under blue and green paint layers remain invisible. However, infrared reflectography can be used where such layers are present. An example of how an under-drawing was revealed using infrared reflectography is illustrated in Figure 3.1,

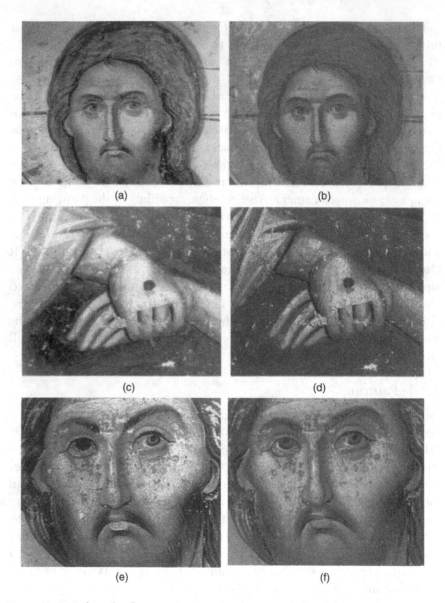

Figure 3.1 Infrared reflectograms (a,c,e) and corresponding photographs (b,d,f) of a Greek Byzantine wall painting. Reprinted from Journal of Cultural Heritage, Vol 1, Daniilia *et al.*, 91–100, 2000 with permission from Elsevier

in which photographs and the corresponding infrared reflectograms of sections of a Byzantine wall painting from Greece are shown [Daniila *et al.*, 2000]. The details of the carbon black used in the original drawing are clearly seen.

3.2.2 Written Material

The legibility of manuscripts and documents can be considerably improved with the use of infrared techniques [Mairinger, 2000a, 2004]. By employing infrared photography, it is possible to differentiate pigments and inks, for example brown writing fluids such as sepia and iron gall ink, which appear identical in visible light. When such documents have become illegible, it is possible to improve visibility using an infrared approach. Infrared reflectography can be somewhat limited for examining such objects because the sensitivity of the technique in the spectral range of inks tends to be very low. Thus, the choice of filter is important. Drawings produced using iron gall ink will vanish at 0.9 μm and inks containing soot remain visible up to 0.2 μm [Mairinger, 2000a].

3.3 ULTRAVIOLET TECHNIQUES

Ultraviolet (UV) radiation can be used in several ways to examine objects. Ultraviolet reflectography involves irradiating the object of interest with UV radiation and recording the reflected radiation [Mairinger, 2000b, 2004]. As with infrared reflectography, the reflected UV light may be recorded using photographic or electronic devices. Ultraviolet reflectography is useful for providing information regarding the surface details of objects, such as paper and documents, as surface roughness and stains can be enhanced by the technique.

A more widely used UV technique for the examination of paintings and documents is UV fluorescence [Dorrell, 1989; Hain 2003; Messinger, 1992; Mairinger, 2000b; Mairinger, 2004]. Fluorescence is a form of luminescence: when molecules absorb radiation in electronic transitions to form excited states, the excited states lose energy through the emission of radiation. Fluorescence ceases immediately the exciting radiation is removed. Like ultraviolet reflectography, UV fluorescence involves irradiating the sample of interest with ultraviolet radiation. However, fluorescence radiation of lower wavelength emerging from the sample is measured. The shift in wavelength means that the reflected light occurs in the visible region of the electromagnetic spectrum and the colour produced provides an indicator for the type of material being examined.

An UV fluorescence apparatus consists of an UV radiation lamp [Mairinger, 2000b, 2004]. Film and video cameras can be used to record the image. Filters are used in front of the UV source to remove visible

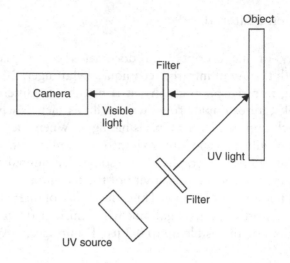

Figure 3.2   UV fluorescence apparatus

light and in front of the camera to remove ultraviolet radiation. The
layout of an UV fluorescence apparatus is illustrated in Figure 3.2.

3.3.1 Paintings

Ultraviolet fluorescence is applicable to the examination of paintings [Mairinger, 2000b], since many of the components of paint exhibit
characteristic coloured fluorescence. The organic and inorganic pigments
that are used in paint produce characteristic fluorescence, examples of
which are listed in Table 3.1. As small amounts of impurities can produce
strong fluorescence, care should be taken with the interpretation of the
colours produced. Also, certain substances can quench the fluorescence
of other substances. For example, verdigris can quench the fluorescence
of resins like mastic or dammar. As described, the varnish layers of
a painting can produce a strong fluorescence making it impossible to
examine the underlying layers. However, ultraviolet fluorescence can
still be used to identify the pigments in water colours, tempera and wall
paintings.

3.3.2 Written Material

Ultraviolet photography has proved useful for examining graphic documents [Mairinger, 2000b, 2004]. For the study of faded iron gall inks,

Table 3.1 Fluorescence properties of pigments

Pigment	Fluorescence properties
Blue	
Azurite	Dark blue
Cerulean blue	Lavender blue
Cobalt blue	Red
Egyptian blue	Purple
Indigo	Dark purple
Phthalocyanine blue	No fluorescence
Prussian blue	No fluorescence
Smalt	Light purple
Green	
Green earth	Bright blue
Phthalocyanine green	No fluorescence
Verdigris	No fluorescence
Viridian	Bright red
Red	
Alizarin	No fluorescence
Cadmium red	Red
Madder	Yellow
Red lead	Dark red
Red ochre	No fluorescence
Vermillion	Red
White	
Chalk	Dark yellow
Gypsum	Violet
Lithopone	Orange–yellow
White lead	Brown–pink
Zinc white	Light green
Yellow	
Cadmium yellow	Light red
Chrome yellow	Red
Naples yellow	Light red
Orpiment	Light yellow
Zinc yellow	Bright red

this approach provides better results than UV fluorescence. Iron gall ink absorbs UV light without producing fluorescence. Also, the effects of biodeterioration on paper can be observed using UV photography. Damage caused by bacteria or fungi that is not visible to the eye appears as grey in the ultraviolet.

3.4 RADIOGRAPHY

X-ray radiography is a non-destructive technique in which an object is irradiated with X-rays of wavelength 10^{-7}–10^{-11} m [Mairinger, 2004].

As X-rays are of a shorter wavelength than visible and UV light, they are able to penetrate materials that are opaque to such radiation. X-rays will be absorbed or pass through a material, depending on the composition of the material. When X-ray photons interact with a material, some of the photons are transmitted, some absorbed and some scattered from their path of incidence. As a result, the incident beam is attenuated. A shadow image is generated behind the object being studied.

An apparatus used for X-ray radiography employs an X-ray source, generally an X-ray tube where electrons are accelerated onto a target material (Figure 3.3). The instruments operate over a range of voltage and the voltage employed in an experiment depends on the object being examined. For instance, materials such as paintings, paper, textiles and wood are examined using a lower voltage (<100 kV), while metallic objects are examined with a higher voltage (>100 kV). Filters are used in X-ray radiography to reduce excessive subject contrast. Behind the object being studied is a form of detector: photographic film or semiconductor detectors may be used to register X-rays passing through the object under investigation to produce a radiograph.

γ-radiation can be used instead of X-rays when very dense or large objects are to be examined. γ-rays have shorter wavelengths than X-rays and are more energetic. This means that they are able to penetrate an object more deeply.

Computerised axial tomography (CAT) can be used to produce two- or three-dimensional images of an object. CAT scans, more familiar in medical applications, are used to focus X-rays at a point within an

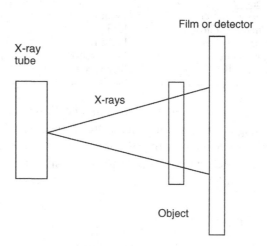

Figure 3.3   X-ray radiography apparatus

object. X-ray and γ-ray CAT provides information about the spatial distribution of chemical species and density within objects at a micron level [Mairinger, 2004].

3.4.1 Paintings and Written Material

X-ray radiography has been widely used to study paintings and illuminated manuscripts [Mairinger, 2004]. Radiographs provide information about the pigments employed and the nature of the canvas and support used. Pigments will absorb X-rays differently depending on the atomic weight and the density. For instance, a pigment containing lead or mercury will absorb more X-rays than a pigment containing chromium or cobalt [Ianna, 2001]. The canvas of a painting can be examined—in most cases a clear impression of the ground layer will be visible. The structure and diameter of the textile yarn will be visible, which is important as these factors can be characteristic of a particular artistic period. Wood used in paintings may also be characterised even with paint on both sides. In gilded panels the gold or silver foil is invisible in a radiograph.

X-ray radiography can, significantly, provide information regarding the painting technique and layer structure. Radiographs will present details not visually observed, since they provide a summation of all the absorbing layers. Variations in paint thickness will also affect the radiograph produced.

X-ray radiography may be employed to determine changes that have occurred to a painting such as compositional, dimensional, ageing, damage or later additions. For example, cracks in a paint layer appear black in a radiograph, so if they appear white the cracks must have been over painted. However, the loss of glazes is difficult to detect using X-ray radiography, so a microscopic examination is more suitable in such cases [Mairinger, 2004].

3.4.2 Metals

Radiography is commonly used to examine metal objects [Cronyn, 1990; Ianna, 2001]. The differences in densities of metals are readily compared using radiography. This approach is particularly useful for studying corrosion. A metal is more dense to X-rays than its corrosion products. For example, radiography has been used to examine the metal embedded within concrete objects recovered from the HMS Pandora shipwreck and located at the Queensland Museum in Australia [Ianna, 2001]. This technique enables conservators to determine the best approach to the treatment of such objects, including the method of cleaning.

3.4.3 Sculptures

To date, radiography has not been used to study sculptures as widely as it has for paintings [Alfrey and James, 1986; Hatziandreov and Ladopoulos, 1981; Ianna, 2001; Mairinger, 2004; Vitali *et al.*, 1986], but should become a more employed technique with the wider use of CAT scanners. Sculptures, including stone and marble, can be examined using radiography to reveal repair work, evidence of deterioration and the presence and shape of internal armature. For wooden sculptures, radiography can be used to provide information about the construction method, the presence of cavities or restoration work [Mairinger, 2004].

3.5 REFRACTOMETRY

When light travels into a transparent material it experiences a change in velocity and is bent at the interface. This phenomenon is known as refraction. The refractive index (or index of refraction) (RI) of a material is the ratio of the velocity in a vacuum to the velocity in some medium. The RI is temperature dependent. The RI is a useful property that can be used in the identification of materials; Table 3.2 lists the RI values of some common materials.

A refractometer is an optical instrument that is used to determine the RI of a material. In a refractometer, a sample is sandwiched between a measuring prism and a cover plate. Light travels to the sample and either passes through a reticle (wires placed in the focal plane) or is totally internally reflected. A shadow line is then formed between the light and dark areas and it is at the point that this shadow line crosses a scale where a reading can be viewed through a magnifying eyepiece. There are various refractometers available for measuring liquids, gases and transparent and translucent solids. There are hand-held analog and digital refractometers. There are also Abbe refractometers that are designed for use on a bench and have higher precision. Such instruments require circulating water to control the instrument temperature.

3.6 OPTICAL MICROSCOPY

Optical (or light) microscopes are used to magnify small objects and can provide information about the structure and characteristics of a sample.

Table 3.2   Refractive indices

Material	RI
Natural materials	
Amber	1.546
Asphalt	1.635
Colophony	1.548
Copal	1.528
Gum arabic	1.480
Ivory	1.539–1.541
Mastic	1.535
Paraffin	1.433
Minerals and stones	
Amethyst	1.543–1.554
Emerald	1.576–1.582
Garnet	1.72–1.94
Haematite	2.94–3.22
Lapis lazuli	1.5
Obsidian	1.482–1.492
Opal	1.44–1.46
Quartz	1.544–1.553
Ruby	1.77
Sapphire	1.762–1.778
Topaz	1.619–1.627
Turquoise	1.61–1.65
Glass	
Silica glass	1.458
Borosilicate glass	1.47
Soda–lime glass	1.51
Polymers	
PE	1.51
PMMA	1.49
PP	1.49
PS	1.60
PTFE	1.35

Optical microscopy (OM) involves the interaction of light with a sample and a magnification of the sample from × 2 to × 2000 is attainable. A resolution of about 0.5 μm is possible, depending on the limits of the instrument and the nature of the sample being examined. Optical microscopy is a quick method for identifying a broad range of materials including minerals, metals, ceramics, fibres, hair and paint.

There is a variety of light microscopy techniques that may be used to examine materials. Samples may be examined with transmitted light, reflected light and via stereomicroscopy where a three-dimensional image is obtained. There are also different imaging modes that may be used. Bright field is the normal mode of operation in OM. With transmitted light, the contrast is based on variations of colour and optical density in the material. With reflected light, if the bright field mode does not provide adequate contrast, then the dark field mode may be employed. This mode excludes unscattered light, the background light becomes approximately zero and a high image contrast is possible.

Polarised light microscopy (PLM) is often the first technique used to analyse the structure of an object [McCrone, 1994]. PLM can be used to study oriented samples such as fibres and crystalline materials. This technique involves the study of samples using polarised light and a standard microscope can be fitted with polarising filters. Anisotropic materials, which include 90 % of all solid materials, have optical properties that vary with the orientation of the incident light with the crystallographic axes. Such materials demonstrate a range of refractive indices depending on the propagation direction of light through the material and on the vibrational plane coordinates. Anisotropic materials act as beam splitters and divide light rays into two parts. Polarising microscopy exploits the interference of the split light rays as they are reunited along the same optical path to extract information about anisotropic materials. By contrast, isotropic materials (gases, liquids, unstressed glasses, cubic crystals) show the same optical properties in all directions. Such materials have only one refractive index and there is no restriction on the vibrational direction of light passing through.

There is a range of sampling techniques available in OM. Powders may be observed by dispersion on a microscope slide. The sampling of small quantities of material, such as paint, on the nanogram scale can be achieved using a fine tungsten needle or a microscalpel. If there are difficulties removing small samples from the needle a liquid agent, such as amyl acetate in the case of paints, may be used. Small samples, such as fibres, may need to be embedded in a suitable material such as an acrylic or epoxy resin. Samples that are examined in thin sections may be stained to improve contrast and osmium tetraoxide is commonly used. Thin sections may be obtained using a microtome, and for materials such as polymers melt pressing or solvent casting can be used to produce thin layers suitable for transmitted light observation. Metal surfaces are examined in a reflecting mode and the surface is often treated with an appropriate chemical reagent to reveal the microstructure (known as etching).

When using OM, it is sometimes necessary to carry out reactions on the sample under study to aid in identification [Chamot and Mason, 1983; McCrone, 1986]. Pigments, corrosion products, minerals or metals can be identified using a single drop of a test substance and reagent. The test produces a precipitate with a characteristic crystal formation.

3.6.1 Textiles

Light microscopy provides a simple identification method for textile fibres [McCrone, 1994; Peacock, 1996; Perry, 1985; Ryder and Gabra-Sanders, 1985]. A visual examination of the structure and colour of a fibre under the microscope can provide a quick confirmatory test. Even, if not always sufficiently conclusive for an exact identification, this approach can certainly enable the fibre group to be identified. Natural fibres such as cotton, wool and silk, have characteristic features. Examination of fibres commonly involves a longitudinal view of the fibre mounted using water or liquid paraffin. Cotton has characteristic twists in its separated fibres and wool is of variable diameter but shows characteristic scales which allow its fibres to interlock with other fibres. Flax and hemp show a bamboo-like structure and silk has very thin fibres. Man-made fibres may be readily differentiated from natural fibre. Rayon, for example, which is produced from cellulose, has a perfectly smooth shaft of fibres enabling it to be differentiated from cotton. Sometimes striations may be observed depending on the extrusion process used to make the fibres. Information about the fibre can also be obtained by examining a cross-section of a 2–3 mm sample obtained by cutting a fibre embedded in an epoxy or acrylic resin with a microtome. Fibres have a characteristic appearance in the cross-sectional view: cotton is flat and elongated; viscose rayon is uniformly round or serrated depending on the extrusion process; and wool shows a variable diameter and is oval to circular in appearance.

The refractive index and birefringence of fibres from PLM may also be used to differentiate samples. The birefringence (Δn) can be calculated by determining the difference between the refractive index parallel to the axis along the fibre ($n_{||}$) and the refractive index perpendicular to the axis across the fibre ($n_{\perp}$) ($\Delta n = n_{||} + n_{\perp}$). The refractive indices and birefringence values of some common fibres are listed in Table 3.3 [Perry, 1985]. When a birefringent fibre is examined between crossed polars under a polarising microscope at a 45 °C angle, specific coloured interference bands are observed. This approach is useful for synthetic fibres. The colours observed for some common fibres when a first order red plate is used in the lens between the crossed polars are listed in Table 3.4 [Perry, 1985].

3.6.2 Written Material

The structural properties of paper can be characterised using OM. The technique has been used to analyse ancient and modern papyrus

Table 3.3   Refractive indices of fibres

Fibre	$n_{||}$	$n_{\perp}$	Δn
Cellulose diacetate	1.476	1.473	0.003
Cellulose triacetate	1.469	1.469	0.000
Acrylic	1.511	1.514	−0.003
Nylon 11	1.553	1.507	0.046
Nylon 6	1.575	1.526	0.049
Nylon 6,6	1.578	1.522	0.056
Polyester	1.706	1.546	0.160
PP	1.530	1.496	0.034
PE	1.574	1.522	0.052
Viscose	1.542	1.520	0.022
Wool	1.557	1.547	0.010
Cotton	1.577	1.529	0.048
Silk	1.591	1.538	0.053
Flax	1.58–1.60	1.52–1.53	0.06

Table 3.4   Fibre colours from polarising light microscopy

Fibre	$n_{||}$	$n_{\perp}$
Viscose	Green, Yellow	Dark green, Yellow
Acetate	Blue	Orange
Acrylic	Yellow–Orange	Blue
Polyester	Yellow, Green	Yellow, Green
Nylon	Yellow, Orange, Green	Green, Red, Yellow

papers [Franceschi *et al.*, 2004]. The samples were stained with both acid fluoroglucine and toluidine blue to identify regions containing lignin. Sections were prepared for analysis by embedding them in resin and cutting with a microtome. The acid fluoroglucine turns the lignified parts red and toluidine turns then blue–green. It is also possible to detect calcium oxalate crystals and starch granules within the plant tissue using PLM. A section of papyrus sheet stained with toluidine blue with non-polarised and polarised light is illustrated in Figure 3.4.

3.6.3 Paintings

Polarised light microscopy provides an excellent tool for examining the pigments contained in paintings [Burgio *et al.*, 2003; McCrone, 1982, 1994]. Where pigments may appear similar by eye, they show very different crystal properties at a microscopic level. An examination of the shape, size, colour and the measurement of the refractive index and birefringence can be made to confirm the identity of pigment

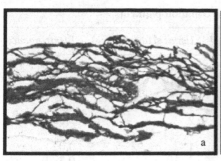 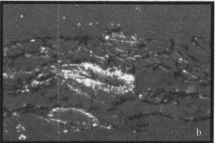

Figure 3.4 Optical micrographs of papyrus with (a) non-polarised light and (b) polarised light. Reprinted from Thermochimica Acta, Vol 418, Franceschi *et al.*, 39–45, 2004 with permission from Elsevier

structures. The degree of birefringence can be estimated by reference to a Michel–Levy chart. Some of the microscopic properties of some common paint pigments [McCrone, 1982] are summarised in Table 3.5. Microchemical tests can also be carried out on the particles under the microscope as confirmatory tests. Such tests involve the use of specific reagents to precipitate easily recognised crystals that are characteristic of specific ions [Chamot and Mason, 1983; McCrone, 1971].

3.6.4 Metals

Optical microscopy is commonly used to examine metal surfaces [Scott, 1992]. This approach is useful for examining corrosion in metals, with the nature of the corrosion and the corrosion products able to be observed [De Ryck *et al.*, 2004]. An example of the use of OM to study metal surfaces is the characterisation of patinas on coins. An investigation of the surface coatings on a series of 18th century silver Mexican coins using OM was able to characterise the patina on these coins [Rojas-Rodriquez *et al.*, 2004]. Patinas are often the result of chemical changes, but in some cases they are deposits resulting from external sources such as cleaning processes. From the morphological appearance of the surface determined using OM and the non-uniform distribution of its features, it was established that the patina originated from a chemical process and had been similarly observed on authenticated coins made during the same period.

3.6.5 Stone

The use of PLM to examine mineral-containing materials is known as thin-section petrography by geologists. Such an approach may also

Table 3.5   Microscopic properties of some common pigments

Colour	Pigment	Usual size (μm)	Birefringence
White	Chalk	1–10	0.172
	Gypsum	5–50	0.009
	Lead white	1–50	0.15
	Titanium white	<1	0.28
	Zinc white	<2	0.02
Yellow	Cadmium yellow	1	0.023
	Chrome yellow	1–10	0.37
	Lead–tin yellow	1–2	0.00
	Massicot	5–50	0.20
	Naples yellow	1–5	0.00
	Orpiment/Auripigmentum	1–30	0.6
	Yellow Ochre	5–50	0.00
	Zinc yellow	1–2	0.10
Red	Cadmium red	1	0.00–0.023
	Haematite	1–40	0.21
	Red lead	1–50	0.01–0.05
	Red ochre	1–3	0.21
	Vermillion	1–30	0.33
Blue	Azurite	1–50	0.016
	Cerulean blue	1–10	0.00
	Cobalt blue	1–50	0.00
	Prussian blue	0–5	0.00
	Smalt	1–50	0.00
	Ultramarine	1–50	0.00
Green	Chromium oxide	1	>0.05
	Emerald green	1–10	0.07
	Malachite	1–50	0.25
	Verdigris	1–30	0.03
Black			
	Charcoal black	1–100	Opaque
	Ivory black/Bone black	1–20	Opaque

be used for the examination of cultural objects [Reedy, 1994]. Thin sections of stone can be studied to identify the mineral composition and its source. This makes it possible to group objects such as sculptures, identify the geological origin and study the manufacturing technology. It is also possible to study any deterioration appearing in stone materials and monitor the effects of conservation treatments. Some of the stone sculpture materials that have been characterised using thin-section studies are sandstone, limestone, marble, granite and gypsum [Kempe and Harvey, 1983; Newman, 1992; Reedy 1994].

Stone samples are often mounted in thin sections of the order of 30 μm at magnifications ranging from × 16 to × 400 [Reedy, 1994]. By investigating a variety of optical properties, such as colour, transparency, opaqueness, refractive index and birefringence, it is possible to characterise minerals. Statistical analysis can be applied to carry out quantitative studies of samples [Reedy and Reedy, 1994].

3.6.6 Ceramics

As with the microscopic studies of stone, ceramics have been successfully examined using thin-section petrography [Alaimo *et al.*, 2004; Kempe and Harvey, 1983; Reedy, 1994; Reiderer, 2004]. By comparing the mineral components of a ceramic to potential geological sources it is possible to locate the source of manufacture. It is also possible to relate objects that might have a common source. Additionally, an examination of the quality and structure of components such as clays provides information about the manufacturing process used to produce a ceramic.

3.6.7 Glass

Optical microscopy provides a helpful approach to the identification of the structural properties of historic glasses [Bertoncello *et al.*, 2002; Corradi *et al.*, 2005; Dal Bianco *et al.*, 2004]. For instance, the appearance of air bubbles and inclusions on a μm scale can be observed. Additionally, PLM is useful for examining regions with different optical properties, such as crystal structures.

3.7 TRANSMISSION ELECTRON MICROSCOPY

Transmission electron microscopy (TEM) parallels OM, but instead of using light and lenses, high speed electrons and electromagnetic fields are used [Adriaens and Dowsett, 2004; Bulcock, 2000; Jose-Yacaman and Ascencio, 2000]. The technique exploits the diffraction of electrons. The wave characteristics of electrons are used to obtain pictures of very small objects.

In TEM it is possible to produce images of objects that cannot be seen with light microscopes. According to the laws of optics, it is possible to form an image of an object that is smaller than half the wavelength of the light used, which means that for visible light of wavelength at 400 nm the smallest that may be observed is 2×10^{-5} m. Charged electrons also have the advantage that they may be focused by applying an electric or magnetic field. In TEM a ×200 000 magnification and 0.5 nm resolution may be obtained. De Broglie's equation relates the wave and particle properties of a particle:

$$\lambda = \frac{h}{mv} \tag{3.1}$$

where λ is the wavelength, h is the Planck constant, m is the particle mass and v is the particle velocity. Equation 3.1 shows that the wavelength of an electron is inversely proportional to its velocity. By accelerating electrons at very high velocities, it is possible to obtain wavelengths as short as 0.004 nm.

A typical layout of a transmission electron microscope is shown in Figure 3.5. An electron gun provides a source of electrons. Magnetic fields can be produced with gradients that act as convex lenses for the electron waves. In TEM the detector is mounted behind the sample as electrons are transmitted through a thin section of the sample. Magnification is achieved by using lenses underneath the sample to project the image formed by the transmitted electrons onto a recording device.

Transmission electron microscopy is a useful technique for characterising morphology and structure, but in recent years analytical TEM has developed. Scanning TEM (STEM) involves scanning a sample in

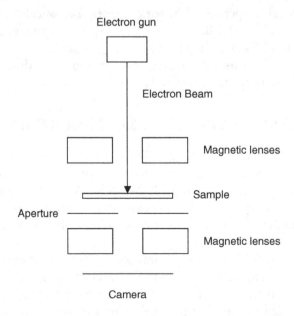

Figure 3.5 Schematic diagram of a transmission electron microscope (TEM)

a raster manner with a small electron probe. It is possible to combine STEM with other analytical techniques such as energy dispersive X-ray spectroscopy (EDS). EDS can be used for the chemical analysis of a sample. The signal from the sample is converted into a voltage and sent to a converter. The data is displayed in the form of a spectrum with intensity versus emission energy. The spectrum provides a fingerprint of the specific elements present in the sample. Electron energy loss spectroscopy (EELS) is also a technique used in TEM and can be used to provide similar information to EDS. The energy loss resulting from a beam of electrons reflected from a surface is measured and is characteristic for a material. High resolution and sensitivity are attained, even for light elements.

In TEM diffraction patterns can be used to identify crystal structures [Jose-Yacaman and Ascencio, 2000]. If the sample is a single crystalline phase and a spot pattern is obtained, it is then necessary to tilt the sample to obtain the different axes information. A table of interplanar distance values is then obtained from the pattern. Combining the information with an EDS analysis to identify the elements present narrows down the list of potential compounds. The parameters obtained are compared with a crystallographic database to identify the compound.

An important aspect of TEM is sample preparation [Jose-Yacaman and Ascencio, 2000]. Very thin sections of the order of 50–100 nm are required – an ultramicrotome may be used for this purpose. For paint fragments, the fragment is first dehydrated by heating and then embedded in an epoxy resin [San Andres et al., 1997]. The embedded sample is then cut using a microtome to the desired thickness. Sample preparation is relatively straightforward when the sample is in the form of a powder with a small grain size. Such a powder is dissolved in acetone or an alcohol and then a drop of the solution is placed on a metal mesh grid for analysis. Bulk metal samples can be examined in TEM by electrolytically thinning the sample with an acid solution. For samples that are difficult to grind, such as fibres, polymers or ceramics, ion milling is a good method. In ion milling, the sample is bombarded with an ion or neutral atom beam that produces neutral and/or charged species in the sample.

3.7.1 Paintings

TEM can be a suitable technique for examining paint layers [Barla et al., 1995; Bulcock, 2000; San Andres et al., 1997]. It can have an advantage over scanning electron microscopy (see Section 3.8) in that elemental and crystallographic information may be obtained simultaneously and the experiment can be carried out at a greater operating voltage. The spatial resolution in the microanalysis and diffraction modes is of the order of nanometres and so enables the components in a separate layer to be analysed. EDS can be used to identify the elements in the sample and electron diffraction can be used to determine the crystal structure of each particle of paint.

An example of the application of TEM to the identification of a paint sample is a study of a 15th century Venetian panel painting entitled 'The Garden of Love' and attributed to Antonio Vivrami [Bulcock, 2000]. The painting was studied for the National Gallery of Victoria, Australia, because it had been through a number of restorations and analysis was required. To distinguish the original material and the restoration material, a sample 200 μm × 20 μm in size was removed from a region of the painting which appeared uncharacteristic of the period in which it was originally painted. The sample was dispersed in ethanol using an ultrasonic bath before being deposited on a nickel TEM grid and drying in a vacuum oven. One of the grains in the paint sample was identified using TEM analysis. The EDS spectrum collected for this grain showed four major peaks at energies of 1.74, 3.69, 7.47 and 8.04 keV. These peaks were attributed to silicon, calcium, nickel and copper, respectively. The nickel was associated with scattering from the grid. The electron diffraction pattern collected for the same grain showed a square symmetry with identical spacings in both directions, which is indicative of a tetragonal structure. The dimensions of the structure were measured to be 3.6 Å. A comparison of the pattern with compounds containing silicon, calcium and copper on a database was used to identify the compound as Egyptian blue.

An example of the use of TEM to characterise different paint layers is provided by a study of a 14th century Spanish work of tempera on canvas, the artist unknown [San Andres et al., 1997]. This painting showed evidence of overpainting. The TEM of a ultra-thin section from the painting embedded in epoxy resin is illustrated in Figure 3.6. Electron diffraction and EDS analysis were carried out on the different layers. Region A in the figure is a white primer and B is the original blue-coloured pictorial layer consisting of coarse granules of azurite and white lead. On top of this layer is a dark coat, region C, which is the varnish of the original work. It can be seen that immediately on top of the varnish is the overpainted layer, which was dated much later due to the darkening of the varnish. The overpainted layer shows characteristics very similar

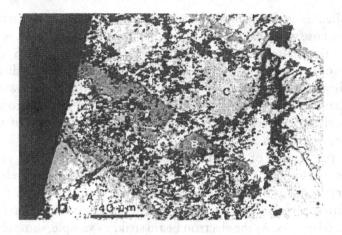

Figure 3.6 TEM of a thin section of tempera paint on canvas [San Andres *et al.*, 1997]

to the original layer in that it contains coarsely ground azurite and white lead. However, there are some green malachite crystals in the top layer. The differing proportions of malachite in each layer suggests that the azurite used in the overpainting was of a different origin.

3.7.2 Ceramics

TEM can be used to examine ceramic materials intact to provide information about composition, crystal structures and textures [Mata *et al.*, 2002]. For example, because of the scale of analysis in TEM, closely related clay minerals such as smectite, illite and muscovite, which occur as grains of the order of nanometres in pottery, may be differentiated in TEM. It is possible with TEM to characterise both the source raw materials and products of the firing process used to produce ceramics, thus allowing provenance studies to be carried out.

3.8 SCANNING ELECTRON MICROSCOPY

As with TEM, in scanning electron microscopy (SEM) the image of an object is created using a beam of electrons rather than traditional visible light [Goodhew *et al.*, 2001; Henson and Jergovich, 2001; Jose-Yacaman and Ascencio, 2000; Skoog and Leary, 1992]. However, SEM has the advantage that the surface of a sample is studied and the preparation of the thin sections used in TEM can be avoided. A scanning electron microscope can magnify objects of the order of 100 000 times and detailed three-dimensional images can be produced. Such a technique is clearly of use when examining very small regions of artefacts. Scanning electron microscopy may also be combined with energy EDS in order to carry out elemental analysis of very small samples. Additionally, the applicability of electron microscopy to material conservation has been expanded in recent times with the development of environmental scanning electron microscopy (ESEM), which can be carried out at much higher pressures than SEM.

When the surface of a sample is scanned with a beam of energetic electrons a number of signals are produced. Back-scattered electrons, secondary electrons, Auger electrons and X-ray fluorescence are possible. Back-scattered and secondary electrons are the basis of SEM and X-rays are detected in EDS. As the electron beam strikes a sample, some electrons are elastically scattered by the sample atoms without a significant loss of energy. These electrons retain slightly less energy than the beam and are known as back-scattered electrons. Other electrons produce an inelastic collision and may cause the sample atom to ionise. These electrons lose more energy and are known as secondary electrons. Before the back-scattered electrons leave the sample they may result in inelastic collisions and generate secondary electrons. Electrons from outer energy atomic shells fill the holes produced by the emission of secondary electrons from the inner energy shells. During this process, energy is released in the form of a characteristic X-ray. EDS analysis involves the analysis of the X-rays and allows elements to be identified.

A scanning electron microscope operates by producing a beam of electrons in a vacuum [Henson and Jergovich, 2001; Jose-Yacaman and Ascencio, 2000; Skoog and Leary, 1992]. A schematic diagram of a scanning electron microscope is shown in Figure 3.7. An electron beam is generated by an electron gun consisting of a heated tungsten filament, a Wehnelt cylinder and an anode. A large voltage of up to 50 kV is applied to the filament and Wehnelt cylinder. Electrons are released from the filament by applying an appropriate current which results in emission. The anode is earthed and the electrons are accelerated towards the anode. Part of the electron beam passes through a hole in the anode and travels towards the sample. Electromagnetic lenses are used to focus the beam as it passes through the instrument. The beam is swept back and forth (rastered) over a selected area of the sample. The interaction of the electrons with the sample causes electrons to be dislodged from the atoms within the sample. The electrons generated are detected and amplified. The most common detector used in SEM is an Everhart–Thornley detector and can be used to detect both secondary and back-scattered

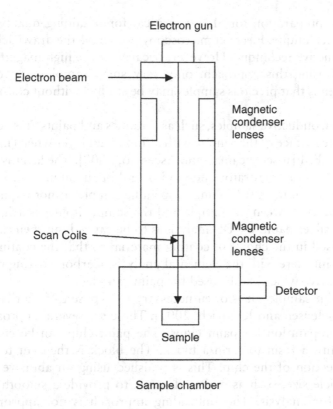

Figure 3.7 Schematic diagram of a scanning electron microscope (SEM)

electrons. The topography of a sample can be illustrated by examining the back-scattered electrons.

An environmental scanning electron microscope enables samples to be examined at higher pressures than those used for conventional SEM [Danilatos, 1997]. In traditional SEM samples are commonly coated with a metal film as many samples are poor conductors. If a nonconductive sample is examined, secondary electrons leaving the sample generate an excess of positive charge of the surface. This will deflect electrons travelling to the detector and result in a blurred image. However, ESEM can be used to avoid coating the samples and for sensitive samples that may not be stable in a vacuum. It uses a moderate pressure (0.75–150 kPa) of a gas such as water in the sample chamber instead of a sample coating to remove the excess charge. A series of apertures and vacuum pumps between the sample chamber and the filament source makes this technique feasible. The use of the ionised gaseous environment enables wet samples to be investigated.

Sample preparation for SEM is critical for obtaining quality results from the technique. Electron microscopy does have the drawback that it is a destructive technique. However, because of the high magnifications obtained using this approach, only very small samples are required, which means that precious samples may be studied without considerable damage.

For nonconductive samples, such as ceramics and paints, it is necessary to coat the surface of the sample with a thin metal layer when employing standard SEM [Jose-Yacaman and Ascencio, 2000]. The layer is applied using a vacuum evaporator and gold or gold-palladium are commonly used deposition metals. Coating of metallic samples is not required and good contact between the sample and the sample holder is achieved by applying silver paint. If EDS analysis is to be carried out then care must be exercised in the choice of coating material so that the coating metal does not interfere with the elemental analysis. Carbon coating provides an alternative which can be used for paint samples.

For paint samples, it is often necessary to expose different layers for analysis [Hensen and Jergovich, 2001]. There are several approaches to sample preparation for paint chips. The paint chip can be embedded as it is into a resin to form a block. The block is then cut to expose a cross-section of the chip. This is polished using an abrasive of very fine particle size, such as diamond paste, to provide a smooth surface suitable for analysis. The embedding approach is not appropriate if sample availability is limited—it is difficult to retrieve the sample from the resin for further analysis with other techniques. Alternatively, the paint chip may be mounted on its edge to expose the cross-section to the SEM beam. Samples can be mounted on aluminium or carbon stubs. Carbon or aluminium paints and carbon tape or cement can be used to adhere the sample to the stub. There is also the stair-step preparation approach, but this approach requires experience. The paint chips are prepared by exposing each layer in a stair-step manner using a scalpel or other sharp cutting tool. Thin peels of each paint layer may also be separately removed and mounted using a sharp tool.

3.8.1 Paintings

Scanning electron microscopy has proved a popular means of examining the materials that make up paintings [Burnstock and Jones, 2000; Doehne and Bower, 1993; Feller, 1986; Roy, 1993; West Fitzhugh, 1997]. ESEM and the use of EDS particularly lend themselves to these sorts of materials [Bower *et al.*, 1994]. SEM can be used to identify inorganic pigment particles and EDS is helpful where pigments are similar in appearance. The surface characteristics of paintings may be investigated using scattered and back-scattered electron imaging in SEM. Scattered electron images of paintings can provide information about surface texture such as fine cracking, the nature of the relationship between the pigment and the binding medium at the surface, paint drying defects and surface pores. Back-scattered electron images provide information regarding the atomic number contrast in the sample and can be used prior to EDS analysis. As an example, the surface topography and composition of samples from paintings by Rembrandt have been studied using scattered and back-scattered electron images [Groen, 1997].

SEM also proves useful for examining surface coatings. However, care must be exercised when studying partially dry organic coatings, such as linseed oil, using conventional SEM [Burnstock and Jones, 2000]. Coatings which retain solvent or moisture tend to produce artefacts, observed structures not due to the actual material. SEM can also be useful when cleaning paintings. ESEM has been used to check the cotton swabs used to clean the Sistine Chapel in Rome [Doehne, 1997; Doehne and Stulik, 1990]. The swabs were dark grey to black in appearance and there was interest in determining whether pigment particles were being removed during the cleaning process in addition to soot particles. Pigment particles appear much denser than soot particles and so can be readily differentiated using SEM. The electron micrographs produced showed cotton fibres with only a few dense particles appearing on the fibres. The EDS spectrum of particles showed peaks due to calcium, silicon, sulfur and iron. The particles were identified as salt particles, silicate dust and air pollution particles.

3.8.2 Written Material

SEM can be employed to examine the materials used in heritage manuscripts [Cappitelli et al., 2006; Doehne, 1997; Doehne and Stulik, 1990; Pinzari et al., 2006; Remazeilles et al., 2005; Wagner et al., 2001]. This approach is particularly helpful for ascertaining the state of deterioration of such materials. ESEM has been used to examine a fragment of the Dead Sea Scrolls to assess their condition [Doehne, 1997; Doehne and Stulik, 1990]. The fragment was examined non-destructively (undried and uncoated) and the relatively high partial pressure of water vapour (~ 100 Pa) in the sample chamber provided long term dimensional stability throughout the experiment. The micrographs produced clearly showed the layered fibrous structures of collagen as well as the almost structureless layer of gelatin above and beneath the collagen. A comparison with a micrograph of modern parchment demonstrated the dramatically different features of the ancient parchment. Dynamic experiments were also carried out on Dead Sea Scroll fragments to study the response of the material to changes in moisture. These experiments revealed the presence of bromide and potassium chloride salts, soil and calcite salts. When the humidity fluctuates, the salts dissolve and recrystallise and produce stresses in the parchment which can result in delamination. The salts are believed to originate from the original processing of the animal skins into parchment using water from the Dead Sea, which contains high levels of bromine and have probably contributed to the protection of the parchment from microorganisms.

3.8.3 Metals

While it does not provide information about the low level elemental composition of metals to the same level as some X-ray techniques, SEM can provide valuable information about the microstructure of metals. SEM-EDS can be used to provide information about the nature of the metal without the need for time-consuming dissolution procedure [Adriaens and Dowsett, 2004]. The shape and size of the grains may assist in determining whether an object was cast into a mould or was worked by hammering and annealing. SEM has been successfully applied to the study of the development of copper-based alloys in Asia and Europe: the chemical composition and phase analysis of metallic remains have been examined [Klein and Hauptman, 1999; Ryndina et al., 1999; Shalev and Northover, 1993].

SEM may also be used to identify corrosion products seen on metal surfaces in order to assist in the development of conservation procedures [Doehne, 1997; De Ryck et al., 2004; Doehne and Stulik, 1990; Ingo et al., 2004; Linke et al., 2004a; 2004b]. As formaldehyde emitted from particleboard and other sources is implicated in the corrosion of lead objects in museum collections, an understanding of the corrosion products is important for conservators. ESEM has been used to study the kinetics of corrosion product formation on a lead sample [Doehne, 1997; Doehne and Stulik, 1990]. The formation of corrosion products was monitored for a lead wire exposed to formaldehyde vapours at different time intervals. Different crystal formations are clearly observed in the micrographs at different time intervals. The composition of the corrosion product was confirmed to be lead formate by X-ray diffraction (XRD).

3.8.4 Stone

ESEM is particularly helpful for stone conservation studies [Gaspar et al., 2003; Doehne, 1997; Doehne and Stulik, 1990; Rao et al., 1996; Stulik and Doehne, 1991]. This approach allows stone structures to be investigated at the microscopic level in their natural state. The morphology and chemical constituents of samples can be characterised and samples with protective coatings may also be examined to investigate the effects of consolidation. ESEM can be used to study the environmental factors that cause weathering in stone structures, such as salt crystallisation. For instance, sodium sulfate has been exposed to hydration/dehydration and dissolution/recrystallisation cycles and studied using ESEM [Doehne, 1994, 1997]. The micrographs produced in this study showed the rapid crystallisation of highly porous particles of sodium sulfate decahydrate implicated in the degradation resulted from the hydration cycling processes.

ESEM–EDS has been used to investigate the weathering of Sydney heritage sandstones to gain a better understanding of the degradation processes [Friolo et al., 2003]. The samples required no pre-treatment. The EDS data for weathered and unweathered sandstone are shown in Figure 3.8. The results indicate that there is a significant change to the elemental composition of the stone upon weathering: an increase in the concentration of iron and a decrease in the concentrations of both silicon and aluminium are observed on weathering. The results support the proposition, based on complimentary data, of Fe^{3+} substitution

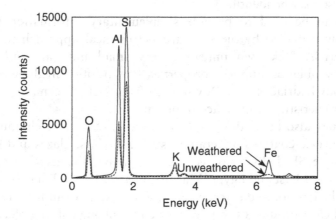

Figure 3.8   ESEM–EDS data for weathered and unweathered sandstones. Reprinted from Journal of Cultural Heritage, Vol 4, Friolo et al., 211–220, 2003 with permission from Elsevier

into the kaolinite structure within the stone, replacing the Si^{4+} and Al^{3+} ions.

ESEM combined with EDS has been used to characterise semi-precious stones in ancient jewellery [Derrick et al., 1994; Doehne and Stulik, 1990]. Such samples can be examined without the risk of a surface coating damaging the artefact. A study of an uncoated gold pendant with a dark red stone was carried out. The stone facets and polishing markings were visible in the micrograph and the EDS spectrum produced peaks due to magnesium, aluminium, silicon, carbon and iron. This enabled the stone to be identified as an almandine garnet, which has the structure $Fe_3Al_2(SiO_4)_3$ with substitution of magnesium and calcium for iron in the crystal lattice.

3.8.5 Ceramics

SEM has proved valuable for examining ceramics, particularly for identifying the materials and the production techniques used [Adriaens and Dowsett, 2004; Alaimo et al., 2004; Freestone, 1982; Froh, 2004; Tite et al., 1982]. The composition of the raw materials used to produce a ceramic can be characterised – SEM can be used to identify clay minerals by their shape and size in low fired ceramics. As many ceramics are fired above the temperatures at which clay minerals start breaking down, the composition of the original raw materials can no longer be directly identified using SEM. However, the newly formed silicates present in the glass phase may be identified.

SEM can be used to provide supplementary information to that commonly provided by the standard petrological approach to ceramics materials. Unknown minerals, very small minerals and minerals with ambiguous optical properties can be identified by electron microscopy [Adriaens and Dowsett, 2004]. Such information can be very characteristic for the place of origin.

SEM may also be used to identify the resources and techniques used for the surface coatings of ceramics such as paints, glazes and washes. For example, SEM–EDS has been used to analyse the glazes of ceramics from the Near East and Egypt [Mason and Tite, 1997]. Polished sections of both the interior and exterior of the vessels were examined and back-scattered electron images were generated to distinguish the phases. The chemical composition of tin-opacified glazes was determined using EDS. The analyses were able to show that tin was first used in glazes during the 8th century AD in Iraq.

3.8.6 Glass

As for ceramics, the origin of glass objects can be identified by SEM, provided that well-authenticated glasses are available for comparison. SEM–EDS can be used in conjunction with other analyses to provide information about a glass. SEM is more widely used to investigate the deterioration of glass [Adriaens and Dowsett, 2004; Bertoncello *et al.*, 2002; Corradi *et al.*, 2005; Dal Bianco *et al.*, 2004, 2005; Janssens *et al.*, 1996; Jembrih *et al.*, 2000; Melcher and Schreiner, 2005; Schreiner *et al.*, 1999; Schreiner, 2004; Orlando *et al.*, 1996; Woisetschlager *et al.*, 2000]. SEM provides information about the structure of the corrosion layer and the extent to which leaching has occurred. An example of the examination of weathered glass using SEM is illustrated in Figure 3.9 [Melcher and Schreiner, 2005]. The secondary electron images of the surfaces of glass of medieval composition exposed to an outdoor environment are shown. Figure 3.9(a) shows the image of glass exposed in a sheltered position and weathering products, including gypsum, are clearly visible. Where the samples have been exposed to unsheltered conditions (Figure 3.9(b)-(d)), weathering products are not visible and it is assumed that they have been washed away by rain. The degree of deterioration is visible in the samples shown in Figure 3.9 (c) and Figure 3.9 (d), where cracks and pits are observed.

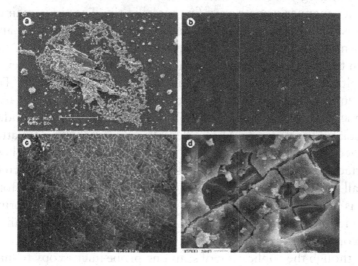

Figure 3.9 SEM of glass samples exposed under shelter (a) and unsheltered (b)–(d) conditions. Reprinted from Journal of Non-Crystalline Solids, Vol 351, Melcher and Schreiner, 1210–1225, 2005 with permission from Elsevier

3.8.7 Textiles

SEM can be used to examine the surfaces of heritage textile fibres, and is suited to the investigation of deterioration. This technique has been utilised to examine the effect of water and burial in soil on model fabrics made of linen, cotton, silk and wool [Peacock, 1996]. The fabrics were mounted on aluminium stubs with copper tape and sputter coated with gold-palladium. Effects including shrinkage and splitting were observed.

3.9　SCANNING PROBE MICROSCOPY

Scanning probe microscopes were developed in the 1980s and are used to examine the surface topography of samples at the atomic level [Birdi, 2003; Bonnell, 2001; Skoog and Leary, 1992; Weisendanger, 1994]. The two most widely used scanning probe microscopies are scanning tunnelling microscopy (STM) and atomic force microscopy (AFM). The techniques are based on the idea that an electron in an atom has a small probability of existing far from the nucleus and, in certain circumstances, it can move, or tunnel, and end up closer to another atom. The tunnelling electrons create a current that can be used to image the atoms of an adjacent surface.

In a scanning tunnelling microscope, the sample surface is scanned in a raster manner using a very fine metallic tip. The tip is maintained at a constant distance above the surface throughout the scan and the movement of the tip reflects the topography of the surface. A current between the tip and the sample surface is held constant and is generated by a voltage that is applied between the tip and the sample. The tip movement is controlled by a piezoelectric transducer.

Atomic force microscopy is able to be used to examine both conducting and insulating surfaces. In AFM a flexible cantilever with a tip attached is scanned in a raster pattern over a sample surface. The force between the surface and the cantilever causes the cantilever to be deflected and the small movements are optically measured. During a scan the force on the tip is held constant by the tip motion and topographic information is provided by the experiment. The schematic layout of an atomic force microscope is illustrated in Figure 3.10.

Even though the application of scanning probe microscopy to material conservation has been limited to date there is plenty of scope for these techniques. The surfaces of a variety of materials may be examined using STM and AFM. Surface studies at the atomic level can be provide

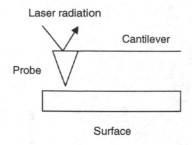

Figure 3.10   Atomic force microscope

detailed information regarding chemical reactions at the surface, such as degradation processes.

3.9.1. Metals

AFM has been used in a number of investigations into the corrosion of bronzes [Calliari *et al.*, 2001; Chiavari *et al.*, 2006; Wadsak *et al.*, 2002; Wang *et al.*, 2006]. One study used AFM to compare the corrosive behaviour of a traditionally used alloy with a new silicon-containing bronze [Chiavari *et al.*, 2006]. The AFM images of the bronzes after 30 days exposure to 1 ppm nitrogen oxide and 1 ppm sulfur dioxide atmosphere in a climate chamber are shown in Figure 3.11. The traditional bronze clearly shows a thicker and more irregular layer of corrosion products.

3.9.2. Glass

AFM has been used to study the initial stages of the weathering of medieval potash–lime–silica glass [Schreiner *et al.*, 1999; Schreiner 2004]. Topographic changes are detected for glass exposed to humid environments.

3.9.3. Ceramics

AFM shows potential as a technique for examining the surfaces of ceramics. This technique has been used to characterise ceramic lustre [Roque *et al.*, 2005]. AFM provided information about the nature of the reaction occurring during lustre formation, demonstrating that the lustre layer

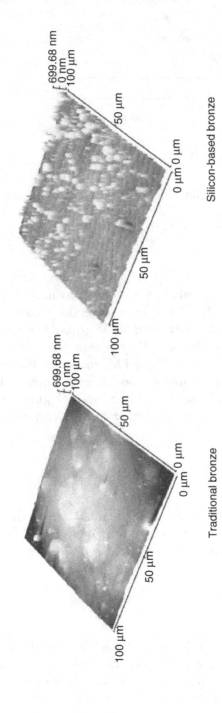

Figure 3.11 AFM images of a traditional bronze and a silicon-based bronze after exposure to a polluted atmosphere. Reprinted from Materials Chemistry and Physics, Vol 95, Chiavari et al., 252–259, 2006 with permission from Elsevier

does not appear as a superimposed layer on top of the glaze, but as a surface roughness due to the growth of crystals inside the glassy matrix.

3.9.4. Stone

A useful application of AFM to the examination of heritage stonework is the study of mortars [Armelao *et al.*, 2000; Boke *et al.*, 2006]. The nature of the topography observed at the interfaces can provide useful information about the effectiveness of the adhesion between mortar and stone.

3.9.5 Written Material

AFM can be used to monitor the degradation of paper [Piantanida *et al.*, 2005, 2006]. The technique can be used to study the degradation of cellulose fibres on the surface of paper and can provide qualitative and quantitative information about ageing processes and deterioration due to fungus. This technique can produce a topographical map revealing the structure at a molecular level. AFM was used to characterise the effects of chemical reactions induced by accelerated ageing at 80 °C and 65 % relative humidity [Piantanida *et al.*, 2005]. The study showed that with ageing there was increased fragmentation of the fibre surface involving changes to the height, width and length. The bundles of cellulose fibres appear to unravel with increased ageing.

3.9.6. Textiles

An AFM examination of unaged and light aged silk fibres was used to characterise the changes to the silk structure as a result of degradation [Odlyha *et al.*, 2005]. Areas of 5×5 μm of the fibres were examined and clearly showed the disappearance of fibrils of the order of 1–2 μm. Such fibrils reduce to smaller segments after exposure to ageing and changes of this order can be monitored using AFM.

REFERENCES

A. Adriaens and M.G. Dowsett, Electron microscopy and its role in the cultural heritage studies, in *Non-Destructive Microanalysis of Cultural Heritage Materials*, (eds S.K. Janssens and R. van Grieken), Elsevier, Amsterdam (2004), pp 73–128.

R. Alaimo, G. Bultrini, J. Fragala, *et al.*, Microchemical and microstructural characterisation of medieval and post-medieval ceramic glaze coatings, *Applied Physics A*, **79** (2004), 263–272.

G.F. Alfrey and K. James, The gamma-ray radiography of decorative plasterwork, *Studies in Conservation*, **31** (1986), 70–76.

L. Armelao, A. Bassan, R. Bertoncello, *et al.*, Silica glass interaction with calcium hydroxide: a surface chemistry approach, *Journal of Cultural Heritage*, **1** (2000), 375–384.

C. Barla, M. San Andres, L. Peinadio, *et al.*, A note on the characterisation of paint layers by transmission electron microscopy, *Studies of Conservation*, **40** (1995), 194–200.

R. Bertoncello, L. Milanese, U. Russo, *et al.*, Chemistry of cultural glasses: the early medieval glasses of Montselice's Hill (Padova, Italy), *Journal of Non-Crystalline Solids*, **306** (2002), 249–262.

K.S. Birdi, *Scanning Probe Microscopies: Application in Science and Technology*, CRC Press, Boca Raton (2003).

H. Boke, S. Akkurt, B. Ipekoglu and E. Ugurlu, Characteristics of brick used as aggregate in historic brick-lime mortars and plasters, *Cement and Concrete Research*, **36** (2006), 1115–1122.

D. Bonnell (ed.), *Scanning Probe Microscopy and Spectroscopy: Theory, Techniques and Application*, 2nd ed, Wiley–VCH, New York (2001).

N.W. Bower, D.C. Stulik and E. Doehne, A critical evaluation of the environmental scanning microscope for the analysis of paint fragments in art conservation, *Fresenius Journal of Analytical Chemistry*, **348** (1994), 402–410.

S. Bulcock, Transmission electron microscopy and its use for the study of paints and pigments, in *Radiation in Art and Archaeometry* (eds. D.C. Creagh and D.A. Bradley), Elsevier, Amsterdam (2000), pp 232–254.

L. Burgio, R.J.H. Clark, L. Sheldon and G.D. Smith, Pigment identification by spectroscopic means: evidence consistent with the attribution of the painting 'Young Woman Seated at a Virginal' to Vermeer, *Analytical Chemistry*, **77** (2003), 1261–1267.

A. Burnstock and C. Jones, Scanning electron microscopy techniques for imaging materials from paintings, in *Radiation in Art and Archaeometry* (eds D.C. Creagh and D.A. Bradley), Elsevier, Amsterdam (2000), pp 202–231.

I. Calliari, M. Dabala, G. Brunoro and G.M. Ingo, Advanced surface techniques for characterisation of copper-based artefacts, *Journal of Radioanalytical and Nuclear Chemistry*, **247** (2001), 601–608.

F. Cappitelli, C. Sorlini, E. Pedemonte, *et al.*, Effectiveness of graft synthetic polymers in preventing biodeterioration of cellulose-based materials, *Macromolecular Symposia* **238** (2006), 84–91.

E.M. Chamot and C.W. Mason, *Handbook of Chemical Microscopy*, 4th ed, John Wiley & Sons Inc., New York (1983).

C. Chiavari, A. Colledan, A. Frignani and G. Brunoro, Corrosion evaluation of traditional and new bronzes for artistic castings, *Materials Chemistry and Physics*, **95** (2006), 252–259.

A. Corradi, C. Leonelli, P. Veronesi, *et al.*, Ancient glass deterioration in mosaics of Pompeii, *Surface Engineering*, **21** (2005), 402–405.

J.M. Cronyn, *Elements of Archaeological Conservation*, Routledge, London (1990).

B. Dal Bianco, R. Bertoncello, L. Milanese and S. Barison, Glasses on the seabed: surface study of chemical corrosion in sunken Roman glasses, *Journal of Non-Crystalline Solids*, **343** (2004), 91–100.

B. Dal Bianco, R. Bertoncello, L. Milanese and S. Barison, Glass corrosion across the Alps: a surface study of chemical corrosion of glasses found in marine and ground environments, *Archaeometry*, **47** (2005), 351–360.

S. Daniilia, S. Soiropoulou, D. Bikiaris, *et al.*, Panselinos' Byzantine wall paintings in the Protaton Church, Mount Athos, Greece: a technical examination, *Journal of Cultural Heritage*, **1** (2000), 91–110.

G.D. Danilatos, Environmental Scanning Electron Microscopy, in *In-Situ Microscopy in Materials Research*, (ed P.L. Gai), Kluwer, Boston (1997), pp 13–44.

M. Derrick, E. Doehne, A.E. Parker and D.C. Stulik, Some new analytical techniques for use in conservation, *Journal of the American Institute for Conservation*, **33** (1994), 171–184.

I. De Ryck, E. van Biezen, K. Leyssens, *et al.*, Study of tin corrosion: the influence of alloying elements, *Journal of Cultural Heritage*, **5** (2004), 189–195.

E. Doehne and D.C. Stulik, Applications of the environmental scanning electron microscope to conservation science, *Scanning Microscopy*, **4** (1990), 275–286.

E. Doehne and N. Bower, Experimental conditions for semi-quantitative SEM/EDS of painting cross sections using the environmental scanning microscope, *Microbeam Analysis*, **2** (1993), 539–540.

E. Doehne, In situ dynamics of sodium sulfate hydration and dehydration in stone pores: Observations at high magnification using the environmental scanning electron microscope, in *The Third International Symposium on the Conservation of Monuments in the Mediterranean basin in Venice, Italy* (eds V. Fassina, H. Ott and F. Zezza) (1994), pp 143–150 (1994).

E. Doehne, ESEM development and application in cultural heritage conservation, in *In-Situ Microscopy in Materials Research*, ed P.L. Gai, Kluwer, Boston (1997), pp 45–62.

P. Dorrell, *Photography in Archaeology and Conservation*, Cambridge University Press, Cambridge (1989).

M. Faries, Analytical capabilities of infrared reflectography: an art historian's perspective, in *Scientific Examination of Art: Modern Techniques in Conservation and Analysis*, National Academies Press, Washington DC (2005), pp 87–104.

R.L. Feller (ed), *Artists' Pigments. A Handbook of their History and Characteristics*, Vol. 1, Cambridge University Press, Cambridge (1986).

E. Franceschi, G. Luciano, F. Carosi, L. Cornara and C. Montanari, Thermal and microscope analysis as a tool in the characterisation of ancient papyri, *Thermochimica Acta*, **418** (2004), 39–45.

I.C. Freestone, Applications and potential of electron micro-probe micro-analysis in technological and provenance investigations of ancient ceramics, *Archaeometry*, **24** (1982), 99–116.

K.H. Friolo, B.H. Stuart and A. Ray, Characterisation of weathering of Sydney sandstones in heritage buildings, *Journal of Cultural Heritage*, **4** (2003), 211–220.

J. Froh, Archaeological ceramics studied by scanning electron microscopy, *Hyperfine Interactions*, **154** (2004), 159–176.

P. Gaspar, C. Hubbard, D. McPhail and A. Cummings, A topographical assessment and comparison of conservation cleaning treatments, *Journal of Cultural Heritage*, **4** (2003), 294s–302s.

P.J. Goodhew, J. Humphreys and R. Beanland, *Electron Microscopy and Analysis*, 3rd ed, Taylor and Francis, London (2001).

K. Groen, Investigation of the use of the binding medium by Rembrandt, *Kunsttechnology und Konservierung*, **11** (1997), 207–227.

M. Hain, J. Bartl and V. Jacko, Multispectral analysis of cultural heritage artefacts *Measurement Science Review* **3** (2003), 9–12.

L. Hatziandreov and G. Ladopoulos, Radiographic examination of the marble statue of Hermes at Olympia, *Studies in Conservation*, **26** (1981), 24–28.

M.L. Henson and T.A. Jergovich, Scanning electron microscopy and energy dispersive X-ray spectrometry (SEM/EDS) for the forensic examination of paints and coatings, in *Forensic Examination of Glass and Paint* (ed B. Caddy, Taylor and Francis), London (2001), pp 243–272.

C. Ianna, Non-destructive techniques used in materials conservation, Proceedings of the 10th Asia–Pacific Conference on Non-Destructive Testing (2001) (online publication).

G.M. Ingo, E. Angelini, T. De Caro, *et al.*, Combined use of GDOES, SEM + EDS, XRD and OM for the microchemical study of the corrosion products on archaeological bronzes, *Applied Physics A*, **79** (2004), 1999–2003.

K. Janssens, A. Aerts, L. Vincze, *et al.*, Corrosion phenomena in electron, proton and synchrotron X-ray microprobe analysis of Roman glass from Qumtan, Jordan, *Nuclear Instrumentation and Methods in Physics B*, **109/110** (1996), 690–695.

D. Jembrih, M. Schreiner, M. Peer, *et al.*, Identification and classification of iridescent glass artefacts with XRF and SEM/EDX, *Mikrochimica Acta*, **133** (2000), 151–157.

M. Jose-Yacaman and J.A. Ascencio, Electron microscopy and its application to the study of archaeological materials and art preservation, in *Modern Analytical Methods in Art and Archaeology* (eds E. Ciliberto and G. Spoto), John Wiley & Sons Inc., New York (2000), pp 405–444.

D.R.C. Kempe and A.P. Harvey (eds), *The Petrology of Archaeological Artefacts*, Clarendon Press, Oxford (1983).

S. Klein and A. Hauptman, Iron age leaded tin bronzes from Khirbet Edh–Dharih, Jordan, *Journal of Archaeological Science*, **26** (1999), 1075–1082.

R. Linke, M. Schreiner and G. Demortier, The application of photon, electron and proton induced X-ray analysis for the identification and characterisation of medieval silver coins, *Nuclear Instruments and Methods in Physics Research B*, **226** (2004), 172–178.

R. Linke, M. Schreiner, G. Demortier, *et al.*, The provenance of medieval silver coins: analysis with EDXRF, SEM/EDX and PIXE, in *Non-Destructive Microanalysis of Cultural Heritage Materials* (eds K. Janssens and R. van Grieken), Elsevier, Amsterdam (2004), pp 605–663.

F. Mairinger, The infrared examination of paintings, in *Radiation in Art and Archaeometry* (eds D.C. Creagh and D.A. Bradley), Elsevier, Amsterdam (2000a), pp 40–55.

F. Mairinger, The ultraviolet and fluorescence study of paintings and manuscripts, in *Radiation in Art and Archaeometry* (eds D.C. Creagh and D.A. Bradley), Elsevier, Amsterdam (2000b), pp 56–75.

F. Mairinger, UV-, IR- and X-ray imaging, in *Non-Destructive Microanalysis of Cultural Heritage Materials* (eds S.K. Janssens and R. van Grieken), Elsevier, Amsterdam (2004), pp 15–71.

R.B. Mason and M. Tite, The beginnings of tin-opacification of pottery glazes, *Archaeometry*, **39** (1997), 41–58.

M.P. Mata, D.R. Peacor and M.D. Gallart-Marti, Transmission electron microscopy (TEM) applied to ancient pottery, *Archaeometry*, **44** (2002), 155–176.

W.C. McCrone, Ultramicrominaturisation of microchemical tests, *Microscope*, **19** (1971), 235–241.

W.C. McCrone, The microscopical identification of artists' pigments, *Journal of the International Institute for Conservation – Canadian Group*, **7** (1982), 11–34.

W.C. McCrone. Solubility, recrystallisation and microchemical tests on nanogram single particles, *The Microscope*, 34 (1986), 107–118.

W.C. McCrone, Polarised light microscopy in conservation: a personal perspective, *Journal of the American Institute of Conservation*, 33 (1994), 101–114.

M. Melcher and M. Schreiner, Evaluation procedure for leaching studies on naturally weathered potash-lime-silica glass with medieval composition by scanning electron microscopy, *Journal of Non-Crystalline Solids*, 351 (2005), 1210–1225.

J.M. Messinger, Ultraviolet-fluorescence microscopy of paint cross sections, *Journal of the American Institute of Conservation* 31 (1992) 267–276.

R. Newman, Applications of petrography and electron microprobe analysis to the study of Indian stone sculpture, *Archaeometry*, 34 (1992), 163–174.

M. Odlyha, Q. Wang, G.M. Foster, *et al.*, Thermal analysis of model and historic tapestries, *Journal of Thermal Analysis and Calorimetry*, 82 (2005), 627–636.

A. Orlando, F. Olmini, G. Vaggelli and M. Bacci, Medieval stained glasses of Pisa Cathedral (Italy): Their composition and alteration products, *Analyst*, 121 (1996), 553–558.

E.E. Peacock, Biodegradation and characterisation of water-degraded archaeological textiles created for conservation research, *International Biodeterioration and Biodegradation*, 38 (1996), 49–59.

R. Perry, *Identification of Textile Materials*, 7th ed., The Textile Institute, Manchester (1985).

G. Piantanida, M. Bicchieri and C. Coluzza, Atomic force microscopy characterisation of the ageing of pure cellulose paper, *Polymer*, 46 (2005), 12313–12321.

G. Piantanida, F. Pinzari, M. Montanari, *et al.*, Atomic force microscopy applied to the study of Whatman paper surface deteriorated by a celluloytic filamentous fungus, *Macromolecular Symposia* 238 (2006), 92–97.

F. Pinzari, G. Pasquariello and A. De Mico, Biodeterioration of paper: a SEM study of fungal spoilage reproduced under controlled conditions, *Macromolecular Symposia* 238 (2006), 57–66.

S.M. Rao, C.J. Brinker and T.J. Ross, Environmental microscopy in stone conservation, *Scanning*, 18 (1996), 508–514.

C.L. Reedy, Thin-section petrography in studies of cultural material, *Journal of the American Institute of Conservation*, 33 (1994), 115–129.

T.J. Reedy and C.L. Reedy, Statistical analysis in conservation science, *Archaeometry*, 36 (1994), 1–23.

C. Remazeilles, V. Rochen-Quillet, J. Bernard, *et al.*, Influence of gum Arabic on iron-gall ink corrosion. Part II: Observation and elemental analysis of originals, *Restaurator*, 26 (2005), 118–133.

J. Riederer, Thin section microscopy applied to the study of archaeological ceramics, *Hyperfine Interactions*, 154 (2004), 143–158.

I. Rojas-Rodriquez, A. Herrar, C. Vazquez-Lopez, *et al.*, On the authenticity of eight Reales 1730 Mexican silver coins by X-ray diffraction and by energy dispersion spectroscopy techniques, *Nuclear Instruments and Methods in Physics Research B*, 215 (2004), 537–544.

J. Roque, T. Pradell, J. Molera and M. Vendrell-Saz, Evidence of nucleation and growth of metal Cu and Ag nanoparticles in lustre: AFM surface characterisation, *Journal of Non-Crystalline Solids*, 351 (2005), 568–575.

A. Roy (ed), *Artists' Pigments. A Handbook of their History and Characteristics*, Vol. 2, Oxford University Press, Oxford (1993).

M.L. Ryder and T. Gabra-Sanders, The application of microscopy to textile history, *Textile History*, **16** (1985), 123–140.

N. Ryndina, G. Indenbaum and V. Kolosova, Copper production from polymetallic sulfide ores in the Northeastern Balkan neolithic culture, *Journal of Archaeological Science*, **26** (1999), 1059–1068.

M. San Andres, M.I. Baez, J.L. Baldonedo and C. Barba, Transmission electron micro scopy applied to the study of works of art: sample preparation methodology and possible techniques, *Journal of Microscopy*, **188** (1997), 42–50.

D.A. Scott, *Metallography and Microstructure in Ancient and Historic Metals*, Getty Conservation Institute, Los Angeles (1992).

S. Shalev and P. Northover, The metallurgy of the Nahal Mishmar Hoard reconsidered, *Archaeometry*, **35** (1993), 35–47.

M. Schreiner, G. Woisetschlager, I. Schmidt and M. Wadsak, Characterisation of surface layers formed under natural environmental conditions on medieval stained glass and ancient copper alloys using SEM, SIMS and atomic force microscopy, *Journal of Analytical and Atomic Spectroscopy*, **14** (1999), 395–403.

M. Schreiner, Corrosion of historic glass and enamels, in *Non-Destructive Micro Analysis of Cultural Heritage Materials* (eds K. Jannsens and R. Van Grieken), Elsevier, Amsterdam (2004), pp 713–754.

D.A. Skoog and J.J. Leary, *Principles of Instrumental Analysis*, Harcourt Brace College Publishers, Fort Worth (1992).

D.C. Stulik and E. Doehne, Applications of environmental scanning electron microscopy in art conservation and archaeology, in *Materials Issues In Art and Archaeology II*, Vol. 185 Materials Research Society Proceedings (eds P. Vandiver, J. Druzik and G.S. Wheeler) (1991), pp 23–29.

M.S. Tite, I.C. Freestone, N.D. Meeks and M. Bimson, The use of scanning electron microscopy in the technological examination of ancient ceramics, in *Archaeological Ceramics*, (eds J.S. Olin and A.O. Franklin), Smithsonian Institution Press, Washington (1982), 109–120.

J.R.J. van Asperen de Boer, Infrared reflectography: A method for the examination of paintings, *Applied Optics*, **7** (1968), 1711–1714.

V. Vitali, J. Darcovich and W. Williams, Construction of a fudo–myoo sculpture: an X-radiographic study, *Studies in Conservation*, **31** (1986), 185–189.

M. Wadsak, T. Aastrup, I. Odnevall Wallinder, *et al.*, Multianalytical in situ investigation of the initial atmospheric corrosion of bronze, *Corrosion Science*, **44** (2002), 791–802.

B. Wagner, E. Bulska, A. Hulanicki, *et al.*, Topographical investigation of ancient manuscripts, *Fresenius Journal of Analytical Chemistry*, **369** (2001), 674–679.

J. Wang, C. Xu and G. Lu, Formation processes of copper chloride and regenerated copper crystals on bronze surfaces in neutral and acidic media, *Applied Surface Science*, **252** (2006), 6294–6303.

R. Weisendanger, *Scanning Probe Microanalysis and Spectroscopy: Methods and Appli-cations*, Cambridge University Press, Cambridge (1994).

E. West Fitzhugh (ed), *Artists' Pigments. A Handbook of their History and Characteris-tics*, Vol. 3, Oxford University Press, Oxford (1997).

G. Woisetschlager, M. Dutz, S. Paul and M. Schreiner, Weathering phenomena on naturally weathered potash–lime–silica–glass with medieval composition studied by scanning electron microscopy and energy dispersive microanalysis, *Mikrochimica Acta*, **135** (2000), 121–130.

4
Molecular Spectroscopy

4.1 INTRODUCTION

Spectroscopy is the analysis of electromagnetic radiation absorbed, emitted or scattered by molecules or atoms as they undergo transitions between energy levels. Quantum theory shows that molecules and atoms exist in discrete states known as energy levels. The frequency (ν) of the electromagnetic radiation associated with a transition between two energy levels (ΔE) is given by:

$$\Delta E = h\nu \qquad (4.1)$$

where h is the Planck constant. Different types of spectroscopic techniques using radiation from different regions of the electromagnetic spectrum are used to investigate different magnitudes of energy level separation. Atomic spectroscopy provides information about the electronic structure of an atom and is described in the Chapter 5. Molecular spectroscopy is more complex and involves rotational, vibrational and electronic transitions.

The characteristic spectral frequencies associated with each molecule mean that spectroscopy can be used to identify specimens. Infrared spectroscopy utilises infrared radiation to cause the excitation of vibrations of bonds in molecules. This type of spectroscopy is widely used and can be applied to a range of material types in conservation. Raman spectroscopy is another form of vibrational spectroscopy, but measures the scattering of radiation by a molecule. Raman spectroscopy can also be used to investigate a variety of materials and has been particularly successful for identifying paint pigments. Radiation in the UV and visible

Analytical Techniques in Materials Conservation Barbara H. Stuart
© 2007 John Wiley & Sons, Ltd

regions are used in UV-visible spectroscopy. The technique involves transitions in the electronic energy levels of the bonds of a molecule and tends to be restricted to the study of compounds such as dyes. Photoluminescence spectroscopy examines the fluorescent and phosphorescent behaviour of molecules. Spectroscopic techniques that examine changes in spin in molecules are nuclear magnetic resonance and electron spin resonance spectroscopies. In addition, Mössbauer spectroscopy is a technique that is used to investigate the electronic environment with γ-rays, but is limited to the study of specimens containing specific elements.

4.2 INFRARED SPECTROSCOPY

Infrared spectroscopy is a technique based on the vibrations of the atoms of a molecule [Derrick et al., 1999; Stuart, 2004]. An infrared spectrum is commonly obtained by passing infrared radiation through a sample and determining what fraction of the incident radiation is absorbed at a particular energy. The energy at which any peak in an absorption spectrum appears corresponds to the frequency of a vibration of a part of a sample molecule. For a molecule to show infrared absorptions it must possess a specific feature: an electric dipole moment of the molecule must change during the vibration. This is the selection rule for infrared spectroscopy. Vibrations can involve either a change in bond length (stretching) or bond angle (bending). Some bonds can stretch in-phase (symmetrical stretching) or out-of-phase (asymmetric stretching). If a molecule has different terminal atoms then the two stretching modes are no longer symmetric and asymmetric vibrations of similar bonds, but will have varying proportions of the stretching motion of each group known as coupling. There will be many different vibrations for even fairly simple molecules. The complexity of an infrared spectrum arises from the coupling of vibrations over a large part of or over the complete molecule and such vibrations are called skeletal vibrations. Bands associated with skeletal vibrations are likely to conform to a pattern or fingerprint of the molecule as a whole, rather than a specific group within the molecule.

Fourier transform infrared (FTIR) spectroscopy is the most common technique and is based on the interference of radiation between two beams yielding an interferogram [Griffiths and de Haseth, 1986; Stuart, 2004]. An interferogram is a signal produced as a function of the change of path length between the two beams. The two domains of distance and frequency are interconvertible by the mathematical method of Fourier transformation. In a FTIR spectrometer the radiation

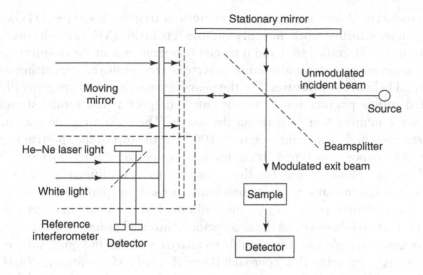

Figure 4.1 A Michelson interferometer used in FTIR spectrometry

emerging from a source is passed through an interferometer to the sample before reaching a detector. FTIR spectrometers use a Globar or Nernst source for the mid-infrared region or a high-pressure mercury lamp for the far-infrared region. For the near-infrared, tungsten-halogen lamps are used as sources. The data are converted to a digital form by an analog-to-digital converter and transferred to the computer for Fourier transformation. The most common interferometer used in FTIR spectrometry is a Michelson interferometer, and consists of two perpendicularly plane mirrors, one of which can travel in a direction perpendicular to the plane (Figure 4.1). A semi-reflecting film, the beamsplitter, bisects the planes of these two mirrors. If a collimated beam of monochromatic radiation of wavelength λ cm is passed into an ideal beamsplitter, 50 % of the incident radiation will be reflected to one of the mirrors and 50 % will be transmitted to the other mirror. The two beams are reflected from these mirrors, returning to the beamsplitter where they recombine and interfere. Fifty percent of the beam reflected from the fixed mirror is transmitted through the beamsplitter and 50 % is reflected back in the direction of the source. The beam which emerges from the interferometer at 90° to the input beam is called the transmitted beam and this is the beam detected in FTIR spectrometry. The moving mirror produces an optical path difference between the two arms of the interferometer. There are two commonly used detectors used in the mid-infrared region. The normal detector for routine use is a pyroelectric device incorporating deuterium tryglycine sulfate (DTGS); for more sensitive work, mercury cadmium telluride (MCT) can be used, but has to be cooled to liquid nitrogen temperatures. In the far-infrared germanium or indium-antimony detectors are employed, operating at liquid helium temperatures. For the near-infrared detectors are generally lead sulfide photoconductors. The infrared spectra that results shows a wave number (cm^{-1}) scale on the x-axis. The ordinate scale may be presented in % transmittance with 100 % at the top of the spectrum or in absorbance from 0 to 1. It comes down to personal preference which of the two modes to use, but the transmittance is traditionally used for spectral interpretation, while absorbance is used for quantitative work.

Transmission spectroscopy is the traditional infrared sampling method. The method is based upon the absorption of infrared radiation as it passes through a sample and it is possible to analyse samples in liquid, solid or gaseous form using this approach [Derrick et al., 1999; Stuart, 2004]. There are several different types of transmission solution cells available. Fixed path length sealed cells are useful for volatile liquids, but cannot be taken apart for cleaning. Semi-permanent cells are demountable so that the windows can be cleaned. The spacer is usually made of PTFE and is available in a variety of thicknesses. Variable path length cells incorporate a mechanism for continuously adjusting the path length and a vernier scale allows accurate adjustment. All of these cell types are filled using a syringe and the syringe ports are sealed with PTFE plugs before sampling. If quantitative analysis of a sample is required it is necessary to use a cell of known path length. An important consideration in the choice of infrared cells is the type of window material. The material must be transparent to the incident infrared radiation and normally alkali halides are used in transmission methods. The least expensive material is sodium chloride, but other materials used are potassium bromide, calcium fluoride and caesium bromide. In choosing a solvent for a sample the following factors need to be considered: it has to dissolve the compound; it should be as non-polar as possible to minimise solute–solvent interactions; and it should not strongly absorb infrared radiation. Liquid films also provide a quick method of examining liquid samples. A drop of liquid may be sandwiched between two infrared plates that are then mounted in a cellholder. Liquid samples might be exudates on the surface of an object or a solvent. Such samples can be collected with a capillary tube or a cotton swab [Derrick et al., 1999; Derrick, 1995].

There are three general methods for examining solid samples in transmission infrared spectroscopy: alkali halide discs, mulls and films. The use of alkali halide discs involves mixing a solid sample with a dry alkali halide powder (around 2 to 3 mg of sample with about 200 mg of halide). The mixture is usually ground with an agate mortar and pestle and subjected to a pressure of about 1.6×10^5 kg m^{-2} in an evacuated die. This sinters the mixture and produces a clear transparent disc. The most commonly used alkali halide is potassium bromide, which is completely transparent in the mid-infrared region. The mull method for solid samples involves grinding the sample then suspending (about 50 mg) in 1–2 drops of a mulling agent (usually Nujol, liquid paraffin). This is followed by further grinding until a smooth paste is obtained. Films can be produced by either solvent casting or by melt casting.

Gases have densities several orders of magnitude less than liquids, so the path lengths must be correspondingly greater, usually 10 cm or longer. The walls are of glass or brass with the usual choice of windows. The cells can be filled by flushing or by using portable infrared monitors.

Reflectance techniques may be used for samples where destruction of the sample needs to be avoided or when the surface properties are of interest. Attenuated total reflectance (ATR) spectroscopy utilises the phenomenon of total internal reflection (Figure 4.2). A beam of radiation entering a crystal will undergo total internal reflection when the angle of incidence at the interface between the sample and crystal is greater than the critical angle. The critical angle is a function of the refractive indices of the two surfaces. The beam penetrates a fraction of a wavelength beyond the reflecting surface and when a material that selectively absorbs radiation is in close contact with the reflecting surface, the beam loses energy at the wavelength where the material absorbs. The resultant attenuated radiation is measured and plotted as a function of wavelength by the spectrometer and gives rise to the absorption spectral characteristics of the sample. The depth of penetration in ATR

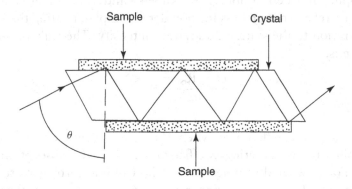

Figure 4.2   ATR spectroscopy

is a function of wavelength, the refractive index of the crystal and the angle of incident radiation. As a consequence, the relative intensities of the infrared bands in an ATR spectrum will appear different to those observed in a transmission spectrum; the intensities will be greater at lower wave number values. The crystals used in ATR cells are made from materials that have low solubility in water and are of a very high refractive index. Such materials include zinc selenide (ZnSe), germanium (Ge) and thallium/iodide (KRS-5).

Specular reflectance occurs when the reflected angle of radiation equals the angle of incidence. The amount of light reflected depends on the angle of incidence, the refractive index, surface roughness and absorption properties of the sample. For most materials the reflected energy is only 5–10 %, but in regions of strong absorptions, the reflected intensity is greater. The resultant data appears different from normal transmission spectra, as derivative-like bands result from the superposition of the normal extinction coefficient spectrum with the refractive index dispersion (based upon Fresnel's relationships). However, the reflectance spectrum can be corrected using a Kramers–Kronig transformation (or K–K transformation). The corrected spectrum appears like the familiar transmission spectrum.

In external reflectance, the energy that penetrates one or more particles into a sample is reflected in all directions and this component is called diffuse reflectance. In the diffuse reflectance technique, commonly called DRIFT, a powdered sample is mixed with potassium bromide powder. The DRIFT cell reflects radiation to the powder and collects the energy reflected back over a large angle. Diffusely scattered light can be collected directly from material in a sampling cup or from material collected using an abrasive sampling pad. DRIFT is particularly useful for sampling powders or fibres. The Kubelka and Munk theory describes the diffuse reflectance process for powdered samples, relating the sample concentration to the scattered radiation intensity. The Kubelka–Munk equation is:

$$\frac{(1 - R_\infty)^2}{2R_\infty} = \frac{k}{s} \tag{4.2}$$

where R_∞ is the reflectance when the layer's depth can be said to be infinite, s is a scattering coefficient and k is the absorption coefficient. For powdered samples diluted in a non-absorbing matrix, the Kubelka–Munk function is linear with concentration. An alternative relationship between the concentration and the reflected intensity is now widely used in near-infrared diffuse reflectance spectroscopy:

$$\log\left(\frac{1}{R}\right) = k'c \qquad (4.3)$$

where k' is a constant and c is the concentration.

Photoacoustic spectroscopy (PAS) is a non-invasive reflectance technique with penetration depths in the range from micrometres down to several molecular monolayers. Gaseous, liquid or solid samples can be measured using PAS and the technique is particularly useful for highly absorbing samples. The photoacoustic effect occurs when intensity modulated light is absorbed by the surface of a sample located in an acoustically isolated chamber filled with an inert gas. A spectrum is obtained by measuring the heat generated from the sample due to a re-absorption process. The sample absorbs photons of the modulated radiation, which have energy corresponding to the vibrational states of the molecules. The absorbed energy is released in the form of heat generated by the sample, which causes temperature fluctuations and, subsequently, periodic acoustic waves. A microphone detects the resulting pressure changes that are then converted to an electrical signal. Fourier transformation of the resulting signal produces a characteristic infrared spectrum.

Optical fibres may be used in conjunction with infrared spectrometers to carry out remote measurements. Fibre optics allow light to be propagated over long distances by utilising total internal reflection. The fibres transfer the signal to and from a sensing probe and are made of materials that are flexible and infrared transparent. Mid-infrared chalcogenide fibre optics has been combined with FTIR benches, making it possible to perform non-invasive reflectance measurements of objects [Vohra et al., 1996].

For some samples, dipole moment changes may be in a fixed direction during a molecular vibration and, as such, can only be induced when the infrared radiation is polarised in that direction. Polarised infrared radiation can be produced by using a polariser consisting of a fine grating of parallel metal wires. This approach is known a linear infrared dichroism [Buffeteau and Pezolet, 2002].

It is possible to combine an infrared spectrometer with a microscope to study very small samples [Derrick, 1995; Humecki, 1999; Katon and Sommers, 1992; Katon, 1996; Messerschmidt and Harthcock, 1998; Sommer, 2002]. In recent years there have been considerable advances in FTIR microscopy with samples of the order of micrometres being characterised. In FTIR microscopy, the microscope sits above the FTIR sampling compartment. Infrared radiation from the spectrometer is focused onto a sample placed on a standard microscope x–y stage. After passing through the sample, the infrared beam is collected by a Cassegrain objective that produces an image of the sample within the barrel of the microscope. A variable aperture is placed in this image plane. The radiation is then focused on a small area MCT detector by another Cassegrain condenser. The microscope also contains glass objectives to allow visual inspection of the sample. In addition, by switching mirrors in the optical set-up, the microscope can be converted from transmission mode to reflectance mode.

If a microscope is not available, there are other special sampling accessories available that allow examination of microgram or microlitre amounts. This is accomplished using a beam condenser so that as much as possible of the beam passes through the sample. Microcells are available with volumes of around 4 μl and path lengths up to 1 mm. A diamond anvil cell (DAC) uses two diamonds to compress a sample to a thickness suitable for measurement and increase the surface area. This technique can be used at normal atmospheric pressures, but it may also be applied to study samples under high pressures and improve the quality of the spectrum of trace samples. Alternatively, a multiple internal reflectance cell may be used as this technique can produce stronger spectra.

Infrared imaging using FTIR microspectroscopic techniques has emerged as an effective approach to studying complex specimens [Kidder et al., 2002; Krishnan et al., 1995; Mansfield et al., 2002]. The technique can be used to produce a two- or three-dimensional picture of the properties of a sample. This is possible because, instead of reading the signal of only one detector as in conventional FTIR spectroscopy, a large number of detector elements are read during the acquisition of spectra. This is possible due to the development of focal plane array (FPA) detectors. Currently, a step-scanning approach is used, which means that the moving mirror does not move continuously during data acquisition but waits for each detector readout to be completed before moving onto the next position. This allows thousands of interferograms to be collected simultaneously and then transformed into infrared spectra. The infrared beam from a Michelson interferometer is focused onto a sample with a reflective Cassegrain condenser. The light transmitted is collected by a Cassegrain objective and focussed onto a FPA detector. The data are collected as interferograms with each pixel on the array having a response determined by its corresponding location on the sample. Each point of the interferogram represents a particular moving mirror position and the spectral data is obtained by performing a Fourier transform for each pixel on the array. Thus, each pixel (or spatial location) is represented by an infrared spectrum.

There are a number of techniques available to assist in the interpretation of infrared spectra [Stuart, 2004]. Difference spectroscopy may be carried out simply by subtracting the infrared spectrum of one component of the system from the combined spectrum to leave the spectrum of the other component. If the interaction between components results in a change in the spectral properties of either one or both of the components, the changes will be observed in the difference spectra. The changes may manifest themselves via the appearance of positive or negative peaks in the spectrum. Spectra may also be differentiated. Resolution is enhanced in the first derivative since changes in the gradient are examined and the second derivative gives a negative peak for each band and shoulder in the absorption spectrum. Deconvolution is the process of compensating for the intrinsic linewidths of bands in order to resolve overlapping bands. The technique yields spectra that have much narrower bands and is able to distinguish closely-spaced features. The instrumental resolution is not increased, but the ability to differentiate spectral features can be significantly improved. Curve-fitting may also be applied for quantitative analysis.

Infrared spectroscopy, and a number of the techniques described elsewhere in this book, generate data which can be used to provide a quantitative analysis. Data is often collected in order to identify or characterise a single specimen such as a possible fraud. Data is also collected to characterise a group of specimens such as a collection of bronze objects. There are a number of statistical methods that can be used for the analysis of cultural objects [Baxter and Buck, 2000; Brown, 2000]. Multivariate analytical techniques can be classified as either supervised or unsupervised methods, depending on whether the sample grouping is known in advance. Unsupervised methods do not require information for classification and individual samples are clustered based on the similarity amongst the sample data. One of the most popular unsupervised methods is principal component analysis (PCA), which computes a few linear combinations of the original variables, so minimising the data with a minimal loss of information. Cluster analysis (or hierarchical clustering) is another popular unsupervised technique. In this approach, the similarity of samples based on the data is defined and an algorithm from grouping samples is specified. The data are usually presented in the form of a dendrogram. Supervised methods include linear discriminant analysis (LDA) and artificial neural networks (ANNs). Such approaches use prior assumptions about the existence of groups, such as known provenances of the samples. These methods are capable of comparing a large number of variables within a data set.

An infrared spectrum can be divided into three main regions: the far-infrared (<400 cm^{-1}), the mid-infrared (4000–400 cm^{-1}) and the near-infrared (13 000–4000 cm^{-1}). Many infrared applications employ the mid-infrared region, but the near- and far-infrared also provide important information about certain materials. The mid-infrared spectrum can be divided approximately into four regions and the nature of a group frequency may generally be determined by the region in which it is located. The regions are generalised as: the X–H stretching region (4000–2500 cm^{-1}), the triple bond region (2500–2000 cm^{-1}), the double bond region (2000–1500 cm^{-1}) and the fingerprint region (1500–600 cm^{-1}). The absorptions observed in the near infrared region (13 000–4000 cm^{-1}) are overtones or combinations of the fundamental stretching bands that occur in the region 3000–1700 cm^{-1}. The bands involved are usually due to C–H, N–H or O–H stretching. The far-infrared region is more limited than the mid-infrared for spectra-structure correlations, but does provide information regarding the vibrations of molecules containing heavy atoms, molecular skeleton vibrations, molecular torsions and crystal lattice vibrations. In Table 4.1 the main infrared bands for common materials, are summarised.

4.2.1 Natural Materials

The appearance of the mid-infrared spectra of different oils is similar as they are all are mixtures of triglycerides [Derrick *et al.*, 1999]. The main infrared bands produced by oils are listed in Table 4.2 and the infrared spectrum of linseed oil is shown in Figure 4.3. Infrared spectroscopy

Table 4.1 Common infrared bands of organic molecules

Wave number (cm^{-1})	Assignment
3700–3600	O–H stretching
3400–3300	N–H stretching
3100–3000	Aromatic C–H stretching
3000–2850	Aliphatic C–H stretching
2300–2050	C≡C stretching
2300–2200	C≡N stretching
1830–1650	C=O stretching
1650	C=C stretching
1500–650	Fingerprint region (bending, rocking)

does not readily provide the specific identity of an oil as does, say, chromatography, but the technique may certainly be used to identify this class of compound. However, multivariate analysis has been applied to differentiate the mid-infrared spectra of vegetable oils [Rusak *et al.*, 2003].

The infrared spectra of waxes are characterised by bands due to the methylene chain [Birshtein and Tulchinskii, 1977a; Derrick *et al.*, 1999; Newman, 1979, 1998]. Table 4.2 lists the infrared bands in the spectra of waxes. As well as strong C–H stretching bands in the $3000-2800$ cm^{-1} region, there are confirmatory doublets for waxes at $1470-1460$ cm^{-1}

Table 4.2   Major mid-infrared bands of natural materials

Material	Wave numbers (cm^{-1})
Oils	3600–3200, 3000–2800, 1750–1730, 1480–1300, 1300–900, 750–700
Waxes	3000–2800,1470–1460,730–720
Carbohydrates	3600–3200, 3000–2800, 1650, 1480–1300, 1300–900
Proteins	3400–3200, 1660–1600,1565–1500, 1480–1300
Resins	3600–3200, 3100–2800, 2700–2500,1740–1640, 1650–1600,1480–1300

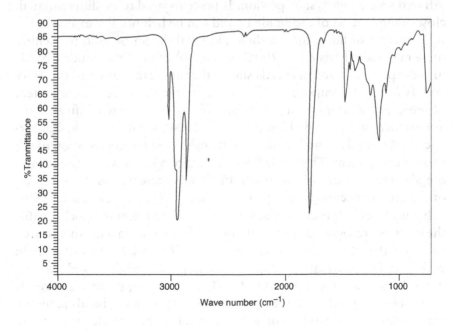

Figure 4.3   Infrared spectrum of linseed oil

and 730–720 cm^{-1}. Beeswax shows additional bands due to C=O and C–O stretching of the ester groups.

Carbohydrates such as gums, starch and cellulose consist of polysaccharides that contain a large amount of O–H groups. Polysaccharides show strong broad bands near 3300 cm^{-1} due to O–H stretching and near 1080 cm^{-1} due to C–O stretching (Table 4.2) [Birshtein and Tulchinskii, 1977b; Derrick et al., 1999; Newman, 1998]. These sugars also show a band at 1620 cm^{-1} associated with intermolecular water and a carboxyl group. The C–H stretching bands are relatively weak in the spectra of polysaccharides.

The identification of natural materials containing proteins, such as animal glues, casein, egg, horn and ivory, is aided by the characteristic bands due to the protein [Birshtein and Tulchinskii, 1983; Fabian, 2000; Fabian and Mantele, 2002; Stuart, 2004; Paris et al., 2005]. Proteins exhibit infrared bands associated with their characteristic amide group and are labelled as amide bands. The characteristic protein infrared bands are also listed in Table 4.2. Proteins are usually characterised by the appearance of the amide I and II bands. The N–H stretching band near 3350 cm^{-1} may also be used to confirm the presence of an amide group. There are many types of proteins with differing amino acid composition, but such differences are usually too subtle to differentiate the infrared spectra on first inspection. If proteins need to be differentiated, closer examination of the amide I band can be helpful. For example, the ATR spectra of horn, tortoiseshell and galalith can be used to identify these materials [Paris et al., 2005]. Deconvolution of the amide I bands of these protein-based materials shows that an examination of the 1639 and 1614 cm^{-1} components allows the materials to be differentiated. However, denaturation can affect identification as this will influence the appearance of the amide I band [Stuart, 2004]. Changes to the appearance of the amide I and II bands can be used to show degradation of the protein component. The amide II band shifts to a lower wave number and the absorbance ratio of the amide I/II bands increases as denaturation of the protein occurs [Brodsky-Doyle et al., 1975; Derrick et al., 1999].

Natural resins produce characteristic infrared bands which enable them to be recognised and each type of resin produces an infrared spectrum that shows differentiating features [Derrick et al., 1999, 1989; Feller, 1954; Newman, 1998]. The major infrared bands shown by natural resins are listed in Table 4.2. The cyclic ring structure of resins of plant origin produces a spectrum with C–H stretching bands at higher wave numbers than oils. These bands tend to be broader than those observed for oils and waxes because of the presence of more methyl

(CH_3) end groups and a variety of molecular environments for the methylene (CH_2) groups. Resins also show a distinctive broad band in the 2700–2500 cm^{-1} range due to O–H stretching of a dimerised carboxyl group. These compounds also produce carbonyl stretching bands near 1700 cm^{-1}. In order to differentiate resins, the fingerprint region may be used.

Shellac, a resin of insect origin, also shows distinctive infrared bands. The C–H stretching bands appear at similar wave numbers to those of oils and there is a small olefinic band at 1636 cm^{-1}. The C=O stretching band of shellac is a doublet with peaks at 1735 and 1715 cm^{-1} due to the ester and the acid, respectively. The ester, acid and alcohol groups produce C–O stretching bands at 1240, 1163 and 1040 cm^{-1}, respectively. A flowchart has been published to aid in the identification of natural resins and is illustrated in Figure 4.4 [Derrick, 1989]. The infrared spectra of

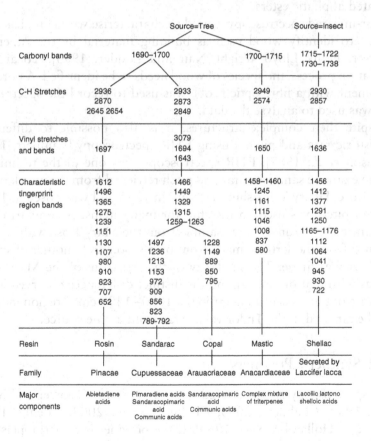

Figure 4.4 Flowchart for the identification of natural resins based on infrared spectra [Derrick, 1989]

resins may also be used to investigate the deterioration of these materials. The wave numbers in the spectra remain the same with deterioration of a sample, but the band shapes and the relative intensities do change. In particular, the carbonyl stretching band broadens and decreases in intensity when a resin degrades and oxidises [Derrick *et al.*, 1999].

Infrared spectroscopy is an established and widely used method for characterising ambers [Angelini and Bellintani, 2005; Beck *et al.*, 1964, 1965; Beck, 1986; Langenheim and Beck, 1965; Mosini and Munziante Cesaro, 1996]. Apart from confirming that a material is amber, infrared spectroscopy has been used to differentiate ambers of different origin. For example, a particular type of Baltic amber, succinite, shows characteristic bands in the 1250–1110 cm^{-1} region of the spectrum. Succinite shows a broad shoulder between 1250 and 1175 cm^{-1} and a sharp band at 1150 cm^{-1}, which have been attributed to the C–O stretching of saturated aliphatic esters.

Near-infrared spectroscopy is used to characterise wood and has been applied to identify wood used as building material by the American architect, Frank Lloyd Wright [Nair and Lodder, 1993]. To aid the restoration process, the species of wood needs to be identified. A portable instrument with a fibre optic probe was used to record the spectra and PCA was used to analyse the data.

Despite their complex structures, it is also possible to differentiate historic tars and pitches using FTIR spectroscopy [Lambert, 1997; Robinson *et al.*, 1987]. FTIR spectroscopy was one of the techniques used to examine samples of tar and pitch retrieved from the wreck of the Mary Rose, Henry VIII's ship that sank in 1545 and was raised in 1981. Tar and pitch was used on the ship for many items. A comparison of the infrared spectrum of a tar sample from the Mary Rose with spectra obtained for a modern tar made from pine wood and another of spruce wood tar was made. The similarity of the spectrum of the Mary Rose tar was with that of pine tar and the distinct differences observed in the spectrum of the spruce wood tar in the 1800–1400 cm^{-1} region indicate that the tar used in the Tudor vessel was from a pine source.

4.2.2 Synthetic Polymers

FTIR spectroscopy is widely used to study polymers [Bower and Maddams, 1989; Chalmers, 2000; Chalmers *et al.*, 2002; Koenig, 1999; Siesler and Holland-Moritz, 1980]. If the polymer is a thermoplastic it may be softened by warming and pressed in a hydraulic press into a thin film. Alternatively, the polymer may be dissolved in a volatile solvent and the solution allowed to evaporate to a thin film on an alkali halide plate. Some polymers, such as cross-linked synthetic rubbers, can be cut into thin slices with a blade using a microtome. A solution in a suitable solvent is also a possibility. As non-destructive sampling techniques are preferred, reflectance and microspectroscopic techniques are readily applied to polymer samples.

As most polymers are organically based, the well-established spectral assignments made for organic molecules are helpful when interpreting the infrared spectra of polymers. There are useful reference spectra collections and spectral databases for polymers and their additives available, making the identification of these materials straightforward [Hummel and Scholl, 1981; Pouchert, 1995]. Correlation tables provide a useful summary to assist in identification; a typical table is illustrated in Figure 4.5 [Sandler et al., 1998; Stuart, 2004]. A flowchart can also be established as a schematic aid in identifying the type or class of polymer present. An example of a flowchart that will aid in identifying polymers from particular infrared bands is illustrated in Figure 4.6 [Derrick et al., 1999]. The infrared bands listed in Table 4.1 may be used to identify synthetic polymers commonly encountered by conservation scientists.

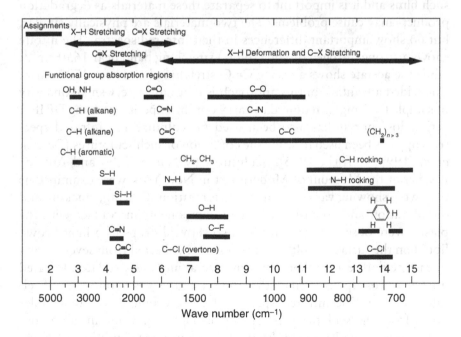

Figure 4.5 Correlation table for the infrared bands of polymers. Reprinted from Polymer Synthesis and Characterisation: A Laboratory Manual, ISBN 012618240X, Sandler et al., page 99, 1998 with permission from Elsevier

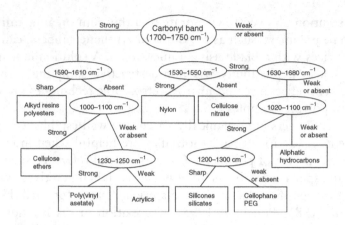

Figure 4.6 Flowchart for the identification of polymers based on infrared spectra. Reproduced by permission from Infrared Spectroscopy in Conservation Science, M R Derrick, D Stulick & J M Landry, 1999

An example of the use of infrared spectroscopy to identify polymeric materials is the study of photographic negative materials [Keneghan, 1998]. Cellulose acetate and cellulose nitrate have been used to produce such films and it is important to separate these materials as degradation products may cause problems. The two materials are physically similar, but do show important differences in their infrared spectra. The nitrate shows a strong nitrogen dioxide (NO_2) stretching band near 1650 cm^{-1}, while the acetate shows a strong C=O stretching band near 1730 cm^{-1}. The added advantage of this approach is that only a very small quantity of sample (~ 2 mg) is required to carry out the experiment using DRIFT.

Cellulose nitrate has also been used in sculpture and infrared spectroscopy has been used to study the condition of such sculptures [Derrick *et al.*, 1999; Derrick, 1995]. Sculptures by Naum Gabo and Antoine Pevsner at the Museum of Modern Art in New York were examined as they were showing various signs of deterioration. Crazing, cracking and discoloration and corrosion of metal components in contact with the plastic were observed. Two pieces also showed drops of a light brown liquid on the surface. Multiple samples were collected from several areas of each sculpture as broken fragments, scrapings and exudates. Infrared microspectroscopy confirmed that the sculptures were composed of cellulose nitrate. The infrared spectra of three cellulose nitrate samples taken from the sculptures are shown in Figure 4.7: one in good condition, one in moderate condition and one in poor condition [Derrick, 1995]. The spectra of camphor and liquid nitrate salt are also shown. Camphor was used as a plasticiser in the cellulose nitrate and nitrate

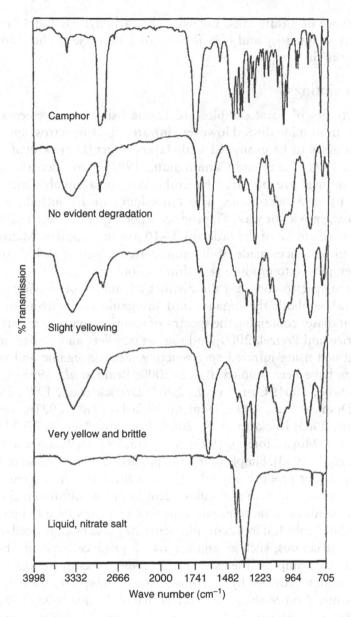

Figure 4.7 Infrared spectra for cellulose nitrate samples taken from sculptures [Derrick, 1995]

salts can be formed by the reaction of nitric acid, a degradation product, with fillers, stabilisers or colourants such as zinc oxide. Inspection of the cellulose nitrate spectra in Figure 4.7 reveals an increase in the intensity of the band at 1340 cm^{-1} in the more degraded samples; this is due to the presence of nitrate salts. Likewise the carbonyl band at 1735 cm^{-1} decreases in intensity and is indicative of a decrease in the amount of camphor in the sample.

4.2.3 Paintings

Many studies of paint samples are transmission measurements using potassium bromide discs. However, infrared microspectroscopy allows paint samples to be examined with layers intact [Derrick et al., 1999; Derrick, 1995; Tsang and Cunningham, 1991]. Cross section samples from paintings are prepared by embedding in a suitable media and cutting using a microtome. The embedding media include acrylics, polyesters, epoxies or wax. To produce a good quality spectrum, a paint sample needs to be of the order of 1–10 μm in thickness. Microtomes with cutting devices made of diamond, steel, tungsten carbide or glass can be employed to produce very thin sections.

Infrared spectroscopy has the distinct advantage of providing information about both the organic and inorganic components in paint. FTIR databases containing the spectra of painting components are available [Price and Pretzel, 2000] and pigments, fillers and binders may all be identified using infrared spectroscopy. Both inorganic and organic pigments have been studied [Bacci, 2000; Bruni et al., 1999; Casadio and Toniolo, 2001; Castro et al., 2003; Derrick et al., 1999; Derrick, 1995; Desnica et al., 2003; Domenech-Carbo et al., 2001b; Genestar and Pons, 2005; Kuckova et al., 2005; Low and Baer, 1977; Marengo et al., 2005; Moya Moreno, 1997; Newman, 1979; Salvado et al., 2005; Shearer et al., 1983]. Simple inorganic pigments, such as metal oxides or sulfides, do not produce any vibrations in the mid-infrared region; the lattice vibrations of such molecules occur in the far-infrared region. The infrared bands of some common inorganic pigments found in paintings are listed in Table 4.3. Inorganic pigments may also be composed of more complex structures, such as anions, which produce bands in the mid-infrared. An example of the use of infrared spectroscopy for identifying a blue pigment is provided by the comparison of the spectra for azurite and ultramarine. Azurite shows distinctive bands in the 1600–1300 cm^{-1} range due to carbonate (CO_3^{2-}) stretching groups, while ultramarine shows a strong band in the 1100–900 cm^{-1} region due to Si–O–Si and Si–O–Al stretching bands. Fillers (or extenders) that are also composed of inorganic compounds are included in Table 4.3 as well.

Organic pigments show more complex infrared spectra than inorganic pigments. Consideration must also be given to the fact that such pigments are more likely to change in structure due to ageing. Degradation

Table 4.3 Major mid-infrared bands of some common inorganic pigments and fillers

Compound	Wave number cm^{-1}
Azurite	3425,1490, 1415, 1090, 952, 837
Barium sulfate	1185, 1128–1120, 1082, 639, 614, 200
Cadmium sulfide	250
Calcite	1492–1429, 879, 706
Calcium sulfate	1140–1080, 620, 3700–3200
Chrome yellow	887
Chromium oxide	632, 566
Cinnabar	347, 285, 130
Gypsum	3500–3400, 1700–1600, 1150–1100, 700–600
Indigo	3400–3200, 3100-2800, 1700–1550, 1620–1420
Kaolinite	3700–3200, 1100–1000, 910–830
Lead chromate	905, 860, 830
Lead white	3535–3530, 1400, 1047–1045, 693–683
Litharge	375, 295
Malachite	3400, 3320, 1500, 1400, 1095, 1045
Naples yellow	666, 408
Orpiment	305, 183, 139
Prussian blue	3500–3000, 2083
Realgar	373, 367, 343, 225, 170
Red lead	530, 455, 320, 152, 132
Silica	1100–1000
Ultramarine	1150–950
Viridian	3500–3000, 1600, 555, 481
Zinc sulfide	290
Zinc white	400–500

produces products that will produce their own infrared spectra. The infrared bands of some common organic functional groups to aid in the identification of organic pigments are summarised in Table 4.1.

Paint binding media have been characterised using infrared spectroscopy in a number of studies [Bacci, 2000; Birshtein and Tulchinskii, 1977a; Birshtein and Tulchinskii, 1977b; Cappitelli, 2004; Cappitelli and Koussiaki, 2006; Danillia *et al.*, 2002; Domenech-Carbo *et al.*, 1996, 1997, 2001a, 2001b; Felder-Casagrande and Odlyha, 1997; Jurado-Lopez and Luque de Castro, 2004; Kuckova *et al.*, 2005; Learner, 2005, 2004, 1996; Learner, 1998; Meilunas *et al.*, 1990; Moya Moreno, 1997; Odlyha, 1995; Ortega-Aviles *et al.*, 2005; Salvado *et al.*, 2005]. Protective varnishes, composed of waxes or resins, may also be investigated. Table 4.2 list the infrared bands of some common varnish and binder components including oils, resins and proteins. The infrared spectra of some polymers that are used in paint binders, including acrylics, alkyds and PVA, are illustrated in the Appendix. There will clearly be overlap between a number of these components, which needs to be considered when interpreting spectra. For example, there can be difficulty in detecting egg yolk in the presence of an oil, such as linseed, since both these components contain triglycerides. However, egg yolk also contains protein, so the characteristic protein bands can be used to identify the egg.

It is important to note that ageing can cause significant changes to the infrared bands of certain binding media. For example, the ageing of linseed oil is a result of the oxidation of the triglycerides and the differences will be observed in the spectra [Meilunas et al., 1990]. Before ageing the oil shows C–H stretching bands in the region 3000–2800 cm^{-1}, a C=O stretching band at 1746 cm^{-1}, CO–O–C stretching bands in the region 1250–1100 cm^{-1} and cis–C=CH bands at 3010, 1652 and 722 cm^{-1}. After heating at 120 °C for 24 hours, the resulting infrared spectrum of linseed oil shows significant changes. The bands at 3010 and 722 cm^{-1} disappear, while a band at 970 cm^{-1} appears. These changes indicate a loss of cis double bonds and the appearance of trans double bonds. Broad carbonyl bands also appear in the 1850–1600 cm^{-1} region and O–H stretching bands in the 3600–3000 cm^{-1} as a result of the formation of oxidation products.

FTIR imaging has emerged as an excellent new analytical approach to the study of paint cross sections [van der Weerd et al., 2002; van der Weerd, 2002; van Loon and Boon, 2004]. The problem of grinding samples and opaqueness are avoided. This technique provides information about the spatial distribution of characteristic functional groups in a paint cross-section. A false colour plot is used to represent the distribution of a particular absorption band across the surface of the different colours. A FTIR imaging study of Rembrandt's 'Portrait of a Standing Man' was able to identify the characteristics of the different layers making up the painting [van der Weerd et al., 2002; van der Weerd, 2002]. Silicates were identified in a red ground layer, lead white and lead carboxylates in a white ground layer, calcium carbonate (CaCO$_3$) in a paint layer and various organic functional groups were located in the varnish layers. Although this technique is currently limited to a spatial resolution of about 7 μm, improvements in technology should see this improve.

Fibre optic sampling in the mid-infrared is also possible for the nondestructive analysis of paint layers [Bacci et al., 2001; Dupuis et al., 2002; Fabbri et al., 2001]. However, it is important to be aware that such measurements can show large distortions in the spectrum, affecting

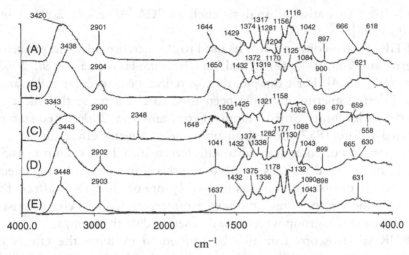

(A) Tang Dynasty; (B) Song Dynasty (C) Xi Xia Dynasty;
(D) Late Qing Dynasty; (E) Modern Era

Figure 4.8 FTIR spectra of rice paper of five different eras. Reprinted from Talanta, Vol 64, Na *et al.*, 1000–1008, 2004 with permission from Elsevier

the band shape and frequency. Thus, it is difficult to compare reflectance spectra with those collected in transmission.

4.2.4 Written Material

FTIR spectroscopy has been used in a number of studies of historic paper. The ATR and diffuse reflectance sampling techniques are suitable for non-destructive examination of such materials. Paper shows a characteristic infrared spectrum in the mid-infrared and near-infrared regions, depending on the source and age of the sample and this information can be used to categorise paper [Ali *et al.*, 2001; Lang *et al.*, 1998; Na *et al.*, 2004; Sistach *et al.*, 1998]. Figure 4.8 illustrates the diffuse reflectance spectra of rice paper from calligraphies held by the National Library of China from different eras: the Tang Dynasty, the Song Dynasty, the Xi Xia Dynasty, the Late Qing Dynasty and the modern era [Na *et al.*, 2004]. The spectrum for each era is characteristic of that period, based on band positions and peak height ratios. However, the spectra of samples D and E belong to samples of eras nearer in time and show more similarity than to the other spectra. As there are often subtle differences between the FTIR spectra of paper samples, this data lends itself to statistical analysis such as PCA [Ali *et al.*, 2001; Lang *et al.*, 1998].

FTIR spectroscopy can also be used to characterise inks and pigments [Ferrer and Vila, 2006; Ferrer and Sistach, 2005; Havermans *et al.*, 2005; Na *et al.*, 2004]. For example, diffuse reflectance, specular reflectance and infrared microscopy have been used to characterise the ink used on the 19th century British stamps [Ferrer and Vila, 2006]. Red stamps printed before 1857 showed an unintended blueing effect due to the incorporation of potassium hexacyanoferrate into the ink. During 1857 the formula of red ink was changed and this stopped the blueing effect, so whether or not a stamp was printed before or after 1857 affects the value of the stamp. Potassium hexacyanoferrate shows a characteristic band for the CN group which can be used to date the stamps.

FTIR spectroscopy can also be utilised to examine the effects of cleaning treatments, such as plasma or laser cleaning [Kolar *et al.*, 2003; Laguardia *et al.*, 2005]. FTIR can be used to monitor any changes in composition of the paper.

Infrared spectroscopy can be employed to characterise parchment and be used to detect signs of deterioration. During the 1980s, the Getty Conservation Institute began a study of the optimum storage and display conditions for the Dead Sea Scrolls. Infrared spectroscopy has been used to evaluate their state of degradation [Derrick, 1995; Derrick *et al.*, 1999]. Fragments (2 mm × 1 mm) were examined using step-scan analysis by infrared microspectroscopy. The fragments were embedded in polyester-styrene and cut with a microtome to produced 10μm sections. A linear spectral map was produced across the width of the sample in 15μm increments. As collagen is the major component of parchment and can be converted to gelatin with time, an examination of the changes to the collagen spectrum provided information about the degree of deterioration. Collagen denaturation is monitored using infrared spectroscopy by observing a shift in the amide II band position from 1550 to 1530 cm^{-1} and changes in the relative intensities of the amide I and II bands [Brodsky-Doyle *et al.*, 1975]. The spectral map of a Dead Sea Scroll sample section is illustrated in Figure 4.9 [Derrick *et al.*, 1999]. The window was stepped from the flesh side of the parchment (scan 1) to the skin side of the parchment (scan 12). The shift in the amide II band and the change in the amide I and II relative intensities are localised in the exterior parts of the sample. The denaturation of the collagen in this sample occurs mainly in the outer 20–50μm and the skin side of the sample shows more degradation, based on the collagen denaturation.

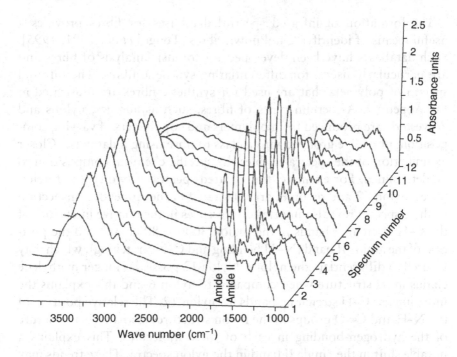

Figure 4.9 Linear infrared spectral map of a Dead Sea Scroll cross-section sample. Reproduced by permission from Infrared Spectroscopy in Conservation Science, M R Derrick, D Stulick & J M Landry, 1999

4.2.5 Textiles

Infrared spectroscopy is a useful technique for the characterisation of natural and synthetic fibres [Derrick *et al.*, 1999]. Fibres for analysis may be collected from objects such as textiles with fine forceps. Loose fibres from a surface may be collected using adhesive tape, but thought must be given to the possibility of contamination of the surface as a result of this method of collection. A scalpel can be used to slice thin sections from a fibre.

A variety of sampling techniques may be used to investigate fibres. If a fibre is not too thick, a conventional transmission method may be used. The fibre can be examined by taping the ends across a hole in a metal disk or by simply laying it on an infrared window [Tungol *et al.*, 1991, 1995]. Infrared microscopy is the most useful for examining small samples of fibres and dyes. Thicker fibres can be flattened to increase the surface area and reduce the thickness: a diamond anvil cell is an easy method for flattening fibres. Generally fibres are too thick for transmission techniques and reflection methods are used. ATR spectroscopy is particularly useful for analysis larger samples such as fabrics.

The formation of infrared spectral databases for fibres provides a useful means of identifying unknown fibres [Tungol et al., 1991, 1995]. Such databases have been developed for forensic analysis of fibres and are particularly useful for differentiating synthetic fibres. The infrared spectra of polymers that are used for synthetic fibres are illustrated in the Appendix. As certain types of fibres, such as acrylics, nylons and polyesters, are produced from polymers and copolymers of varying composition, fibres are grouped into a class of fibre using a database. Closer examination of the spectra enables the specific chemical composition to be determined. For example, the infrared spectra of two types of nylon fibres, nylon 12 and nylon 6, are illustrated in the Appendix. Inspection of these spectra reveals important differences in the relative intensities of the C–H stretching bands in the region $3000–2800$ cm^{-1} and the position of the amide II band (N–H bending and C–N stretching), which may be used to differentiate the nylons. Nylon 12 possesses longer methylene chains in its structure when compared to nylon 6 and this explains the more intense C–H stretching bands for nylon 12. The relative position of the N–H and C=O groups in the nylon structures also affects the nature of the hydrogen-bonding in each of these molecules. This explains a notable shift in the amide II band in the nylon spectra. These trends may also be applied to other nylons and be used to identify unknown nylons.

The infrared spectra of natural textile fibres show characteristic protein bands. Although the spectra can appear similar, the relative intensities of the bands will vary according to the type of protein present. For example, the infrared spectra of wool and silk fibres are illustrated in Figure 4.10. Wool is made of keratin, while silk is composed of fibroin and the intensities of the infrared bands are different due to the different composition of these proteins.

Changes in the crystalline properties of historic textile fibres can be used as an indication of degradation. Such changes are readily identified using polarised infrared spectroscopy [Cardamone, 1989; Chen and Jakes, 2001; Garside et al., 2005; Garside and Wyeth, 2002]. Various parts of the spectrum have been used to monitor degradation, including the carbonyl band as an indication of oxidative degradation of cellulose [Cardomone, 1989]. Different peaks are chosen, depending on the type of material, to calculate an infrared crystallinity ratio. For example, one study of historic cotton fibres used the relative peak intensities at 2900 and 1372 cm^{-1} to calculate the ratio [Chen and Jakes, 2001].

It is also possible to analyse dyes on historic textile fibres using FTIR microscopy [Gillard et al., 1993, 1994]. A single fibre can be sufficient to obtain a signal if the quantity of dye is adequate. As the technique

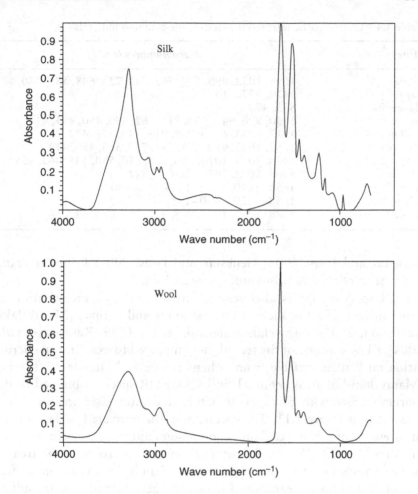

Figure 4.10   Infrared spectra of wool and silk

is non-destructive, this approach is a useful step before, for instance, destructive chromatographic techniques are used. A reference library of the infrared spectra of suitable dyes is required so an unknown may be identified.

4.2.6 Stone

Naturally occurring minerals found in stone have been well characterised using infrared spectroscopy and comprehensive databases and hand-books are available [Farmer, 1974; Ferraro, 1982; Gadsen, 1975; Price et al., 1998; van der Marel and Beutelspacher 1976]. The main infrared bands for some common minerals are listed in Table 4.4. FTIR has also been widely used to characterise historic mortars [Bakolas et al., 1995;

Table 4.4   Characteristic infrared bands of some common minerals

Mineral	Wave numbers (cm^{-1})
Albite	1096, 1032, 990, 784, 762, 742, 723, 648, 588, 530, 425
Calcite	1420, 877, 714
Haematite	535, 475
Illite	1030, 990, 948, 905, 815, 762, 490, 460, 431, 414
Kaolinite	1117, 1033, 1010, 938, 915, 540, 472, 432
Muscovite	1062, 1022, 990, 935, 754, 727, 553, 480, 412
Orthoclase	1120, 1040, 1010, 770, 728, 650, 580, 535, 463, 428
Quartz	1160, 1082, 797, 778, 695, 512
Weddelite	1650–1620, 1330–1315, 790–780
Whewellite	1650–1620, 1330–1315, 790–780
Wollastonite	1088, 1064, 1023, 968, 930, 905, 684, 647, 563

Genestar and Pons, 2003; Genestar and Pons, 2005; Genestar *et al.*, 2005; Silva *et al.*, 2005; Biscontin *et al.*, 2002].

FTIR spectroscopy is also very effective for the characterisation of stone surface deposits such as encrustations and patinas [Maravelaki-Kalaitzaki, 2005; Maravelaki-Kalaitzaki *et al.*, 1999; Rampazzi *et al.*, 2004]. FTIR was one of the techniques employed to examine an encrustation on Pentelic marble from Athens before and after laser cleaning [Maravelaki-Kalaitzaki *et al.*, 1999]. The FTIR spectra obtained of the marble surface with black crust before and after laser treatment are illustrated in Figure 4.11. The spectrum of the untreated surface shows the presence of calcium carbonate, potassium nitrate and calcium sulfate dihydrate ($CaSO_4.2H_2O$). By comparison, the spectrum of the treated surface shows a decrease in intensity of the bands due to calcium sulfate dihydrate and the appearance of bands characteristic of calcium sulfate hemihydrate ($CaSO_4.0.5H_2O$) and calcium sulfate anhydrite ($CaSO_4$). The hemihydrate and anhydrite result from the loss of water molecules from gypsum as a result of treatment. Being able to monitor the transformation of gypsum by laser treatment is of importance to conservators as the goal is to remove surface layers of dirt without damaging the inner gypsum layer and underlying stone.

4.2.7   Ceramics

FTIR spectroscopy is used to characterise the mineral content of ceramic materials [Barone *et al.*, 2003; Barilaro *et al.*, 2005; De Benedetto *et al.*, 2002, 2005; Maritan *et al.*, 2006; Romani *et al.*, 2000]. Samples are generally examined in powder form in a potassium bromide disc or

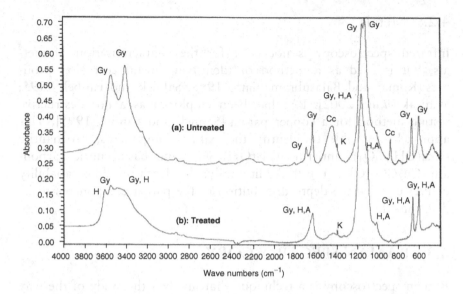

Figure 4.11   FTIR spectra of black crust on marble (a) before and (b) after laser treatment. Reprinted from Applied Surface Science, Vol 148, Maravelaki-Kalaitzaki *et al.*, 92–104, 1999 with permission from Elsevier

by using a diffuse reflectance approach as milligram quantities can be studied. As for stone structures, the minerals used to produce ceramics have been well characterised using infrared spectroscopy [Farmer, 1974; Ferraro, 1982; Gadsen, 1975; Price and Pretzel, 1998; van der Marel and Beutelspacher, 1976] and Table 4.4 lists the characteristic infrared bands of some common minerals that may be found in ceramics. Statistical analysis, such as PCA, of infrared data obtained for groups of ceramics enables samples to be classified [De Benedetto *et al.*, 2005].

4.2.8   Glass

Infrared reflectance spectroscopy can be used to examine glass surfaces for signs of deterioration [Schreiner *et al.*, 1999; Vilarigues and Da Silva, 2004]. This approach has been used to examine structural changes in the surface layer of medieval glass windows. Changes due to the leaching of alkaline and earth alkaline glass components appear as wave number shifts and intensity changes to the non-bridging and bridging Si-O stretching bands at 950 and 1050 cm^{-1}, respectively. FTIR spectroscopy has also been employed in a study of adhesives used in the restoration of archaeological glass [Lopez-Ballester *et al.*, 1999].

4.2.9 Metals

Infrared spectroscopy is not used for the characterisation of metals, but is used as a method of identifying metal corrosion products [Kumar and Balasubramaniam, 1998; Salnick and Faubel, 1995; Wadsak *et al.*, 2002]. PAS has been employed as a non-destructive testing method for a copper patina [Salnick and Faubel, 1995]. This approach was able to identify the patina components, brochantite ($Cu_4(OH)_6SO_4$), antlerite ($Cu_3(OH)_4SO_4$) and basic cupric carbonate ($Cu_2CO_3(OH)_6.H_2O$). Additionally, the depth-profiling capability of PAS enabled the depth distribution of the patina components to be determined.

4.3   RAMAN SPECTROSCOPY

Raman spectroscopy is a technique that involves the study of the way in which radiation is scattered by a sample [Banwell, 1983; Cariati and Baini, 2000; Edwards and Chalmers, 2005; Edwards and de Faria, 2004; Ferraro, 2003; Long, 2005]. The source of radiation used may be in the near-UV, visible or near-infrared regions of the spectrum. When radiation falling on a molecule does not correspond to that of an absorption process, it is scattered. Most of the scattered radiation is unchanged in wavelength and is known as Rayleigh scattering. A small proportion of the scattered light is slightly increased or decreased in wavelength and this is known as Raman scattering. When the wavelength is increased, the process is known as Stokes Raman scattering, while a decreased wavelength is associated with anti-Stokes Raman scattering.

Raman scattering results from the same type of quantised vibrational changes that are associated with infrared spectroscopy and the difference in wavelength between the incident and scattered radiation corresponds to wavelengths in the mid-infrared region. Raman scattering by molecules involves transitions between rotational or vibrational states. In order to be Raman active, a molecular rotation or vibration must cause a change in a component of the molecular polarisability – a measure of the degree to which a molecule can have its positive and negative charges separated. As a result of their different origins, infrared and Raman spectra provide complementary information. Raman spectra are plotted in terms of the wave number shift from the excitation radiation, $\Delta \bar{v}$. However, it is common to see just wave number used on the scale as Stokes Raman bands are shown. The $\Delta \bar{v}$ scale increases to the left. The ordinate scale shows intensity.

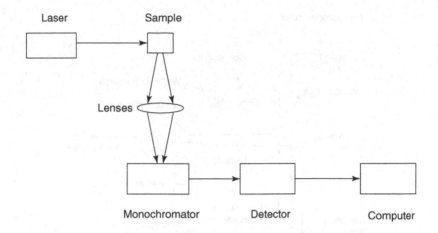

Figure 4.12   Schematic diagram of a Raman spectrometer

The layout of a typical dispersive Raman spectrometer is illustrated in Figure 4.12 [Ferraro, 2003; Stronmen and Nakamoto, 1984]. Helium–neon lasers at a wavelength of 632.8 nm or an argon laser at 514.5 nm are often used as an excitation source. A consideration with the choice of a source is the potential for fluorescence produced by the sample or impurities that can swamp the spectrum. However, fluorescence can be minimised by using a source of higher wavelength near-infrared sources. A neodymium–yytrium aluminium garnet (Nd-YAG) laser at 1064 nm can be used with a Fourier transform Raman spectrometer [Hendra *et al.*, 1991]. An interferometer is required when working with higher wavelengths, as the intensity of the scattered signal is dramatically decreased when the radiation moves to higher values. The interferometer allows for the collection of weak spectra. The scattered light from the sample passes through a series of focusing and collection optics. Optical filters are used for the rejection of Rayleigh scattered light and the light is sent to the detector. A number of detectors are used in Raman spectrometers – photomultiplier tubes, photodiode array detectors, charge-coupled devices and indium gallium arsenide detectors, depending on the type of instrument used.

Raman microscopy is an important analytical technique for a range of conservation problems. This method enables the Raman spectra with 1 μm spatial resolution of samples in the picogram range to be investigated [Corset *et al.*, 1989; Ferraro, 2003; Smith and Clark, 2004]. A typical set up involves coupling a microscope to a spectrometer; a schematic diagram of the layout of a Raman microscope is shown in Figure 4.13. The sample is placed on the stage of the microscope and

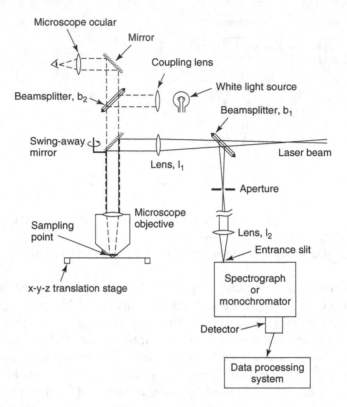

Figure 4.13 Schematic diagram of a Raman microscope. Reprinted from Endeavour, Vol 16, Best *et al.*, 66–73, 1992 with permission from Elsevier

illuminated by white light and brought into focus by adjusting the objective. The illuminator lamp is switched off and the laser radiation is directed to the beamsplitter. The scattered light from the sample is collected by the objective and is sent to the spectrometer. Low laser powers (less than 5 mW) at the sample are used to reduce the possibility of degradation.

There are several special Raman techniques that are available to aid in the recording of spectra. Resonance Raman spectroscopy (RRS) uses incident radiation that nearly coincides with the frequency of an electronic transition of the sample [Edwards and Rull Perez, 2004]. During the process an electron is promoted into an excited electronic state followed by an immediate relaxation to a vibrational level of the electronic ground state. RRS is characterised by a much greater intensity of the scattered radiation. Surface enhanced Raman spectroscopy (SERS) involves obtaining the Raman spectra of samples adsorbed on metal surfaces such as silver or gold. Again, the intensity of certain bands in the spectrum is enhanced for a molecule on a metal surface [Edwards and Rull Perez, 2004; Ferraro, 2003; Long, 2005].

Like infrared spectroscopy, Raman spectroscopy may be used for imaging [Ferraro, 2003; Gardiner and Graves, 1989]. In Raman imaging, the Raman spectrum of the entire illuminated area is measured with a charge-coupled device camera. The data is transferred to a monochromator where a frequency is chosen and the distribution of the component associated with the characteristic frequency is identified throughout the illuminated area.

Raman spectroscopy has the advantage that a variety of sample types can be examined with minimal preparation [Ferraro, 2003]. Solid polymers in the form of powders, films or fibres can be studied: there is a range of appropriate solid cells commercially available. Liquids can be examined using capillaries or ampules depending on the amount of sample available. Temperature cells are also available. If there is a risk of overheating the sample with laser radiation, the sample can be cooled or a rotating cell might be used. Fibre optic sampling is also feasible with Raman spectroscopy as visible or near-infrared radiation can be transmitted over considerable distances through optical fibres, enabling difficult or bulky samples to be examined [Edwards and Rull Perez, 2004].

4.3.1 Paintings

Raman spectroscopy is an excellent tool for the characterisation of paintings and has been used to examine easel, panel and wall paintings, as well as rock art, from a range of periods. While certain paintings may be of a size which enable direct Raman analysis, where larger pieces are to be examined, it may require a small scraping to be sampled for study. With the continual improvement in portable Raman spectrometers, it is becoming increasingly feasible to directly analyse paintings in situ with a fibre optic probe [Vandenabeele *et al.*, 2001a]. An important consideration when carrying out Raman experiments on paintings is the potential for damage through excessive laser power.

Raman spectroscopy is highly applicable when identifying pigments in paintings and many studies have been reported [Bell *et al.*, 1997; Burgio and Clark, 2001; Edwards, 2005a, 2000; Otieno-Alego, 2000; Perardi *et al.*, 2000; Rull Perez, 2001; Vandenabeele and Moens, 2005; and references therein]. Inorganic pigments produce good quality Raman spectra as they are crystalline materials – such materials show sharp Raman bands with lattice vibrations appearing in the low wave number region of the spectrum. However, pigments can often produce considerable fluorescence which can hide the Raman spectrum, but the use of a larger excitation wavelength can avoid this problem. The study of the Raman spectra of synthetic organic pigments is more problematic. There is a large number of 20th century organic pigments which produce strong fluorescence and generally produce much weaker spectra than their inorganic counterparts. However, Raman databases suitable for identifying pigments in modern paintings are being developed [Vandenabeele *et al.*, 2001b, 2000a].

Collections of the Raman spectra of commonly encountered pigments have been published and a collection of the spectra may be accessed online [Bell *et al.*, 1997; Burgio and Clark, 2001; Burrafato *et al.*, 2004; Castro *et al.*, 2005]. A list of the Raman bands observed in the spectra of a number of paint pigments recorded using conventional and FT-Raman methods is given in Table 4.5 [Bell *et al.*, 1997; Burgio and Clark, 2001; Edwards *et al.*, 2000; Edwards, 2005a, 2002; Vandenabeele and Moens, 2004]. The Raman spectra of the pigments identified in prehistoric art encountered in rock art have also been collected [Edwards, 2002, 2005a; Edwards *et al.*, 2000; Bouchard and Smith, 2005]. The Raman bands associated with the minerals used as pigments in prehistoric times are also included in Table 4.5. It should be noted that some of the bands listed can appear weak and may not be easily observed, depending on the method used.

Raman spectroscopy has the added advantage that it may be used to differentiate pigments with the same chemical formula but of different crystalline structures (polymorphs). For example, the white titanium dioxide pigments rutile and anatase each show unique Raman spectra, enabling unambiguous identification. The Raman spectra of pigments of comparable composition may be distinguished and the example of chrome yellow deep and chrome yellow orange, both with the structure $PbCrO_4.PbO$, is illustrated in Figure 4.14 [Bell *et al.*, 1997]. The degradation of pigments can also be investigated using Raman spectroscopy. For example, pararealgar (yellow) is recognised as a polymorph of realgar (orange) as a result of a photodegradation process [Clark and Gibbs, 1997a,1998].

The organic binding media and varnishes of paintings may also be identified using Raman spectroscopy [Burgio and Clark, 2001; Vandenabeele *et al.*, 2000b, 2004]. However, as the composition of a number of these organic compounds varies due to their biological nature, care must be taken when identifying such compounds by comparison with 'reference' spectra. The Raman bands observed for a selection of commonly encountered binding media and varnishes are listed in Table 4.6 [Burgio and Clark, 2001; Vandenabeele *et al.*, 2000b].

Table 4.5 Raman bands of pigments

Pigment	Excitation wavelength (nm)	Wave numbers (cm^{-1})
White		
Anatase	1064	144, 201, 397, 512, 634
Bone white	632.8	431, 590, 961, 1046, 1071
Calcite	1064	154, 282, 712, 1086
	514.5	157, 282, 1088
Gypsum	1064	140, 181, 493, 619, 670, 1007, 1132
	514.5	181, 414, 493, 619, 670, 1007, 1132
Lead white	1064	415, 681, 1051, 1055
		154, 203, 260, 353, 418, 1056
	514.5	665, 687, 829, 1050
Lithopone	1064	348, 453, 462, 617, 647, 988
	514.5	216, 276, 342, 453, 616, 647, 988
Barium white	1064	454, 464, 619, 648, 989, 1087, 1105, 1142, 1168
	632.8	453, 616, 647, 988
Rutile	1064	144, 232, 447, 609, 147, 242, 440, 611
Zinc white	1064	100, 331, 438, 489
	514.5	331, 383, 438
Yellow		
Barium yellow	1064	68, 351, 361, 404, 412, 428, 864, 873, 885, 901
	514.5	352, 403, 427, 863, 901
Berberine	1064	714, 732, 754, 772, 837, 979, 1105, 1120, 1145, 1205, 1237, 1278, 1345, 1365, 1398, 1426, 1449, 1501, 1521, 1570, 1623, 1636, 2849, 2912, 2954, 3001, 3022, 3074
	632.8	1203, 1235, 1276, 1342, 1361, 1397, 1424, 1449, 1501, 1518, 1568, 1626
Cadmium yellow	514.5	304, 609
Chrome yellow light	1064	141, 339, 361, 378, 405, 842, 864
Chrome yellow deep	632.8	336, 358, 374, 401, 838
Chrome yellow-orange	1064	63, 151, 341, 359, 824
	632.8	149, 828
Cobalt yellow	1064	111, 180, 276, 305, 821, 837, 1258, 1327, 1398
	632.8	179, 264, 304, 821, 836, 1257, 1326, 1398, 1215, 1246, 1265, 1330, 1433, 1592, 1633
Gamboge	1064	1224, 1249, 1281, 1333, 1383, 1437, 1594, 1634
Hansa yellow	1064	70, 85, 95, 118, 124, 158, 177, 185, 212, 284, 353, 386, 394, 414, 617, 626, 655, 742, 761, 770, 785, 823, 849, 953, 1001, 1068, 1111, 1141, 1181, 1192, 1257, 1306, 1325, 1336,

Table 4.5 *(continued)*

Pigment	Excitation wavelength (nm)	Wave numbers (cm⁻¹)
Indian yellow	632.8	1360, 1386, 1403, 1451, 1491, 1534, 1561, 1568, 1605, 1619, 1672 484, 610, 631, 697, 772, 811, 877, 1009, 1047, 1097, 1127, 1178, 1218, 1266, 1345, 1414, 1476, 1503, 1599
Lead-tin yellow (type I)	1064	81, 113, 131, 197, 276, 294, 459
	514.5	129, 196, 291, 303, 379, 457, 525
Lead-tin yellow (type II)	514.5	138, 324
Mars yellow	632.8	245, 299, 387, 480, 549
Massicot	1064	72, 87, 142, 285
	632.8	143, 289, 385
	647.1	52, 77, 88, 143, 289, 385
Mosaic gold	1064	314
Naples yellow	1064	74, 88, 144, 289, 342, 345, 465
	632.8	140, 329, 448
Orpiment	1064	70, 107, 137, 155, 180, 183, 193, 203, 221, 294, 308, 353, 361, 369, 384
	632.8	136, 154, 181, 202, 220, 230, 292, 309, 353, 381
Pararealgar	1064	118, 142, 153, 158, 172, 176, 191, 197, 204, 230, 236, 275, 315, 320, 333, 346, 364, 371
	632.8	141, 152, 157, 171, 174, 190, 195, 202, 222, 229, 235, 273, 319, 332, 344
Saffron	1064	1020, 1166, 1210, 1283, 1537, 1613
	514.5	1165, 1210, 1282, 1536
Strontium yellow	1064	339, 348, 374, 431, 859, 867, 890, 895, 917, 931
	514.5	339, 348, 374, 431, 865, 893, 916, 930
Turmeric	1064	964, 1171, 1186, 1249, 1312, 1431, 1438, 1602, 1632
Yellow ochre	632.8	240, 246, 300, 387, 416, 482, 551, 1008
Zinc yellow	632.8	343, 357, 370, 409, 772, 872, 892, 941

Red

Pigment	Excitation wavelength (nm)	Wave numbers (cm⁻¹)
Bright red	1064	75, 99, 149, 247, 298, 347, 386, 431, 442, 454, 463, 528, 573, 619, 681, 725, 731, 746, 813, 968, 989, 1063, 1099, 1109, 1162, 1205, 1231, 1244, 1261, 1282, 1332, 1359, 1376, 1393, 1449, 1463, 1484, 1552, 1580, 1607
Cadmium red	1064	136, 239, 335, 988
Carmine	1064	465, 474, 1108, 1257, 1314, 1440, 1489, 1529, 1645
	780.0	375, 434, 453, 520, 554, 687, 958, 1003, 1076, 1091, 1149, 1225, 1296, 1460, 1572, 1636

Table 4.5   (*continued*)

Pigment	Excitation wavelength (nm)	Wave numbers (cm^{-1})
Cinnabar	1064	252, 282, 345
Cuprite		146, 217, 242
Hansa red		77, 104, 125, 139, 169, 197, 342, 366, 384, 403, 424, 457, 480, 503, 514, 618, 678, 724, 798, 844, 925, 987, 1077, 1084, 1099, 1129, 1159, 1187, 1217, 1252, 1258, 1282, 1309, 1322, 1334, 1397, 1446, 1482, 1496, 1527, 1556, 1576, 1607, 1622
Goethite	1064	118, 203, 241, 299, 393, 533
Haematite	632.8	224, 243, 290, 408, 495, 609
	1064	224, 244, 292, 409, 610
Litharge	1064	83, 147, 289, 339
	632.8	145, 285, 336
Kermes	1064	1451, 1603
Madder	1064	239, 485, 841, 1189, 1221, 1292, 1327, 1354, 1482, 1519, 1577, 1635
Mars red	632.8	224, 291, 407, 494, 610, 660
	1064	294
Purpurin	632.8	953, 1019, 1049, 1091, 1138, 1160, 1229, 1312, 1334, 1394, 1452
Realgar	1064	56, 61, 66, 125, 144, 167, 172, 183, 193, 213, 231, 329, 344, 355, 369, 375
	632.8	142, 164, 171, 182, 192, 220, 233, 327, 342, 354, 367, 375
red lead	647.1	54, 65, 86, 122, 152, 225, 313, 391, 549
	632.8	122, 149, 223, 313, 340, 390, 480, 548
	1064	65, 122, 144, 148, 150, 224, 314, 391, 456, 477, 55
Red ochre	632.8	220, 286, 402, 491, 601
Tyrian purple	1064	110, 126, 190, 308, 386, 693, 760, 1051, 1105, 1212, 1254, 1312, 1366, 1444, 1584, 1626, 1702
Vermillion	647.1	42, 253, 284, 343
	632.8	252, 282, 343
	1064	253, 284, 343
<u>Blue</u>		
Azurite	1064	86, 115, 137, 155, 177, 249, 282, 332, 401, 739, 765, 838, 938, 1096, 1427, 1459, 1578
	514.5	145, 180, 250, 284, 335, 403, 545, 767, 839, 940, 1098, 1432, 1459, 1580, 1623
Cerulean blue	514.5	495, 532, 674
Cobalt blue	514.5	203, 512

Table 4.5   (*continued*)

Pigment	Excitation wavelength (nm)	Wave numbers (cm^{-1})
Egyptian blue	514.5	114, 137, 200, 230, 358, 377, 430, 475, 571, 597, 762, 789, 992, 1012, 1040, 1086
	1064	431, 465, 1086
Indigo	1064	98, 136, 172, 181, 236, 253, 265, 277, 311, 320, 468, 546, 599, 676, 758, 862, 871, 1015, 1149, 1191, 1226, 1248, 1310, 1363, 1461, 1483, 1572, 1584, 1626, 1701
Lazurite	514.5	258, 548, 822, 1096
	1064	378, 549
Phthalocyanine blue	514.5	166, 173, 233, 255, 592, 681, 747, 777, 841, 952, 1007, 1037, 1106, 1126, 1339, 1450, 1470, 1482, 1527, 1591, 1610
Posnjakite	632.8	135, 208, 278, 327, 467, 612, 983, 1092, 1139
Prussian blue	514.5	282, 538, 2102, 2154
Smalt	514.5	462, 917
Green		
Atacamite	1064	106, 121, 361, 419, 446, 475, 51, 818, 846, 912, 975
	514.5	122, 149, 360, 513, 821, 846, 911, 974
Brochantite	514.5	90, 104, 118, 123, 130, 139, 149, 156, 169, 176, 187, 196, 242, 318, 365, 390, 421, 450, 481, 508, 595, 609, 620, 730, 873, 912, 973, 1077, 1097, 1125, 3225, 3258, 3371, 3399, 3470, 3564, 3587
Chromium oxide	514.5	221, 308, 359, 552, 611
Cobalt green	514.5	328, 434, 471, 555
Copper chloride	514.5	95, 106, 118, 140, 149, 213, 362, 416, 477, 511, 575, 799, 819, 845, 869, 893, 910, 928, 971
Emerald green	514.5	122, 154, 175, 217, 242, 294, 325, 371, 429, 492, 539, 637, 685, 760, 835, 951, 1355, 1441, 1558, 2926
Malachite	514.5	155, 178, 217, 268, 354, 433, 509, 553, 558
Phthalocyanine green	1064	689, 706, 742, 777, 1212, 1292, 1341, 1393, 1538
Scheele's green	514.5	136, 201, 236, 275, 370, 445, 495, 537, 657, 780
terre-verte	514.5	145, 399, 510, 636, 685, 820, 1007, 1084
Verdigris 'raw'	514.5	126, 180, 233, 322, 703, 949, 1360, 1417, 1441, 2943, 2990, 3027

Table 4.5 *(continued)*

Pigment	Excitation wavelength (nm)	Wave numbers (cm^{-1})
Verdigris (no. 1)	514.5	139, 181, 231, 328, 392, 512, 618, 680, 939, 1351, 1417, 1441, 1552, 2937, 2988, 3026
Verdigris (no. 2)	514.5	193, 271, 321, 371, 526, 619, 676, 939, 1351, 1424, 1524, 2939, 3192, 3476, 3573
Viridian	514.5	266, 487, 552, 585
Black		
Bone black	1064	1590, 1360, 1070, 964, 670
Graphite	780.0	1315, 1579
	1064	1590, 1360
Lamp black	1064	1590, 1360
	632.8	1325, 1580
Ivory black	632.8	961, 1325, 1580
Magnetite/Mars black	1064	612, 412, 292
Orange/brown		
Mars orange	632.8	224, 291, 407, 494, 608

4.3.2 Written material

Raman spectroscopy has emerged as an extremely important technique for the study of manuscripts, particularly medieval manuscripts [Best et al., 1993, 1992; Clark, 1995a, 1995b, 1999, 2002, 2005; Edwards, 2000; Otieno-Alego, 2000; Vandenabeele and Moens, 2004]. With improvements in non-destructive and in situ sampling Raman techniques, the number of studies of manuscripts has expanded, particularly for the study of illuminated medieval manuscripts. Where a Raman microscope is equipped with a large stage, an item can be placed directly in the laser beam. Often a glass plate is placed directly on top of the manuscript to reduce the risk of damage via too much laser power.

Raman microscopy is recognised as the most effective technique for identifying the pigments and dyes used to illuminate manuscripts. Many Raman studies of illuminated medieval manuscripts have been reported [Best et al., 1993; Bicchieri et al., 2000; Clark, 1995a, 1995b, 1999, 2002, 2005; Vandenabeele and Moens, 2004; and references therein]. The Raman bands exhibited by the pigments likely to be found in a medieval manuscript can be found in Table 4.5. Raman microscopy has proved effective for identifying the components of pigment mixtures, even down to grain sizes of 1 μm [Clark, 1995a]. For example, in

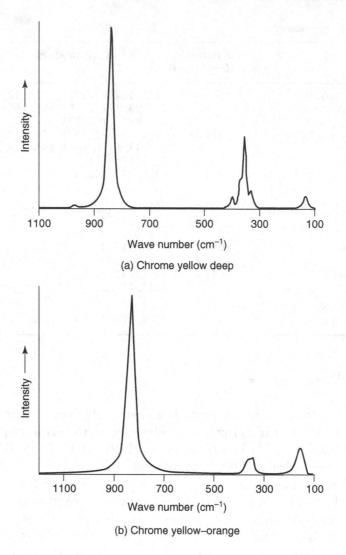

Figure 4.14 Raman spectra of (a) chrome yellow and (b) chrome yellow orange pigments. Reprinted from Spectrochimica Acta A, Vol 53, Bell *et al.*, 2159–2179, 1997 with permission from Elsevier

a study of a 13th century northern Italian choir book using Raman microscopy, layering of lapis lazuli over azurite was detected in a historiated letter [Best *et al.*, 1992; Clark, 1995b]. These blue pigments are readily differentiated by the inspection of their Raman spectra. Such a finding illustrates the practice of hierarchical pigment usage. Lapis lazuli and azurite were both important blue pigments used in the middle ages, but lapis lazuli was scarcer as it was produced through a complex

Table 4.6 Raman bands of paint binding media and varnishes

Material	Excitation wavelength(nm)	Wave number (cm⁻¹)
Beeswax	1064	2881, 2849, 2723, 1460, 1440, 1418, 1369, 1295, 1129, 1062, 891
	780	1735, 1660, 1460, 1439, 1417, 1294, 1171, 1130, 1061, 890
Proteinaceous binders		
Egg white	1064	3059, 2932, 2874, 2728, 1666, 1451, 1004
	780	1666, 1605, 1549, 1519, 1447, 1405, 1338, 1316, 1242, 1206, 1170, 1155, 1124, 1031, 1002, 993, 934, 903, 852, 827, 757, 699, 643, 621, 517, 450, 415
Egg yolk	1064	2933, 2898, 2854, 2730, 1661, 1446, 1304, 1075, 865
Casein	780	1668, 1615, 1549, 1449, 1337, 1247, 1206, 1173, 1155, 1125, 1031, 1002, 936, 879, 851, 829, 757, 643, 621, 540, 490, 405
Gelatin	780	1666, 1451, 1418, 1340, 1316, 1247, 1206, 1162, 1124, 1099, 1031, 1003, 975, 935, 920, 874, 856, 813, 758, 566, 404, 300
Fish glue	780	1669, 1451, 1417, 1321, 1249, 1098, 1029, 999, 921, 883, 855, 815, 718, 618, 528, 422, 406, 278
Carbohydrate binders		
Starch	780	1459, 1376, 1340, 1258, 1207, 1143, 1126, 1105, 1081, 1049, 1023, 941, 865, 769, 718, 616, 577, 526, 477, 440, 409, 359, 303
Gum arabic	1064	2926, 1461, 1419, 1372, 1346, 1269, 1083, 1028, 965, 883, 844, 458
	780	1461, 1340, 1261, 1078, 979, 879, 842, 714, 552, 490, 455, 350
Tragacanth	780	1548, 1453, 1348, 1256, 1226, 1082, 896, 855, 812, 768, 710, 658, 606, 438, 324
Cherry gum	780	1769, 1627, 1460, 1352, 1260, 1138, 1089, 948, 872, 845, 777, 709, 531, 470, 450, 404, 379, 357, 277, 258, 236, 215
Oils		
Linseed	1064	2909, 2855, 1744, 1658, 1443, 1302, 1076, 865

Table 4.6   (*continued*)

Material	Excitation wavelength(nm)	Wave number (cm^{-1})
	780	1744, 1656, 1523, 1429, 1298, 1265, 1157, 1085, 1070, 1021, 971, 942, 917, 866, 843, 727, 600, 542, 427, 294
Poppyseed	780	1746, 1647, 1455, 1439, 1301, 1264, 1078, 971, 910, 875, 842, 768, 724, 601, 545, 464, 402, 372, 314, 217
Walnut	780	1744, 1655, 1439, 1300, 1265, 1082, 970, 913, 867, 840, 725, 601
Sunflower	780	1745, 1657, 1438, 1401, 1367, 1301, 1265, 1110, 1080, 1066, 1043, 1023, 971, 912, 871, 842, 773, 724, 602, 402, 370
Resins		
Shellac	780	1645, 1444, 1300, 1242, 1195, 1133, 1091, 1013, 973, 930, 880, 817, 747, 699, 670, 577, 515, 481, 449, 400, 365, 299
Dammar	780	1655, 1452, 1315, 1198, 1179, 1027, 1001, 954, 938, 918, 830, 800, 761, 674, 601, 582, 556, 529, 464, 412, 318
Colophony	780	1649, 1631, 1611, 1565, 1469, 1442, 1372, 1302, 1256, 1233, 1199, 1133, 1106, 1070, 1050, 970, 950, 926, 882, 741, 706, 556, 529, 460, 371, 311
Sandarac	780	1645, 1450, 1362, 1301, 1244, 1195, 1140, 1093, 1046, 974, 930, 882, 747, 557, 518, 485, 447, 402, 360, 300, 237
copal	780	1644, 1450, 1194, 1014, 930, 749, 699, 484, 366, 302
Amber	780	1660, 1611, 1439, 1307, 1035, 999, 931, 815, 731, 642, 263
Mastic	780	1708, 1639, 1458, 1312, 1175, 1096, 935, 711, 602, 552
Dragon's blood	780	1654, 1600, 1545, 1509, 1450, 1349, 1321, 1284, 1253, 1182, 1157, 1028, 1001, 827, 803, 711, 677, 587, 527, 497, 458
Gamboge	780	1740, 1672, 1633, 1593, 1434, 1379, 1332, 1280, 1247, 1222, 1117, 1050, 1027, 1002, 961, 906, 837, 782, 734, 538, 453, 375, 326

and time consuming process on semi-precious stone only found in what is modern day Afghanistan. Hence, the layering process meant that the more expensive lapis lazuli would go further.

As already outlined, the binding media of paints can be feasibly examined using Raman techniques [Vandenabeele and Moens, 2004]. The main bands of the Raman spectra of the binder materials used in the production of medieval manuscripts, including casein, egg white, egg yolk, starch and gum arabic, are listed in Table 4.6.

Raman spectroscopy can also be used to identify inks used on documents [Bell *et al.*, 2000; Brown and Clark, 2002; Chaplin *et al.*, 2005; Vandenabeele and Moens, 2004; Wise and Wise, 2004]. Medieval iron gall ink has been studied and compared to a synthetic version produced from iron sulfate ($FeSO_4$) and gallic acid in water. Although these components show characteristic Raman bands, it is important to be aware of the complexity of older inks. The Raman spectrum of medieval iron gall ink was more complex than that of the modern version due to complexes formed by the gallnut extract and the crystallinity of the precipitate.

The use of Raman spectroscopy to study the paper used in documents of interest has been limited by the strong fluorescence produced. Resonance Raman techniques need to be used to minimise this problem, particularly when trying to identify dyes on the paper surface. There has been more success with the Raman studies of older writing materials such as parchment and vellum [Edwards, 2001a, 2005b; Edwards and Rull Perez, 2004; Edwards *et al.*, 2001b; Vandenabeele and Moens, 2004]. The Raman spectra of such materials arise from the protein and lipid content [Edwards and Rull Perez, 2004]. Importantly, changes in the Raman bands of these materials can be used to detect deterioration – changes in the bands associated with the protein –CO–NH–structure and the –CS–SC–cysteine bonds provide evidence of changes in the materials. The technique is also able to be used to identify materials used in the original preparation of parchments and vellums. For instance, the presence of sulfates and slaked lime can be detected and their subsequent reactions with carbon dioxide and pollutants such as sulfur dioxide can be monitored using Raman spectroscopy.

4.3.3 Ceramics

The Raman spectra of ceramic bodies, glazes and pigments can provide valuable information regarding identification and the technology of production [Colomban, 2004, 2005; Colomban and Treppoz, 2001; Colomban *et al.*, 2001; Clark *et al.*, 1997a, 1997b]. There are a number of databases containing the Raman spectra of minerals relevant to ceramic studies [Colomban et al., 1999; Clark et al., 1997b; Wang et al., 1994]. An example of the sort of information derived from the ceramic body using Raman spectroscopy is the discrimination between soft- and hard-paste ceramics [Colomban et al., 2001; Colomban, 2005]. Soft-paste ceramics shows characteristic bands due to β-wollastonite ($CaSiO_3$) at 970 and 635 cm^{-1} and/or tricalcium phosphate (β-$Ca_3(PO_4)_2$) bands at 960 and 415 cm^{-1}. These bands are absent for hard-paste ceramics and mullite ($3Al_3O_3.2SiO_2$) or mullite-like glassy phase bands at 480, 960 and 1130 cm^{-1} are observed for this type of ceramic.

Raman spectroscopy can also be used to confirm the processing procedure and the source of raw materials used in ceramics. For instance, in a Raman study of 15th Vietnamese ceramics, anatase was observed in the voids formed inside the glaze [Colomban, 2005; Liem et al., 2002]. As anatase usually transforms into rutile at 800–1100 °C, the presence of the phase indicates that a two-step process was used: the ceramic body was shaped and fired at high temperature and then the glaze was coated with anatase-containing materials and then fired at a lower temperature.

The particularly effective characterisation of inorganic pigments using Raman spectroscopy means that this technique is also very useful for investigating the pigments used in the surface coatings of ceramics [Colomban, 2005; Colomban et al., 2001; Clark et al., 1997a, 1997b]. An example is provided by a Raman study of the blue glaze on medieval pottery excavated in Southern Italy [Clark et al., 1997a]. The study showed for the first time that lapis lazuli is stable at the firing temperature of the glaze and so could be used as a pigment in Italian glaze.

4.3.4 Glass

Glasses may be characterised using Raman spectroscopy because modifications to the silica structure by additives affect the Raman bands of such materials [Colomban, 2005]. It is possible to differentiate glasses based on the Si–O stretching and bending bands that appear near 1000 and 500 cm^{-1}, respectively, in the spectrum. For glasses containing a large quantity of alkali or alkali earth metals, lead or zinc ions that disrupt Si–links, the intensities of the bands in the region of 1000 cm^{-1} are stronger.

The Raman spectra of minerals that are used to pigment stained glass have been reported and aid in the characterisation of these materials [Bouchard and Smith, 2003, 2005; Edwards and Tait, 1998]. For example, a FT–Raman study of 12th and 19th century English stained glass successfully identified red ochre as the red pigment used in the glass.

4.3.5 Stone

Raman spectroscopy has been used widely to study gemstones [Coupry et al., 1997; Calligaro et al., 2002; Chen et al., 2004; Dele-Dubois et al., 1986; Dele et al., 1997; Kiefert et al., 2001,2005; Moroz et al., 2000; Middleton and Ambers, 2005; Reiche et al., 2004; Smallwood et al., 1997; Smith, 2005a, 2005b]. The Raman sampling techniques available, such as fibre optic probes and microscopy, allow precious stones to be examined even when mounted. Additionally, stones may be examined behind glass or plastic. As many precious stones are coloured, fluorescence may be an issue. However, a judicious choice of excitation wavelength can minimise this problem. Collections of the Raman spectra suitable for the identification of gemstones are available [Pinet et al., 1992]. Raman microscopy allows the inclusions in natural stones to be characterised, thus providing information regarding the origin of the stone and, in certain cases, allowing natural and synthetic gems to be characterised. For example, a Raman study of an inclusion in an emerald identified the inclusion as carbon and so was able to ascribe the gemstone to a particular Columbian mine [Sodo et al., 2003]. Additionally, emerald can be readily distinguished from imitations of emerald such as topaz through examination of the Raman spectra.

There is a number of gemstone treatments used to enhance their appearance and such treatments can also be characterised using Raman spectroscopy [Kiefert et al., 2001, 2005; Smith, 2005a]. For instance, for many years pearls have been immersed in silver nitrate to give them a black colour in an attempt to imitate the rare black pearl. Such a treatment is readily identified using Raman spectroscopy because pearls dyed with silver nitrate show a distinct band at 240 cm^{-1} which is not present in natural black pearls [Kiefert et al., 2001, 2005].

Raman spectroscopy also provides a tool for examining of building stones, and is particularly useful for the investigation of surface degradation [Perez-Alonso et al., 2004; Potgieter-Vermaak et al., 2005]. A micro-Raman study of stone from the Cathedral of Seville in Spain, showing a black crust on the surface, was able to characterise the crust [Potgieter-Vermaak et al., 2005]. The study was also used to examine the laser cleaned surface using this technique. Significantly, the Raman examination of the black crust was able to detect the presence of iron oxide (Fe_2O_3), which was not detected by XRD or thermogravimetric analysis (TGA) in the same study.

4.3.6 Natural materials

Raman spectroscopy has proved very effective for examining ivory [Edwards and Farwell, 1995; Edwards *et al.*, 1997c; Edwards *et al.*, 1997a; Edwards *et al.*, 1995; Edwards *et al.*, 1998a; Edwards, 2000; Edwards and Munshi, 2005]. This technique can be used to differentiate genuine and simulated ivory artefacts [Edwards and Farwell, 1995; Edwards *et al.*, 1995]. As the relative compositions of protein and hydroxyapatite are characteristic for ivory, Raman bands associated with these components may be used to identify ivory artefacts. Raman spectroscopy has also been used to differentiate ivory of different species [Edwards *et al.*, 1997c; Edwards *et al.*, 1997a; Edwards *et al.*, 1998a]. Relative band intensities can be used as the amounts of organic and inorganic components are species dependent, albeit a more subtle variation than when a comparison is made with simulated ivory. However, multivariate analysis has proved successful for detecting the small variations in the spectra [Brody *et al.*, 2001a; Shimoyama *et al.*, 2003].

Raman spectroscopy provides an excellent non-destructive tool for characterising a wide range of resins and gums. Detailed assignments of the Raman bands have been made for amber, copal, dammar, sandarac, shellac, rosin, colophony, mastic, frankinscence, myrrh and a variety of gums [Brody *et al.*, 2002; Brody *et al.*, 2001b; Edwards and Farwell, 1996; Edwards and Falk, 1997a; Edwards, 2005b; Edwards *et al.*, 1998b; Edwards and Munshi, 2005; Vandenabeele *et al.*, 2003]. Raman spectroscopy has been useful for the study of ambers, having been used for the identification of the geographical origins of amber jewellery through its succinic acid content and for the detection of modern fakes made from phenolic resins [Brody *et al.*, 2001b; Edwards and Farwell, 1996]. Although the ratio of Raman bands in the 1771–1543 and 1510–1400 cm^{-1} regions of the spectrum and the intensity ratio of the 1646 cm^{-1}/1450 cm^{-1} bands cannot be directly correlated with the geological age of the amber, the I^{1646}/I^{1450} ratio can be used as a broad indicator of the maturity of the sample. For example, where an amber or copal sample contains diterpenoid components, the ratio will approximately 1.5 for immature samples, but has a value of 0.4 for mature samples.

The Raman spectra of keratotic materials, including horn, hoof and tortoiseshell, can be used to characterise these materials and also to differentiate keratotic materials from modern synthetic polymers used to imitate them, such as cellulose nitrate and cellulose acetate [Edwards, 2005b]. The spectra of horn, hoof and tortoiseshell show Raman bands characteristic of keratin. Additionally, there are useful differences between the spectra of keratotic materials which enable them to be specifically identified: differences are observed in the S–S stretching band at 500 cm^{-1} and the C–S stretching band at 640 cm^{-1}. The amide I band at 1650 cm^{-1} and the amide III band at 1260 cm^{-1} can also be used to examine differences in the secondary structures of the keratin in different materials.

The degradation of human mummies has been studied using Raman spectroscopy [Edwards, 2005; Edwards et al., 2002; Edwards and Munshi, 2005]. This technique can be used to examine hair, nail and skin samples from mummies. For instance, the reduced intensities of the protein amide I and III bands in mummified skin are indicative of a loss of protein and changes to the protein secondary structure. Additionally, evidence of embalming treatments can be obtained by examining lipid bands. Hair samples are more difficult to study using Raman spectroscopy because of strong background fluorescence emission. Nail is mainly composed of keratin, so a similar analytical approach to that used for horn, hoof and tortoiseshell can be taken when examining the Raman spectra of such material.

Although wood is a complex substance, Raman spectra may still be used to characterise wooden objects. A number of studies have assigned the Raman bands of different types of wood [Agarwal and Ralph, 1997; Kihara et al., 2002; Marengo et al., 2004; Takayama et al., 1997]. FT-Raman spectroscopy has been used to study the changes to 16th century wood exposed to accelerated ageing. The effects of exposure to sodium hydroxide solution and high temperatures at various time intervals were investigated and the spectra were analysed using PCA. A decrease in the bands associated with lignin and cellulose was clearly observed for the wood samples exposed to ageing.

The assignment of the FT-Raman bands of beeswax and gum arabic have also been reported and can be used to identify these materials in objects [Edwards et al., 1996a]. FT–Raman spectroscopy has been used to investigate the wax component of 19th century paper negatives [Edwards et al., 1996a] and a wax cameo in a collection at the Victoria and Albert Museum in London [Burgio, 2005].

4.3.7 Textiles

Raman spectroscopy has been successfully used to study historic textiles, both for characterising the fibres used and for analysing the dyes on the fibres. A number of studies have reported the Raman spectra of historic fibres including linen, wool and silk and assigned the Raman bands, providing useful reference spectra for identification [Edwards and Wyeth, 2005; Edwards *et al.*, 1997b; Edwards *et al.*, 1996b; Edwards and Falk, 1997b]. The degree of degradation of fibres can also be characterised by examination of their Raman spectra. For instance, the Raman bands of cellulose can be monitored as this component is prone to cleavage of the C–O–C linkage. Changes in the intensity of the ether bands relative to the bands of the CH_2 groups can be utilised to monitor degradation.

Raman spectroscopy has been a popular technique for studying dyes on textiles [Bourgeois and Church, 1990; Coupry *et al.*, 1997; Keen *et al.*, 1998; Schrader *et al.*, 2000; Shadi *et al.*, 2003; Withnall *et al.*, 2005]. Many dye chromophores produce resonance Raman effects so even very low concentrations of dye can be detected on textile fibres. Dyes including indigo, purpurin, alizarin and madder, have been successfully detected using Raman techniques. It has also been possible to characterise indigo from different sources [Coupry *et al.*, 1997; Shadi *et al.*, 2003; Withnall *et al.*, 2005].

4.3.8 Metals

While pure metals and their alloys do not produce Raman spectra, metal oxides and other corrosion products produce characteristic spectra [Bouchard and Smith, 2003; Bouchard and Smith, 2005; Robinet and Thickett, 2005]. Collections of the Raman spectra of the potential corrosion products have been published and aid in the identification of such compounds [Bouchard and Smith, 2003; Bouchard and Smith, 2005; Hayez *et al.*, 2004]. The advantage of Raman microscopy is being able to study a corrosion layer without disturbance. This technique can provide additional information to that obtained using XRD. For example, Raman spectroscopy can be used to distinguish between the iron corrosion products magnetite (Fe_3O_4) and maghernite (γ-Fe_2O_3), while only subtle differences in the two crystal lattices being detected using XRD.

4.3.9 Synthetic polymers

Although the use of Raman spectroscopy to study polymers was limited in the past due to the fluorescence of samples, the emergence of FT–Raman spectroscopy has meant that this technique can be used routinely for the investigation of polymers [Agbenyega *et al.*, 1990; Hendra *et al.*, 1991; Maddams, 1994]. As with infrared spectroscopy, correlation

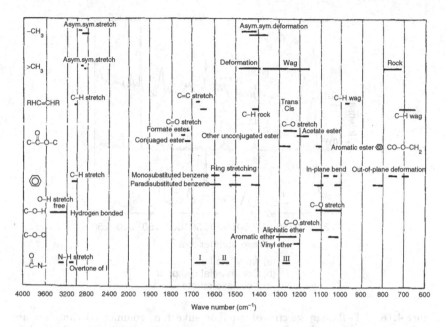

4000 3600 3200 2800   2400 2000   1900 1800 1700   1600 1500   1400 1300 1200   1100 1000   900   800   700   600

Wave number (cm⁻¹)

Figure 4.15 Correlation table for Raman bands of polymers. Reproduced by permission from The Vibrational Spectroscopy of Polymers, D I Bower & W F Maddams, 1989, Cambridge University Press

tables can be used to assign Raman bands to particular functional groups. A correlation table of the major functional groups observed in the Raman spectra of polymers is shown in Figure 4.15 [Bower and Maddams, 1989]. Libraries of the Raman spectra of polymers are also available [Kupstov and Zhizhin, 1998].

It is also possible to use Raman spectra of polymers to differentiate the structures of specific classes of polymers, such as nylons [Hendra et al., 1990; Maddams, 1994]. For instance, the Raman spectra of single number nylons in the series from nylon 4 to nylon 12 can be used for analytical purposes. The differences in the 1700–500 cm⁻¹ region, particularly for those nylons containing short methylene sequences, may be used on an empirical basis for identification. A more quantitative approach is to measure the intensity ratio of the peaks at 1440 cm⁻¹ due to CH_2 bending and the amide I band at 1640 cm⁻¹. This ratio increases linearly with the increasing nylon number.

Raman spectroscopy has been used to characterise the cellulose-based polymeric materials used to produce heritage objects [Paris and Coupry, 2005]. Fans from the Musée de la Mode et du Costume in Paris were examined using FT–Raman. The fans, dating from the late 19th century to the beginning of the 20th century, appeared to be made of a natural

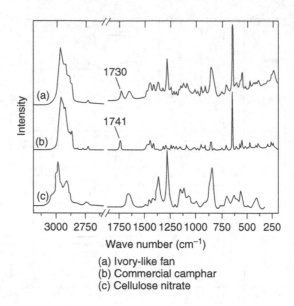

(a) Ivory-like fan
(b) Commercial camphar
(c) Cellulose nitrate

Figure 4.16 FT–Raman spectra of an ivory-like fan, commercial camphor and cellulose nitrate. Journal of Raman Spectroscopy, Paris and Coupry, 2005, John Wiley and Sons Limited with permission

material such as ivory. The Raman spectrum of one of the fans under investigation, dated 1885, is shown in Figure 4.16. This figure also shows the spectrum of commercial camphor, a commonly used plasticiser of the period, and a reference spectrum of cellulose nitrate. The fan's spectrum is a superposition of the two reference spectra. It is noted that the C=O stretching band at 1714 cm^{-1} in the camphor spectrum and at 1730 cm^{-1} in the fan spectrum. This shift can be attributed to hydrogen-bonding between cellulose nitrate and camphor.

Raman spectroscopy has also been used to identify more modern ivory substitutes [Edwards and Farwell, 1995; Edwards, 2005b]. The FT–Raman spectrum of an object, a carved cat is illustrated in Figure 4.17 [Edwards, 2005b]. This artefact was studied because it was, from its appearance, believed to be made of genuine ivory dating from possibly the 17th century. The FT–Raman spectrum of a genuine ivory is also shown in Figure 4.17. Comparison of the spectra demonstrates that the object is not made of ivory. The sample spectrum shows bands at 3060, 1600, 1580 and 1003 cm^{-1} that are characteristic of an aromatic compound and an ester band at 1725 cm^{-1}. A strong band at 1086 cm^{-1} is due to calcite. The bands in the spectrum are notably different to those of genuine ivory and are characteristic of a PMMA/PS copolymer with a calcite additive. The calcite was added in order to

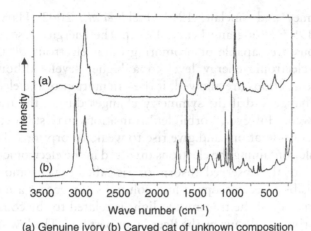

(a) Genuine ivory (b) Carved cat of unknown composition

Figure 4.17 FT–Raman spectra of genuine ivory and a carved cat object of unknown composition. Raman Spectroscopy in Archaeology and Art History, HGM Edwards and J M Chalmers, 263, 2005, reproduced with permission of The Royal Society of Chemistry

simulate the density, texture and appearance of ivory. In addition to not being ivory, the copolymer is a 20th century development, so this object was clearly not 300 years old.

Raman spectroscopy has been used to monitor degradation in plastic objects. One study examined the degradation of cellulose acetate dolls produced in the 1940s–1950s [Edwards *et al.*, 1993]. The dolls were able to be sampled in a non-destructive manner and different regions, those showing signs of deterioration and those not, of the doll could be examined without damage. A comparison of 'diseased' and 'non-diseased' regions of the dolls (apparent by eye) showed significant changes in a number of Raman bands. A mechanism could be proposed for the degradation process – the use of iron hooks to connect the dolls was responsible for catalysing the degradation reaction. Additionally, some regions of the dolls which appeared fine by eye, but the Raman spectra showed evidence of degraded cellulose acetate. Clearly this acts as an early warning sign that an apparently healthy doll is showing evidence of degradation not visible by eye.

4.4 ULTRAVIOLET-VISIBLE SPECTROSCOPY

UV-visible spectrophotometers are used to examine electronic transitions associated with absorptions in the UV (200–400 nm) and visible (400–800 nm) regions of the electromagnetic spectrum [Anderson *et al.*,

2004; Denney and Sinclair, 1991; Field *et al.*, 2002; Harwood and Claridge, 1997; Skoog and Leary, 1992]. The energies associated with these regions are capable of promoting outer electrons of a molecule from one electronic energy level to a higher level. Although many electronic transitions are allowed, if the orientation of the electron spin does not change and if the symmetry changes during the transition it is not allowed. However, 'forbidden' transitions may still occur due to symmetry considerations and give rise to weak absorptions. The part of the molecule containing the electrons involved in the electronic transition responsible for the observed absorptions is called the chromophore.

In UV-visible spectroscopy dilute solutions (or gases) are examined and the intensity of the transmitted light is related to the concentration of the absorbing molecule by the Beer–Lambert law. The law shows that the absorbance (A) of a solution is proportional to the concentration of the absorbing molecule (c) and the path length of the cell (l) containing the solution:

$$A = \varepsilon cl \qquad (4.4)$$

The constant ε is the molar absorptivity of the absorbing molecule. The higher the ε value, the greater the probability of the electronic transition. The bands in the UV-visible spectrum are usually quite broad and the wavelength of the maximum (λ_{max}) is assigned.

UV-visible spectrometers consist of an UV-visible light source, sample cells, a dispersing element and a detector [Field *et al.*, 2002; Harwood and Claridge, 1997; Skoog and Leary, 1992]. Often there are two light sources; a deuterium lamp for UV light and a tungsten lamp for visible light. Single beam UV-visible spectrometers are set-up so that the spectrum of the reference solution is measured first and then that of the sample of interest. The instrument can be set-up to scan the UV-visible light range for the sample, or can be set to measure the absorbance value at a specific wavelength. Double beam UV-visible spectrometers are common, in which the light from the source is separated into two parallel beams passing through two different cells. A schematic layout of a typical double beam UV-visible spectrometer is shown in Figure 4.18. One cell contains the reference solvent and the other the sample of interest. The absorbance is determined by measuring the ratio of the intensity of the light transmitted through the solvent (I_0) and the intensity of light transmitted through the sample (I):

$$A = \log_{10}(I_0/I) \qquad (4.5)$$

Cells for recording the UV-visible spectra of solutions and liquids are commonly made of quartz (usually 1 cm path length). Some common

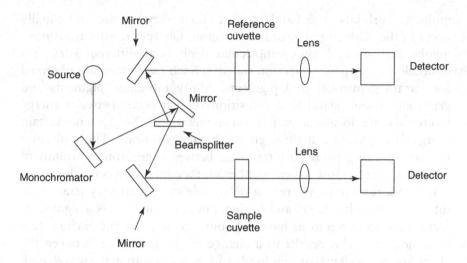

Figure 4.18 Schematic diagram of a UV-visible spectrometer

solvents that may be used in the UV-visible region due to their transparency in this spectral range are ethanol, cyclohexane and hexane.

Samples may also be examined in reflectance mode where reflected light is collected. In recent years, reflection visible light imaging microspectroscopy has been developed [van der Weerd *et al.*, 2003]. A microscope, which enables a homogeneous illumination of the sample, is incorporated into the set-up. A spatial resolution of about 1 μm and a spectral resolution of 4 nm can be achieved.

The development of fibre optics reflectance spectroscopy (FORS) has provided a very useful technique for the examination of culturally important objects [Bacci, 1995, 2000]. For the UV-visible and near-infrared regions, fused silica fibres are used. The optical fibres can be connected to either a spectrophotometer or be part of a spectroanalyser, a transportable instrument.

The types of transitions that result in UV-visible absorptions consist of the excitation of an electron from the highest occupied molecular orbital (usually of non-bonding p or bonding π orbital) to the next lowest unoccupied molecular orbital (an anti-bonding π^* or σ^* orbital). Non-bonding orbitals are represented by n and an asterix is used to represent an anti-bonding orbital. There is a variety of transitions that can be observed between the electronic energy levels of cultural materials [Bacci, 2000]. It is necessary to be aware that some of these transitions can occur at energies that correspond to frequencies in the near-infrared region. Organic pigments and dyes can show transitions between delocalised molecular orbitals. The bands produced are very intense and usually occur in the visible or very near UV region. Charge-transfer transitions involve transitions between molecular orbitals at different sites of a molecule where there is electron transfer. Such transitions are observed for certain gemstones and pigments. Many inorganic pigments and gemstones show ligand field transitions which occur between energy levels that are localised mainly on a metal ion. Metals and certain inorganic pigments can show energy band transitions. The molecular orbitals resulting from an interaction between the atomic orbitals of each atom are so close to one another that a continuum is produced.

It is important to be aware that UV-visible spectra are very sensitive to differences in solvent, pH and conjugation. Solvent type is a significant factor as solvent electrons have the ability to stabilise the excited state of a molecule. This results in a change in the difference between the electronic energy levels of a molecule and, hence, a shift in the wavelength of the associated absorption band. The more conjugation appearing in a molecule, the greater the intensity of the absorption bands of the molecule and the transitions appear at higher wavelengths. Again, the energy difference of the transition is affected. Likewise, the pH of the solvent affects the spectra as the addition or removal of photons in a molecule changes the observed electronic transitions. When there is a shift in the λ_{max} towards longer wavelengths it is known as a red or bathochromic shift. When the shift is towards a shorter wavelength, the change is known as a blue or hypsochromic shift.

4.4.1 Paintings

Ultraviolet-visible spectroscopy can be used to produce characteristic spectra of pigments [Arbizzani et al., 2004; Bacci, 2000; Bacci et al., 1991; Bacci and Piccolo, 1996; Billmeyer et al., 1981; Goltz et al., 2003; Miliani et al., 1998; Orlando et al., 1995; Romani et al., 2006; van der Weerd et al., 2003]. Reflectance techniques have proved particularly useful for studying paint pigments. The bands observed in the reflectance spectra for a number of common pigments are listed in Table 4.7 while absorption bands are listed in Table 4.8 [Bacci et al., 1991; Bacci, 2000]. The presence of soot, the yellowing of varnish and the particle size can all affect the position and shape of reflectance bands. Thus, care must be taken when comparing sample spectra with reference spectra. However, UV-visible spectra of pigments can often be sufficient for identification, particularly for those which show characteristic d-d transitions. Reflectance UV-visible spectroscopy has been used to

Table 4.7 UV-visible reflectance bands for common pigments and dyes (*sh – shoulder)

Colour	Pigment	Bands (nm)
Blue	Azurite	454
	Ultramarine/lapis lazuli	460, >800
	Indigo	411, >800
	Prussian blue	440, >1000
Green	Malachite	518
	Green earth	556, 680, >800
	Verdigris	498
	Viridian	410, 510
Yellow	Yellow ochre	450, 600, 760
	Orpiment	790
Red	Haematite	650 sh, 745
	Cinnabar	>630
	Red lead	>620
Brown	Ochre	460, 610 sh, 770

Table 4.8 UV-visible absorption bands for common pigments and dyes

Colour	Pigment	Bands (nm)
Blue	Smalt	544, 593, 642
	Indigo	599
	Prussian blue	568
	Cobalt violet	531, 594, 860
Green	Viridian	460, 600
Yellow	Chrome yellow	470
Red	Haematite	430, 550, 640, 900

compare the spectra of pure pigments with those used in frescoes at the Brancacci Chapel in Florence [Bacci *et al.*, 1991].

4.4.2 Textiles

UV-visible spectroscopy can be employed to identify the dyes used in heritage textiles [Billmeyer *et al.*, 1981; Cordy and Yeh, 1984; Saltzman, 1992; Taylor, 1983; Wallert and Boynter, 1996]. The UV-visible bands due to some pigments used as dyes are shown in Tables 4.7 and 4.8. In a study of 19th century dyes on flax thread, UV-visible spectroscopy was used as a tool for identifying the dye used [Cordy and Yeh, 1984].

As this study involved the examination of artificially aged fibres, the dye was removed, acid-digested and spectroscopically examined in solution. The method was shown to be effective for the identification of indigo.

4.4.3 Glass

Ultraviolet-visible spectroscopy can be used to characterise the transition metal ions in glass and their state of oxidation. For instance, the technique has been used to study Roman, medieval and renaissance glass [Bianchin *et al.*, 2005; Corradi *et al.*, 2005; Fernandez and la Iglesia, 1999; Garcia-Heras *et al.*, 2005; Green and Hart, 1987; Longworth *et al.*, 1982; Schreurs and Brill, 1984]. The UV-visible-infrared spectra for some samples of Roman glass are shown in Figure 4.19 [Green and Hart, 1987]. Sample (a) is pale green-blue in colour and in its spectrum the strong band centred at 1100 nm is due to Fe^{2+}, the weak sharp bands in the 375–450 nm range and the strong band at 258 nm are due to Fe^{3+}. The Fe^{2+} band in the infrared tails into the visible region

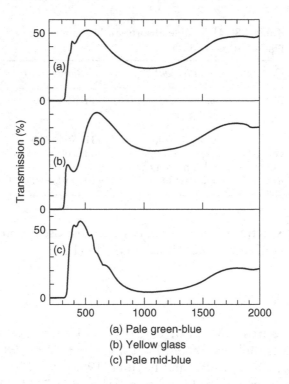

(a) Pale green-blue
(b) Yellow glass
(c) Pale mid-blue

Figure 4.19 UV-visible-infrared spectra of Roman glasses. Reprinted from Journal of Archaeological Science, Vol 14, Green and Hart, 271–282, 1987, with permission from Elsevier

and imparts a pale blue tint. The Fe^{3+} bands produce a pale yellow colour and together with the Fe^{2+} band produces the greenish-blue tint. Sample (b) is a yellow glass and shows iron bands similar to sample (a). However, the glass appears yellow due to the presence of an iron-sulfur band. The intense band which appears at 405 nm is attributed to a Fe^{3+}/S^{2-} species. The sulfur can find its way into the glass during the melting process via a crucible. Sample (c) is a pale mid-blue glass and the spectrum of this sample shows bands due to Fe^{3+} and Fe^{2+}. Bands at 530, 585 and 647 nm are also observed and these are characteristic of tetrahedrally coordinated cobalt ion, Co^{2+}. The cobalt is responsible for the blue colour of the glass.

4.4.4 Stone

Ultraviolet-visible spectroscopy can be used as a tool for the characterisation of gemstones [De Weerdt and Van Royen, 2001; King et al., 2005; Lu et al., 1998; Taran et al., 2003]. For example, a study of Brazilian topazes used UV-visible spectroscopy to determine the origins of the colours which appear in different gems [Taran et al., 2003]. For instance, two bands at 18 000 and 25 000 cm^{-1} were assigned to the d-d transitions of Cr^{3+} ions. These bands are responsible for colours ranging from a light rose to a deep violet colour.

4.5 PHOTOLUMINESCENCE SPECTROSCOPY

When molecules absorb radiation in electronic transitions to form excited states, the latter may lose the acquired energy via several mechanisms. If the energy loss occurs through the emission of radiation, the process is known as luminescence and this is the basis of photoluminescence (PL) spectroscopy [Rendell and Mowthorpe, 1987]. Fluorescence is a form of luminescence and involves emission occurring from the lowest excited single state (S_1) to the singlet ground state (S_0), as illustrated in Figure 4.20. Fluorescence ceases immediately the exciting radiation is removed, with the lifetime usually being of the order of nanoseconds. Phosphorescence is another form of luminescence and involves emission occurring from the lowest excited triplet state (T_1) to the singlet ground state (S_0) (Figure 4.20). Such a process involves inter-system crossing, with the lifetimes in this case ranging from milliseconds to seconds.

Photoluminesence spectroscopy involves an instrument with an excitation light source, a dispersive device such as a grating device, a detector and a recorder [Larson et al., 1991]. Lasers of different wavelengths can

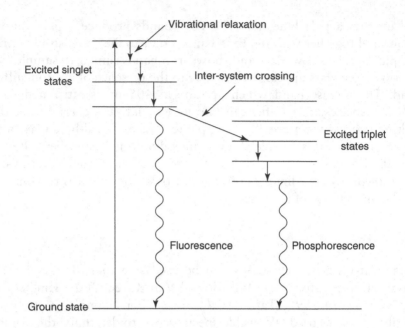

Figure 4.20 Illustration of the process of fluorescence and phosphorescence

be used as an excitation source, depending on the type of sample studied. The sample can be observed at different temperatures by mounting in a holder equipped for heating and cooling. Samples of the order of about 100 μm in diameter are examined, but fibre optics can also be used to examine objects in a non-destructive fashion. There may be the risk of sample decomposition if excessive heating or photochemical reactions occur, so caution should be exercised when irradiating with various laser wavelengths.

An emission or luminescence spectrum is a plot of the intensity of light emitted as a function of wavelength (nm) or energy (cm^{-1}). The spectrum shows either broad bands or bands showing vibronic structure if the vibrational bands in the ground state are resolved. The characteristics of interest in an emission spectrum are the wavelength or wave number of maximum intensity and the full width at half maximum. The overall band shape for a sample with one emitting component consists of a sharp rise in intensity on the low wavelength side of the band, with a gradual decrease on the high wavelength side of the band. Where a sample contains more than one emitting species, the overall band shape will depend on the relative concentrations of the emitting component and their relative quantum yield (the number of quanta emitted per exciting quantum absorbed). The different components may be excited selectively by varying the excitation wavelength.

4.5.1 Natural materials

Amber and other fossil resins have been studied using PL spectroscopy. One study involved the examination of solvent extracts containing aromatic compounds from the resins [Matuszewska and Czaja, 2002]. The compounds found in amber, mainly naphthalenes, phenanthrenes and anthracenes, produce characteristic emission spectra.

4.5.2 Paintings

Photoluminescence spectroscopy can be used to determine the composition of paint layers and pigments, organic binders and resins can all be studied [Ajo *et al.*, 2004; de la Rie, 1982a; de la Rie, 1982b; de la Rie, 1982c; Guineau, 1989; Janes *et al.*, 2001; Larson *et al.*, 1991; Pozza *et al.*, 2000; Miyoshi, 1982; Miyoshi, 1985; Miliani *et al.*, 1998]. One study used this technique to investigate the inorganic pigments Egyptian blue, Han blue and Han purple [Pozza *et al.*, 2000]. Photoluminescence spectra obtained for Egyptian blue and Han blue pigments under 632.8 nm excitation are illustrated in Figure 4.21. The Cu^{2+} ion is responsible for the broad emission band observed in these spectra, as

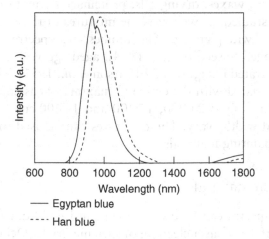

Figure 4.21   PL spectra of Egyptian blue and Han blue pigments. Reprinted from Journal of Cultural Heritage, Vol 1, Pozza *et al.*, 393–398, 2000 with permission from Elsevier

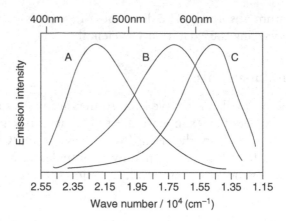

Figure 4.22 Luminescence spectra of (A) casein, (B) egg yolk and (C) linseed oil [Larson *et al.*, 1991]. Reproduced by permission of American Institute for Conservation of Historic and Artistic Works

it is the only optically active species able to absorb energy in the visible range and emit in the region studied. Although the bands observed for Egyptian blue and Han blue share a similar band shape, the band for Han blue appears at a higher wavelength than that observed for Egyptian blue. This is also the case for Han purple and is explained by the change to the ligand-field induced when Ca^{2+} is replaced by the larger Ba^{2+}.

PL spectroscopy has also been used to investigate natural resins and organic binding media used in paintings [Larson *et al.*, 1991]. A range of natural resins, waxes, drying oils, proteinaceous materials and gums have all been studied, as well as some mixtures of these materials with one another and with pigments. The luminescence spectra recorded using 363.8 nm of aged linseed oil (ca. 1934), aged egg yolk (ca. 1937) and casein are illustrated in Figure 4.22 [Larson *et al.*, 1991]. The spectra are clearly differentiated with the emission maxima of casein, egg yolk and linseed oil occurring at 21 900, 17 200 and 14 600 cm^{-1}, respectively. Also, the band widths vary. These features can be used to discriminate among these painting materials.

4.5.3 Written material

Fluorescence spectra can be used to characterise old documents and to examine the effects of laser cleaning of such materials [Ochocinska *et al.*, 2003]. Differences in the structure and shape of emission bands were observed for well preserved and damaged paper samples ranging from the 13th to the 19th century. This approach was also able to indicate that laser cleaning with a 532 nm laser results in less destructive changes to the paper that the use of 266 and 351 nm wavelengths.

4.5.4 Ceramics

Photoluminescence spectroscopy can be used to identify the pigments used on the surface of ceramics as well as coatings such as glazes. A study of fragments of 16th–17th century Italian pottery with lustro decoration was carried out using PL spectroscopy as one of the characterisation techniques [Romani et al., 2000]. Lustro is a decoration made of metal salts or oxides dispersed in clay and superimposed onto glazed ceramics. The PL spectra produced showed that the glaze could be characterised by two maxima at 438 and 465 nm attributable to silicon dioxide, while the lustro showed two emission bands at 380 and in the range 475–580 nm. Significantly, the band in the 475–580 nm range shifts for different groups of samples while the band shape does not significantly change. It was proposed that the shift could be a consequence of the firing procedure.

4.5.5 Stone

Photoluminescence spectroscopy has been successfully employed to characterise gemstones [Carbonin et al., 2001; Hainschwang et al., 2005; Lindblom et al., 2003; Sodo et al., 2003; Taran et al., 2003]. In gemstones, luminescent ions may be present as natural impurities or they may be due to deliberate additions. This enables naturally occurring and synthetic gems to be differentiated using this technique. For example, it is possible to use PL spectroscopy to identify natural and synthetic diamonds [Lindblom et al., 2003]. As synthetic diamonds are grown in a relatively short time and at low temperatures compared to natural diamonds, nitrogen is present in simpler aggregates compared to natural diamonds. Such structural differences manifest themselves in the emission spectra.

4.5.6 Glass

Photoluminescence spectroscopy provides potential as a simple method for the examination of glass samples [Ajo et al., 1998, 1999]. For instance, information regarding the iron oxidation state and concentration can be obtained from PL spectra.

4.5.7 Synthetic polymers

Photoluminescence spectroscopy can be used for the identification of polymers [Beddard and Allen, 1989]. The emission spectra can also be used to detect the presence of additives in polymers as their presence influences the appearance of the spectra. In particular, the fluorescence and phosphorescence spectra of antioxidants and stabilisers have been used for this type of analysis. Additionally, the pigments used in polymers exhibit their own characteristic emission spectra. For example, the different crystalline forms of titanium dioxide may be differentiated by their different characteristic emissions.

4.5.8 Textiles

The potential of fluorescence spectroscopy as a tool for the study of natural dyes used in textiles has been investigated [Clementi *et al.*, 2006]. The development of non-invasive fibre-optic sampling methods for fluorimeters allows textiles to be examined without damage. It is possible to distinguish dyes on different fibre surfaces and identification is feasible if standards are available. The technique is also sensitive to the effects of degradation.

4.6 NUCLEAR MAGNETIC RESONANCE SPECTROSCOPY

Nuclear magnetic resonance (NMR) spectroscopy is a very effective tool for the study of molecular structure, both in solution and in the solid state [Abraham *et al.*, 1988; Bovey, 1989; Derome, 1987; Gunter, 1995; Harris, 1986; McBrierty, 1989; Mirau, 1993]. NMR spectroscopy involves placing a sample in a strong magnetic field and irradiating with radio wave frequency radiation. The absorptions due to the transitions between quantised energy states of nuclei that have been oriented by the magnetic field are observed.

All nuclei carry a charge and in certain nuclei this charge spins on the nuclear axis generating a magnetic dipole. The angular momentum of the spinning charge may be described using spin numbers (I). Each proton and neutron has its own spin and I is the resultant of these spins. NMR spectroscopy utilises nuclei with a $I = 1/2$ such as ^1H, ^{13}C, ^{27}Al, ^{29}Si and ^{31}P. For such nuclei, the magnetic moments align either parallel to or against the magnetic field. The energy difference between the two states is dependent upon the magnetogyric ratio (γ) of the nucleus and the strength (B) of the external magnetic field. The resonance frequency (ν), the frequency of radiation required to effect a transition between the energy states, is given by:

$$\nu = \frac{\gamma B}{2\pi} \qquad (4.6)$$

In theory, a single peak should be observed as a result of the interaction of radio wave frequency radiation and a magnetic field on a nucleus. However, in practice absorptions at different positions are observed. This occurs because the nucleus is shielded to a small extent by its electron cloud, the density of which varies with the environment. For instance, the degree of shielding of a proton and a carbon atom observed using ^{1}H NMR will depend on the inductive effect of the other groups attached to the carbon atom. The difference in the absorption position from that of a reference proton is known as the chemical shift. The chemical shift (δ) is usually expressed in dimensionless units of ppm, determined using:

$$\delta = \frac{\text{frequency of absorption} \times 10^6}{\text{applied frequency}} \qquad (4.7)$$

A common reference compound is tetramethylsilane (TMS). TMS gives single ^{1}H and ^{13}C absorptions corresponding to $\delta = 0$. The appearance of a typical NMR spectrum is illustrated in Figure 4.23 [Sandler et al., 1998]. Each absorption area in a NMR spectrum is proportional to the number of nuclei it represents and these areas are evaluated by an integrator. The ratios of integrations for the difference absorptions are equal to the ratios of the number of the respective nuclei present in the nucleus.

There is another phenomenon, known as spin–spin coupling, which further complicates NMR spectra. Spin–spin coupling is the indirect coupling of nuclei spins through the intervening bonding electrons. This occurs because there is a tendency for a bonding electron to pair its spin with the spin of the nearest $I = 1/2$ nuclei. The spin of the bonding electron influenced will affect the spin of the other bonding electron and so on through to the next $I = 1/2$ nucleus. Where spin–spin coupling occurs, each nucleus will give rise to widely separated absorptions appearing as doublets as the spin of each nucleus is affected slightly by the orientations of the other nucleus through the intervening electrons. The frequency differences between the doublet peaks is proportional to the effectiveness of the coupling and is known as a coupling constant (J) (Figure 4.23).

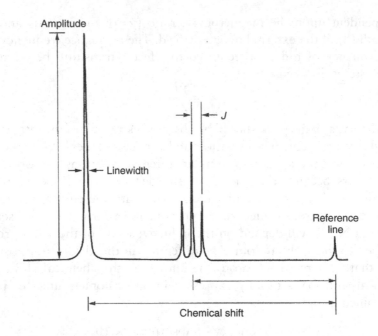

Figure 4.23 Features of a typical NMR spectrum. Reprinted from Polymer Synthesis and Characterisation: A Laboratory Manual, ISBN 012618240X, Sandler *et al.*, page 85, 1998 with permission from Elsevier

While spin–spin coupling is fairly easy to interpret in ^{1}H NMR, in ^{13}C NMR the spin–spin coupling between [1]H–[13]C is significant. This makes the interpretation of ^{13}C spectra difficult as in many cases the coupling is greater than the chemical shift differences between the ^{13}C nuclei absorptions. Hence, it is usual to employ decoupling techniques while recording ^{13}C NMR spectra. Such techniques decouple ^{13}C spins from ^{1}H spins so that the different ^{13}C nuclei give rise to only a single absorption in the spectrum. In addition, two-dimensional (2D) NMR spectra can be recorded, which can represent the different ^{13}C absorptions on one axis and the splitting by [1]H–[13]C spin–spin coupling is shown on a second axis perpendicular to the first.

After the absorption of energy by the nuclei during an NMR experiment there must be a mechanism by which the nuclei can dissipate energy and return to the lower energy state. There are two main processes, spin–lattice relaxation or spin–spin relaxation. Spin–lattice relaxation involves the transfer of energy from the nuclei to the molecular lattice. The spin-lattice relaxation time (T_1) is the time constant for the exponential return of the population of spin states to their equilibrium population. Spin–spin relaxation occurs from direct interactions

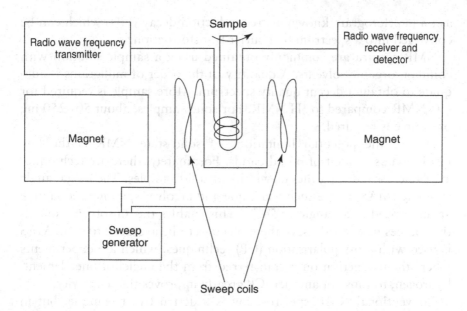

Figure 4.24 Schematic diagram of a NMR spectrometer

between the spins of different nuclei that can cause relaxation without any energy transfer to the lattice. The spin–spin relaxation time (T_2) is the time constant for exponential transfer of energy from one high energy nucleus to another.

A schematic diagram of the layout of a NMR spectrometer is shown in Figure 4.24. NMR spectrometers possess a strong magnet with a homogeneous field, a radio frequency transmitter, a radio wave frequency receiver and a recorder. The instrument also contains a sample holder which positions the sample relative to the magnetic field, the transmitter coil and the receiver coil. The sample holder also spins the sample to increase the apparent homogeneity of the magnetic field. Samples can be studied in solution and solid form. Solutions for ^{1}H and ^{13}C NMR may be prepared using deuterated solvents such as $CDCl_3$ or C_6D_6.

Early NMR spectrometers used permanent magnets or electromagnets (e.g. 60 MHz, 100 MHz) and are known as a continuous wave spectrometer in which a single radio wave frequency is applied continuously. Modern pulsed FT NMR spectrometers use superconducting magnets (e.g. 300 MHz, 500 MHz) and a short pulse of radio wave frequencies is applied to promote the nuclei to the higher energy states. The relaxation of the nuclei back to the lower energy state is detected as an interferogram known as free induction decay (FID) which can be converted into a spectrum by Fourier transformation.

NMR spectra are commonly obtained using a sample solution with common organic solvents. A quantity of the order of milligrams is adequate to obtain a decent quality spectrum. More sample is required for ^{13}C NMR compared to ^1H NMR. For solid samples, about 50–250 mg of sample is required.

One of the potential limitations of solid-state NMR is the low resolution as a result of broad bands. Fortunately, there are techniques available for reducing the linewidths of solid samples. The magic angle spinning (MAS) is one such technique and involves spinning the sample at high speed at an angle of 54.7°. This enables the anisotropy, one of the sources of linewidths, of the molecules to be averaged to zero. MAS is used with cross polarisation (CP) techniques, which are experiments where the magnetisation is transferred from the nuclei of one element, hydrogen, to those of another. CP greatly improves the sensitivity.

Conventional NMR spectroscopy is a destructive technique, but in order to avoid the destruction of precious samples, a non-destructive probe such as the NMR–MOUSE can be employed. The NMR–MOUSE (Mobile Universal Surface Explorer) replaces the magnet and the probe of a regular NMR spectrometer [Blumich et al., 2003]. This sensor can be used with a small spectrometer to provide a portable instrument that can be taken to an object. The spin density, the spin–spin and the spin–lattice relaxation times can be measured using this type of instrument.

Magnetic resonance imaging (MRI) is another non-invasive technique which shows potential in the conservation field [Smith, 1997]. This technique is more commonly associated with medical applications. This technique involves varying the magnetic field linearly across a sample. The protons in different regions will resonate at different frequencies and the intensity of the signal will be proportional to the number of protons at each point, providing a portrayal of the concentration of protons in a solid object.

To aid in the interpretation of NMR spectra extensive correlation tables of chemical shifts are available [Breitmaier, 1993; Field et al., 2002; Pretsch et al., 1989; Silverstein et al., 2005; Williams and Fleming, 1995]. Changes in chemical shifts are mainly the result of intermolecular effects and generally the differences in shifts observed can be related to the structural features within a molecule. The chemical shift regions in which various proton resonances are commonly observed are illustrated in Figure 4.25.

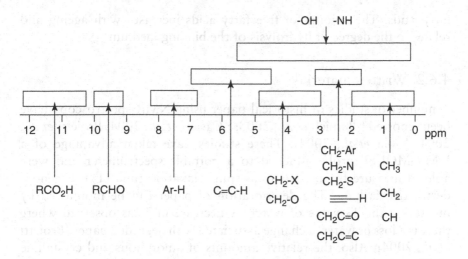

Figure 4.25 Typical proton NMR chemical shift ranges for various environments

4.6.1 Paintings

NMR spectroscopy can be used to identify pigments and dyes used in paintings and textiles [Hvilsted, 1985; Romani *et al.*, 2006]. However, it is the development of non-invasive NMR techniques which holds the most promise as a technique for studying paintings. Relaxation NMR measurements have been used to monitor the state of conservation of a 16th century fresco in Florence [Proietti *et al.*, 2005]. Measurements of the spin–spin relaxation time (T_2) in different regions of the fresco were able to characterise the effect of chemical treatments, cleaning and consolidation. The presence of soluble salts affected the T_2 values.

The binders used in paintings may also be investigated using NMR spectroscopy [Chiantore *et al.*, 2003; Spyros and Anglos, 2004]. One study used one dimensional and two dimensional ^{1}H and ^{13}C NMR techniques to examine the ageing of the binding media of oil paintings from the 17th and 20th centuries [Spyros and Anglos, 2004]. Studies of the solvent-extractable component from model samples of various drying oils, raw oil paints and aged oil paints enabled specific markers of degradation to be identified. During ageing, the triglyceride content of the paint extract decreases based on changes to the NMR spectra. The concentration of the di- and monoglyceride components reach a maximum in the extract followed by a slow decrease with further ageing. These components suffer further hydrolysis to glycerol and free fatty acids. The amount of free fatty acids increases with ageing and relates to the degree of hydrolysis of the binding medium.

4.6.2 Written material

A number of studies of historical paper using NMR spectroscopy have been reported [Blumich et al., 2003; ; Casieri et al., 2004; Proietti et al., 2004; Viola et al., 2004]. These studies have taken advantage of a NMR-MOUSE probe attached to a portable spectrometer and were able to measure the proton spin–spin relaxation time (T_2) in a non-destructive fashion. The deterioration of paper can be monitored by measuring the T_2 value of water. A decrease in T_2 is observed where there is a loss of water, a change associated with degraded paper [Proietti et al., 2004]. Also, the relative amounts of amorphous and crystalline cellulose in paper is an indicator of deterioration and the spin–lattice relaxation time (T_1) value of these components can be estimated using NMR spectroscopy [Casieri et al., 2004]. This approach was used to examine the condition of a 17th century Italian manuscript [Viola et al., 2004]. The ink used was iron-gall ink and was shown by NMR to be responsible for causing deleterious effects on the cellulose fibres in the paper.

4.6.3 Natural Materials

A broad range of natural materials can be characterised using NMR spectroscopy [Ghisalberti and Godfrey, 1998]. Resins, oils, fats, waxes, gums, proteins, bituminous and cellulosic materials can all be studied using this technique. Solution NMR has the advantage that all the components can be analysed with good resolution at the same time without any separation techniques required, and so provides a suitable approach before chromatography and mass spectrometry techniques are used. Of course, the availability of sample that can be removed from an object is a requirement for this sort of experiment. ^{13}C NMR has advantages for organic materials as information regarding the backbone of molecules can be elucidated. Also, the chemical shift range for ^{13}C for most organic compounds is about 200 compared with 10–15 for ^{1}H NMR, resulting in less overlap of the peaks in the ^{13}C spectra. Solid state ^{13}C NMR spectroscopy, particularly ^{13}C CP–MAS NMR, is applicable to a detailed characterisation of natural materials.

NMR spectroscopy has potential as a means of distinguishing natural resins based on their botanical origin [Ghisalberti and Godfrey, 1998].

As particular classes of compounds dominate the composition of resins, the NMR spectra can be used to distinguish between groups. Also, the use of solid-state techniques enables resins that are insoluble in organic solvents to be studied. For example, a solid-state NMR study of an ancient oriental lacquer was used to elucidate an oligomeric structure for the lacquer, information not provided by other techniques [Lambert et al., 1991].

^{13}C NMR has been employed in a number of studies of amber [Cunningham et al., 1983; Ghisalbarti and Godfrey, 1998; Lambert et al., 1988, 1989, 1985, 1990, 1996; Martinez-Richa et al., 2000]. ^{13}C CP–MAS NMR has proved particularly useful for providing information about the C backbone of the polymeric material of amber. NMR provides a means of fingerprinting ambers from different sources. A range of Baltic ambers produced consistent spectra with peaks at shifts of 170–190 (–COO–), 110, 150 ($C=CH_2$), 128, 140 (di- and tri-substituted double bonds), 60–70 (oxygen substituted carbon) and 10–40 (saturated carbon) [Ghisalbarti and Godfrey, 1998]. ^{13}C NMR spectroscopy can also be utilised to identify the plant source of amber. For instance, a study of Dominican amber (20–35 millions years old) showed that it was produced from a different species to that responsible for the production of Mexican amber (20–24 million years old) [Cunningham et al., 1983].

Jet has been characterised using ^{13}C NMR spectroscopy [Lambert et al., 1992]. Chemical shifts demonstrated the presence of phenolic ($\delta \sim$ 150) and quarternary aromatic carbons ($\delta \sim 130$). Saturated aliphatic carbons lacking electronegative groups produced signals at $\delta \sim 10$–50. The spectra were compared to peat and bituminous coal, as well as those materials used to imitate jet, such as glass or onyx, and showed clear differences enabling NMR to clearly identify a jet sample.

NMR is recognised as a useful method for the analysis of wood, particularly as there are NMR sampling methods which do not involve extraction [Cole-Hamilton et al., 1995, 1990; Ghisalbarti and Godfrey, 1998; Viel et al., 2004; Wilson et al., 1993]. NMR spectroscopy was one of the techniques used to analyse wood samples from historic Indian Ocean shipwrecks near Australia, including the Dutch East Indiamen 'Batavia' (1629) [Wilson et al., 1993]. The study investigated the degree of deterioration in the wood using NMR. The spectra showed a greater loss of carbohydrate, measured by the percentage carbon assigned as oxygen – alkyl in the spectrum, for more degraded samples. Samples obtained from a wreck located at an anaerobic site appeared to have undergone little degradation, unlike samples obtained from aerobic sites.

There appeared to be no correlation between the age of the ships and the degradation of the recovered wood.

NMR imaging also provides plenty of promise for the study of historic wood [Cole-Hamilton *et al.*, 1995, 1990]. This approach provides a means of assessing wood and monitoring the effects of consolidants. It is necessary to aware when applying NMR imaging to wood that the quality of the image is affected by the presence of Fe^{2+} and Fe^{3+} salts, species that are often incorporated in wood recovered from shipwrecks. However, NMR imaging of such samples can be used to monitor the removal of such species during the conservation process.

4.6.4 Stone

NMR techniques have proved effective for the investigation of degradation processes and consolidation treatments for heritage stones [Alesiani *et al.*, 2003a; Alesiani *et al.*, 2003b; Appolonia *et al.*, 2001; Blumich *et al.*, 2005; Borgia *et al.*, 2000, 2001, 1999; Cervantes *et al.*, 1999; Cardiano *et al.*, 2005; De Barquin and Dereppe, 1996; Friolo, 2001; Gibson *et al.*, 1997; Rijniers *et al.*, 2005; Sharma *et al.*, 2003]. A traditional approach can be used to characterise degradation products. For example, ^{1}H NMR has been used to determine the degree of hydration of a salt efflorescence formed on the surface of a limestone statue [Gibson *et al.*, 1997]. Another application of NMR is the use of a solid state approach to the study of the pore sizes of stones [Alesiani *et al.*, 2003a; Alesiani *et al.*, 2003b; Friolo, 2001]. In this approach, the sample is immersed in water and the spin–spin relaxation time, T_2, is measured. The pore size is proportional to the T_2 value – the nuclei transfer magnetisation to the pore wall and the smaller the pore size, the quicker the transfer. The relaxation time distribution for a heritage sandstone sample is illustrated in Figure 4.26. The distribution shows the presence of two different populations of pores.

NMR spectroscopy has been used in a number of studies to characterise stone consolidants and to examine the effect of consolidants on the stone structure [Appolonia *et al.*, 2001; Borgia *et al.*, 2000, 1999, 2001; Cervantes *et al.*, 1999; Cardiano *et al.*, 2005; Sharma *et al.*, 2003]. MRI techniques show great potential for determining the spatial distribution of water in a porous material, thus providing information regarding the efficacy of the consolidant treatment.

Solid-state NMR techniques have potential for the examination of precious stones. For instance, ^{29}Si and ^{27}Al NMR techniques have been used to characterise natural opals [Brown *et al.*, 2003].

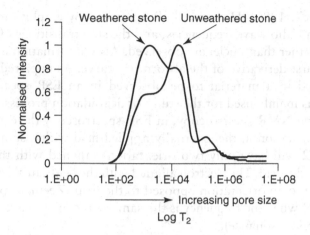

Figure 4.26 NMR relaxation time distribution for weathered and unweathered sandstone

4.6.5 Synthetic Polymers

NMR spectroscopy is commonly used to identify and characterise the structure of polymers [McBriety, 1989; Mirau, 1993; Princi *et al.*, 2005]. [1]H and ^{13}C are the most common nuclei used for the analysis of polymers. NMR spectroscopy has been employed in a characterisation study of a fluorinated polymer to be used as a protective coating for textiles [D'Orazio *et al.*, 2001].

4.6.6 Textiles

Solid-state ^{13}C NMR spectroscopy has been used to examine silk removed from the 12th century coffins of Japanese lords [Chujo *et al.*, 1996]. The mole fractions of glycine residues present were investigated. This quantity was shown to be non-linearly dependent on time, but could be correlated with temperature, providing information about the history of the silk.

4.7   ELECTRON SPIN RESONANCE SPECTROSCOPY

Electron spin resonance (ESR) spectroscopy, also known as electron paramagnetic resonance (EPR) spectroscopy, is based on similar principles to those of NMR spectroscopy [Ikeya, 1993; Ranby and Rabek,

1977; Rabek, 1980]. However, ESR spectroscopy employs microwave rather than radio wave frequencies, and the spin transitions of unpaired electrons rather than nuclei are observed. Also, information is obtained using the first derivative of the absorption curve. As unpaired electrons are required in a material to be observed in an ESR spectrum, this technique is mainly used for the study of degradation processes.

Paralleling NMR spectroscopy, in ESR spectroscopy the electron in a parallel orientation in the externally applied field with a spin quantum number 1/2 will have only two orientations: aligned with the field or opposed to the field. The electron aligned with the field can absorb energy and change to an orientation opposed to the field (Zeeman splitting for an electron) when the frequency is the same as the microwave frequency (electron spin resonance).

The position of an ESR line is usually determined as the point when the derivative spectrum crosses zero. The position is characterised by a g-value, the constant of proportionality between the frequency and the field at which resonance occurs:

$$\Delta E = h\nu = g\beta H_0 \tag{4.8}$$

where ΔE is the energy that will induce transitions, h is the Planck constant, ν is the frequency, g is the g-value (g = 2.002319 for unbound electrons), β is the Bohr magneton and H_0 is the strength of the applied external magnetic field. Generally, the g-value of bound unpaired electrons is not the same as that for unbound electrons.

Splitting in the energy levels of an electron is caused by an increase in the number of energy levels resulting from exposure of a material with unpaired electrons to a magnetic field. The interaction of an electron and the magnetic nucleus is called a hyperfine interaction and splitting in the energy levels is called hyperfine splitting. The hyperfine splitting constant is the separation between two hyperfine lines of the spectrum.

It is also possible to use ESR as a dating method for a range of molecules [Grun, 2000; Jones, 1997]. Ionising radiation produces paramagnetic centres with long lifetimes in a number of materials and the concentration of these centres in a sample is a measure of the radiation dose to which the sample was exposed. The effect can be used to determine the length of time of that exposure. As ESR dating is not as sensitive as thermoluminescence (see Chapter 10) it is not widely used for dating. However, ESR dating has been successfully applied to the dating of archaeological tooth enamel [Grun, 2000].

4.7.1 Synthetic Polymers

ESR spectroscopy has been successfully employed in the study of degradation mechanisms in polymers [Ranby and Rabek, 1977]. This technique is able to detect the presence and nature of radical degradation products. When PVC is exposed to UV light, a free radical, $-[-CH_2-C^\bullet H-]-$, is initially formed. The ESR spectrum of PVC exposed to ultraviolet radiation shows six lines as a result of the interaction of the unpaired electron with the five surrounding protons [Yang et al., 1980].

4.7.2 Stone

ESR spectroscopy provides a useful technique for characterising stone and has been applied to the study of marble and limestone [Armiento et al., 1997; Attanasio et al., 2003; Cordischi et al., 1983; Lazzarini and Antonelli, 2003; Polikreti et al., 2004; Polikreti and Maniatis, 2002]. The ESR spectra of such stones have three regions that may be used to investigate the provenance: peaks due to Mn^{2+} and Fe^{3+} and the region near $g = 2.0000$ [Polikreti and Maniatis, 2002; Polikreti et al., 2004]. The typical Mn^{2+} and Fe^{3+} peaks observed for limestone are illustrated in Figure 4.27 [Polikreti et al., 2004]. Six double lines are associated with the Mn^{2+} ions substituting for Ca^{2+} in the calcium carbonate lattice. This spectral feature is common in polycrystalline carbonates such as marble and calcite. The intensity of these lines depends upon the concentration of Mn^{2+} in the carbonate phase and may be used to discriminate between sources. In the low magnetic field region, four peaks at $g = 14.25, 5.9, 3.7$ and 2.9 are observed for carbonate rocks. These peaks are associated with Fe^{3+} substituting for Ca^{2+} in the calcium carbonate lattice. The peak at $g = 4.3$ is mainly attributed to Fe^{3+} in orthorhombic symmetry.

ESR spectroscopy has potential in the examination of precious stones. For example, natural and cultured pearls can be differentiated using ESR [Haberman et al., 2001]. The Mn^{2+} concentration is determined using this technique and the amount of Mn^{2+} is different in the calcite and aragonite structures of pearls.

ESR spectroscopy has also been used to characterise obsidian from different sources [Duttine et al., 2003; Scrozelli et al., 2001; Tenorio et al., 1998]. The presence of iron in obsidian allows this technique to be used and the complex spectra are mainly due to iron in different states and site location. Each source produces different ESR spectra based on

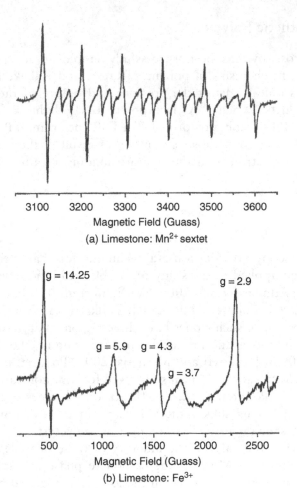

Figure 4.27 ESR spectrum of a limestone: (a) Mn^{2+} sextet and (b) Fe^{3+} peaks. Reprinted from Journal of Archaeological Science, Vol 31, Polikreti *et al.*, 1015–1028, 2004, with permission from Elsevier

the different thermal conditions that occurred during the formation of the obsidian.

4.7.3 Glass

ESR spectroscopy can be used to investigate glass samples, but the spectra are quite difficult to interpret as glass is a disordered system. The ESR signals detected are due to the paramagnetic ions present as impurities or dopants. This technique has been used to study ancient mosaic glass and signals due to Fe^{3+}, Mn^{2+} and Cu^{2+} can be detected and connected to the colouring agents in the glass [Azzoni et al., 2002].

4.7.4 Paintings

ESR spectroscopy has been used to examine the ageing of varnish used for paintings [Deitemann et al., 2001]. Dammar and mastic films were artificially aged by light and heat and the concentration of radicals produced as a result of ageing were detected using ESR. The triterpenoids oxidise, polymerise and decompose during the process. The study also showed that the radical concentration in dammar films stored in darkness is notably less than the concentration for film exposed to light.

4.8 MÖSSBAUER SPECTROSCOPY

In a Mössbauer experiment γ-rays are emitted from a source nucleus and then reabsorbed by the nuclei of the same element present in a solid sample under study [Banwell, 1983; Dickson and Berry, 1986; Maddock, 1997; Wagner et al., 2000; Wagner and Kyek, 2004]. This technique can be employed to investigate the electronic environment of the atoms of the element of interest.

The γ-rays used in a Mössbauer experiment are usually emitted by a radioactive source. The most common source is radioactive ^{57}Co incorporated into a matrix of metallic rhodium, it decays to ^{57}Fe by electron capture with a half-life of 270 days. During this process the first excited state of ^{57}Fe decays to the ground state by emission of a 14.4 keV γ-ray. These rays can excite the ^{57}Fe nuclei in an absorber. Other types of Mössbauer spectroscopy include ^{119}Sn, ^{197}Au and ^{121}Sb. The resonant absorption occurs in a separate sample containing the absorbing nuclei.

Resonant absorption is achieved by mounting the source on a vibrator moving at several mm s^{-1}. This provides a spectral range of energy through modulation of the γ-ray energy by the Doppler effect. The spectrum is measured by recording the count rate at a detector, located behind the absorber as a function of the source velocity. The intensity of the transmitted γ-rays is plotted against a scale of velocity in mm s^{-1}. To record a spectrum with suitable accuracy, one or more days of data collection may be required. Some experiments need to be cooled to liquid nitrogen temperature. A sample of about 200 mg is required, but it is possible to reduce this amount by decreasing the diameter of the absorber at the expense of the count rate. Alternatively, a non-destructive approach can be taken by observing resonantly scattered γ-rays and X-rays. The probing depth is of the order of 10 μm. Conversion electrons emitted after resonant absorption may also be measured in a technique known as conversion electron Mössbauer spectroscopy (CEMS). This examines the surface of an object to within 100–200 nm because of the short range of the low energy conversion electrons. Compared to conventional transmission experiments, these techniques can be used for whole objects without the need to remove samples.

There are three types of hyperfine interactions of the nuclei with their environment in a solid that result in shifts and splittings of the Mössbauer lines observed in the spectra. The isomer or chemical shift is proportional to the product of the electron density at the site of the nuclei and the change of the nuclear charge radius during the Mössbauer transition. The isomer shift causes a displacement of the whole pattern on the velocity scale but not splitting. Electric quadrupole splitting results from the differing spins in the ground and excited states. For instance, ^{57}Fe has a spin $I = 1/2$ unsplit in the ground state and splits the $I = 3/2$ excited state into two substates. This results in the appearance of a doublet, whose separation is called the electric quadrupole splitting. The quadrupole splitting in iron compounds is different for Fe^{3+} and Fe^{2+} and can be used, together with the isomer shift, to distinguish between the two oxidation states. Magnetic hyperfine interaction results from the interaction between the nuclear magnetic dipole moment and the magnetic hyperfine field produced by the electrons at the nucleus and produces Zeeman splitting of the ground and excited state of the nucleus. Such magnetic hyperfine fields exist in the nuclei of magnetically ordered materials. In paramagnetic substances, the magnetic hyperfine interaction normally vanishes.

4.8.1 Metals

Mössbauer spectroscopy can be used to study various metal artefacts, and is particularly useful for characterising corrosion products [Cook, 2004; Hsia and Huang, 2003; Kumar and Balasubramaniam, 1998; Wagner and Kyek, 2004]. This technique can provide supplementary information to that provided by XRD as products may show poor crystallinity and not able to be characterised by XRD. ^{57}Fe Mössbauer spectroscopy can be used to study iron and steel artefacts [Kumar and Balasubramaniam, 1998; Wagner and Kyek, 2004]. This approach can provide a quantitative analysis of the iron phases in steel. Tin and bronze objects can be studied using ^{119}Sn Mössbauer spectroscopy [Hsia and Huang, 2003; Takeda *et al.*, 1977; Wagner and Kyek, 2004; Zhang and Hsia, 1991]. Although the spectra of bronzes do not vary greatly, this approach is useful for detecting oxidation products on the surface and in the interior of bronzes using CEMS experiments. ^{197}Au Mössbauer spectroscopy is used to examine gold artefacts [Kyek *et al.*, 2000; Parish, 1984; Wagner and Kyek, 2004]. The relatively high γ-ray energy requires these experiments to be carried out at liquid nitrogen temperature. A number of Celtic gold coins have been examined Mössbauer spectroscopy [Kyek *et al.*, 2000]. The spectra of gold–silver–copper alloys show broadened single lines with an isomer shift dependent on the alloy composition. Thus, a combination of isomer shift measurements and mass density in the alloys enables the silver and copper content to be determined.

4.8.2 Ceramics

Mössbauer spectroscopy is a particularly suitable technique for the study of ceramic materials, as an analysis of iron compounds provides information regarding the firing technology and the colour of these materials [Guangyong *et al.*, 1989; Hayashida *et al.*, 2003; Hausler, 2004; Hsia and Huang, 2003; Longworth, 1984; Lopez-Arce *et al.*, 2003; Maritan *et al.*, 2006; Murad, 1998; Stievano *et al.*, 2003; Qin and Pan, 1989; Qin and Lin, 1992; Tschauner and Wagner, 2003; Venkateswarlu and Ramshesh, 1984; Wagner *et al.*, 2000; Wagner and Wagner, 2004]. Even though iron is only a minor constituent of ceramics, often in the form of oxides and oxyhydroxides, it undergoes a range of transformations during the firing process. Mössbauer spectra can also be used to classify ceramic materials.

In clay minerals, iron may be present as octahedral Fe^{3+} and Fe^{2+} and tetrahedral Fe^{3+}. These types of iron may be distinguished by their isomer shifts and quadruple splittings. Several broad and overlapping quadruple doublets for octahedral Fe^{3+} and Fe^{2+} are usually observed. Such patterns can be curve-fit with several quadruple doublets with different splittings and isomers shifts and the relative areas of Fe^{3+} and Fe^{2+} can be obtained.

Mössbauer parameters can be used to determine the firing conditions of a ceramic [Hayashida *et al.*, 2003; Hausler, 2004; Maritan *et al.*, 2006; Wagner *et al.*, 2000; Qin and Pan, 1989; Tschauner and Wagner, 2003; Wagner and Wagner, 2004]. By using the intensities, the magnetic

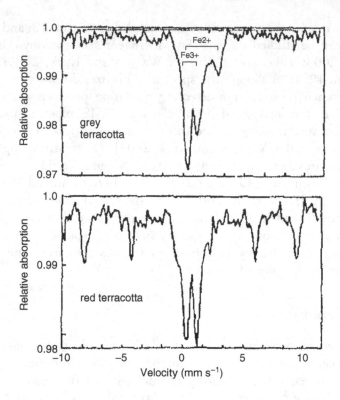

Figure 4.28 Mössbauer spectra of the ceramic material of Qin terracotta warriors and horses. Reproduced by permission of Hyperfine Interactions, Hsia & Huang, 150, 33–50, 2003, Springer

component and the quadruple splitting of Fe^{3+} and Fe^{2+}, the original firing temperature can be estimated. Mössbauer parameters are also a function of the firing conditions – they can be used to determine if the firing process was carried out under oxidising or reducing conditions. For example, a Mössbauer study of terracotta warriors and horses of the Qin Dynasty (221 BC) in China was used to gain insight into the firing process used to produce these artefacts [Hsia and Huang, 2003; Qin and Pan, 1989]. The room temperature Mössbauer spectra of samples taken from two terracotta figures – one grey and one red are illustrated in Figure 4.28. The difference between the grey and red terracotta is clear in the spectra. Analysis of the Fe^{3+} and Fe^{2+} contents based on these spectra reveals that the iron in the grey ceramic is 64 % Fe^{2+}, while in the red ceramic, the iron is predominantly Fe^{3+} (95 %). Ceramic fired in an oxidising atmosphere contains no, or very little, Fe^{2+}, while ceramic fired in a reducing atmosphere contains mostly Fe^{2+}. Thus, it appears that the main difference between the red and grey samples examined is in the atmospheric conditions during the firing process, rather that in the clay itself.

4.8.3 Stone

As described, Mössbauer spectroscopy can be used to characterise the mineral content of ceramic materials. It follows that the clay content of heritage stones can also be characterised, particularly with ^{57}Fe Mössbauer spectroscopy [Nasraoui et al., 2002]. Precious stones, including garnets and amethyst, have also been characterised using Mössbauer spectroscopy [Dedushenko et al., 2004; Zboril et al., 2004].

The provenances of obsidians have been determined using ^{57}Fe Mössbauer spectroscopy [Stewart et al., 2003; Scrozelli et al., 2001]. The spectra of obsidian samples show a broad asymmetrical doublet made up two symmetrical Fe^{2+} quadruple signals with different degrees of distortion and a smaller fraction of Fe^{3+} doublet. These signals were assigned to iron forming part of glass silicates that have different coordination. Different Fe^{3+} concentrations could be used to discriminate obsidians of different Mediterranean sources.

4.8.4 Glass

^{57}Fe Mössbauer spectroscopy of glasses can provide information regarding the colour, composition and the glass making conditions [Brianese et al., 2005; Bertoncello et al., 2002; Casellato et al., 2003; Longworth et al., 1982]. In a study of medieval stained glass obtained mainly from Elgin Cathedral in Scotland, Mössbauer spectroscopy was used to determine the Fe^{2+}/Fe^{3+} ratio. The observed colours could be correlated with particular ratios – green glazes are associated with the presence of both Fe^{2+} and Fe^{3+} ions, while in purple and emerald green glazes the iron incorporated into the glass is only in the ferric oxidation state.

4.8.5 Paintings

Mössbauer spectroscopy can be used to study iron-containing pigments in paintings [Arbizzani et al., 2004; Casellato et al., 2000; Keisch, 1976]. Since the colour of pigments often depends upon the particle size and size is reflected in the Mössbauer spectra, this approach may yield more information on pigments than XRD. Mössbauer spectroscopy has been used to study the pigments in renaissance wall paintings [Casellato *et al.*, 2000]. Commercial iron-containing pigments, red ochre, yellow ochre and green earths, at different temperatures and mixed with binding media, such as egg, milk and animal glue, were also examined using this technique in the study. An important finding of this study was the information gained regarding the oxidation state of the metal ion of the iron-containing pigments and its chemical environment using Mössbauer spectroscopy. The experiments also demonstrated interactions between pigments and binders. In particular, the spectra show a distinct interaction between yellow ochre and various glues, the type and the extent of the interaction depending on the ligand used.

4.8.6 Written Material

^{57}Fe Mössbauer spectroscopy is a suitable technique for investigating the effect of iron gall inks on paper [Bulska and Wagner, 2004; Rouchon-Quillet *et al.*, 2004; Wagner *et al.*, 2004]. Fe^{2+}, contained in the ink, has been implicated in the oxidation of cellulose in paper. Mössbauer spectroscopy can be used to measure Fe^{2+}/Fe^{3+} ratio in a bulk sample so can provide valuable information regarding the corrosion process associated with iron gall ink. In one study, the charge state of iron in iron gall inks was studied by ^{57}Fe Mössbauer spectroscopy [Wagner *et al.*, 2004]. Manuscripts from the 16th century were examined, as well as model samples.

REFERENCES

R. Abraham, J. Fisher and P. Loftus, *Introduction to NMR Spectroscopy*, John Wiley & Sons Ltd, Chichester (1988).

U.P. Agarwal and S.A. Ralph, FT-Raman spectroscopy of wood: identifying contributions of lignin and carbohydrate polymers in the spectrum of black spruce (Picea mariana), *Applied Spectroscopy*, **51** (1997), 1648–1655.

J.K. Agbenyega, G. Ellis, P.J. Hendra, Applications of Fourier transform Raman spectroscopy in the synthetic polymer field, *Spectrochimica Acta Part A*, **46** (1990), 197–216.

D. Ajo, P. Polato and F. De Zuane, Photoluminescence spectroscopy as an attractive technique for quantitative determination of Fe^{3+} in industrial glasses, *Boletin de la Sociedad Espanola de Ceramica y Vidrio*, **37** (1998), 315–318.

D. Ajo, F. De Zuane, G. Caramazza, *et al.*, Spectroscopic techniques for the determination of the iron redox state of glass and their direct application in the glass factory, *Glass Technology*, **40** (1999), 116–120.

D. Ajo, U. Casellato, E. Fiorin, P.A. Vigato, Ciro Ferri's frescoes: a study of painting materials and technique by SEM-EDS microscopy, X-ray diffraction, micro-FTIR and photoluminescence spectroscopy, *Journal of Cultural Heritage*, **5** (2004), 333–348.

M. Alesiani, S. Capuani and B. Maraviglia, NMR study on the early stages of hydration of a porous carbonate stone, *Magnetic Resonance Imaging*, **21** (2003a), 333–335.

M. Alesiani, S. Capuani and B. Maraviglia, NMR applications to low porosity carbonate stones, *Magnetic Resonance Imaging*, **21** (2003b), 799–804.

M. Ali, A.M. Emsley, H. Herman and R.J. Heywood, Spectroscopic studies of the ageing of cellulosic paper, *Polymer*, **42** (2001), 2893–2900.

R.J. Anderson, D.J. Bendell and P.W. Groundwater, *Organic Spectroscopic Analysis*, Royal Society of Chemistry, Cambridge (2004).

I. Angelini and P. Bellintani, Archaeological ambers from Northern Italy: an FTIR-DRIFT study of provenance by comparison with the geological amber database, *Archaeometry*, **47** (2005), 441–454.

L. Appolonia, G.C. Borgia, V. Bortolotti, *et al.*, Effects of hydrophobic treatments of stone on pore water studied by continuous distribution analysis of NMR relaxation times, *Magnetic Resonance Imaging*, **19** (2001), 509–512.

R. Arbizzani, U. Casellato, E. Fiorin, *et al.*, Decay markers for the preventative conservation and maintenance of paintings, *Journal of Cultural Heritage*, **5** (2004), 167–182.

G. Armiento, D. Attanasio and R. Platania, Electron spin resonance study of white marbles from Tharros (Sardinia): a reappraisal of this technique, possibilities and limitations, *Archaeometry*, **39** (1997), 309–319.

D. Attanasio, G. de Marinis, P. Pallecchi, *et al.*, An EPR and isotopic study of the marbles of the Trojan's Arch at Ancona: an example of alleged Hymettian provenance, *Archaeometry*, **45** (2003), 553–568.

C.B. Azzoni, D. Di Matino, C. Chiavari, *et al.*, Electron paramagnetic resonance of mosaic glasses from the Mediterranean area, *Archaeometry*, **44** (2002), 543–554.

M. Bacci, F. Baldini, R. Carla and R. Linari, A colour analysis of the Brancacci Chapel frescoes, *Applied Spectroscopy*, **45** (1991), 26–31.

M. Bacci, Fibre optics applications to works of art, *Sensors and Actuators B*, **29** (1995), 190–196.

M. Bacci and M. Picollo, Non-destructive detection of cobalt (II) in paintings and glass, *Studies in Conservation*, **41** (1996), 136–144.

M. Bacci, UV-VIS-NIR, FT-IR and FORS spectroscopy, in *Modern Analytical Methods in Art and Archaeology* (eds E. Ciliberto and G. Spoto), John Wiley & Sons Inc., New York (2000), pp 321–361.

M. Bacci, M. Fabbri, M. Picollo and S. Porcinai, Non-invasive fibre optic Fourier transform infrared reflectance spectroscopy on painted layers. Identification of materials by means of principal component analysis and Mahalanobis distance, *Analytica Chimica Acta*, **446** (2001), 15–21.

A. Bakolas, G. Biscontin, V. Contardi, *et al.*, Thermoanalytical research on traditional mortars in Venice, *Thermochimica Acta*, **269–270** (1995), 817–828.

C.N. Banwell, *Fundamentals of Molecular Spectroscopy*, 3rd edn, McGraw-Hill, London (1983).

D. Barilaro, G. Barone, V. Crupi, *et al.*, Spectroscopic techniques applied to the characterisation of decorated potteries from Caltagirone (Sicily, Italy), *Journal of Molecular Structure*, **744–747** (2005), 827–831.

G. Barone, V. Crupi, S. Galli, *et al.*, Spectroscopic investigation of Greek ceramic arti-facts, *Journal of Molecular Structure*, **651–653** (2003), 449–458.

M.J. Baxter and C.E. Buck, Data handling and statistical analysis, in *Modern Analytical Methods in Art and Archaeology* (eds E. Ciliberto and G. Spoto), John Wiley & Sons Inc., New York (2000), pp 681–746.

C.W. Beck, E. Wilbur and S. Meret, Infrared spectra and the origin of amber, *Nature*, **201** (1964), 256–257.

C.W. Beck, E. Wilbur, S. Meret, *et al.*, The infrared spectra of amber and the identification of Baltic amber, *Archaeometry*, **8** (1965), 96–109.

C.W. Beck, Spectroscopic investigations of amber, *Applied Spectroscopy Reviews*, **22** (1986), 57–110.

G.S. Beddard and N.S. Allen, Emission spectroscopy, in *Comprehensive Polymer Science*, Vol. 1 (eds C. Booth and C. Price), Pergamon Press, Oxford (1989), pp 499–516.

I.M. Bell, R.J.H. Clark and P.J. Gibbs, Raman spectroscopic library of natural and syn-thetic pigments (pre-~ 1850 AD), *Spectrochimica Acta Part A*, **53** (1997), 2159–2179.

S.E.J. Bell, E.S.O. Bourguignon, A.C. Dennis, *et al.*, Identification of dyes on ancient Chinese paper samples using the subtracted shifted Raman spectroscopy method, *Analytical Chemistry*, **72** (2000), 234–239.

R. Bertoncello, L. Milanese, U. Russo, *et al.*, Chemistry of cultural glasses: the early medieval glasses of Montselice's Hill (Padova, Italy), *Journal of Non-Crystalline Solids*, **306** (2002), 249–262.

S.P. Best, R.J.H. Clark and R. Withnall, Non-destructive pigment analysis of artefacts by Raman spectroscopy, *Endeavour*, **16** (1992), 66–73.

S. Best, R. Clark, M. Daniels and R. Withnall, A bible laid open, *Chemistry in Britain*, **29** (1993), 118–122.

S. Bianchin, N. Brianese, U. Casellato, *et al.*, Medieval and renaissance glass technology in Valdelsa (Florence). Part 3. Vitreous finds and crucibles, *Journal of Cultural Heritage*, **6** (2005), 165–182.

M. Bicchieri, M. Nardone and A. Sodo, Application of micro-Raman spectroscopy to the study of an illuminated medieval manuscript, *Journal Cultural Heritage*, **1** (2000), 5277–5279.

F.W. Billmeyer, R. Kuman and M. Saltzman, Identification of organic colourants in art objects by solution spectrophotometry, *Journal of Chemical Education*, **58** (1981), 307–313.

V.Y. Birshtein and V.M. Tulchinskii, Determination of beeswax and some impurities by IR spectroscopy, *Chemistry of Natural Compounds*, **13** (1977a), 232–235.

V.Y. Birshtein and V.M. Tulchinskii, An investigation and identification of polysaccha-rides isolated from archaeological specimens, *Chemistry of Natural Compounds*, **12** (1977b), 12–15.

V.Y. Birshtein and V.M. Tulchinskii, A study of gelatin by IR spectroscopy, *Chemistry of Natural Compounds*, **18** (1983), 697–700.

G. Biscontin, M. Pellizon Birelli and E. Zendri, Characterisation of binders employed in the manufacture of Venetian historical mortars, *Journal of Cultural Heritage*, **3** (2002), 31–37.

B. Blumich, S. Anferova, S. Sharma, *et al.*, Degradation of historical paper: nondestructive analysis by the NMR-MOUSE, *Journal of Magnetic Resonance*, **161** (2003), 204–209.

B. Blumich, F. Casanova, J. Perlo, *et al.*, Advances of unilateral mobile NMR in nonde-structive materials testing, *Magnetic Resonance Imaging*, **23** (2005), 197–201.

G.C. Borgia, M. Camaiti, F. Cerri, *et al.*, MRI tomography – a new method to evaluate treatment on stone monuments, *La Chimica e l'Industria*, **81** (1999), 729–731.

G.C. Borgia, M. Camaiti,F. Cerri, *et al.*, Study of water penetration in rock materials by nuclear magnetic resonance tomography: hydrophobic treatment effects, *Journal of Cultural Heritage*, **1** (2000), 1–6.

G.C. Borgia, V. Bortolotti, M. Camaiti, *et al.*, Performance evaluation of hydrophobic treatments fro stone conservation investigated by MRI, *Magnetic Resonance Imaging*, **19** (2001), 513–516.

M. Bouchard and D.C. Smith, Catalogue of 45 reference Raman spectra of minerals concerning research in art history or archaeology, especially on corroded metals and coloured glass, *Spectrochimica Acta Part A*, **59** (2003), 227–266.

M. Bouchard and D.C. Smith, Database of 74 Raman spectra of standard minerals of relevance to metal corrosion, stained glass or prehistoric rock art, in *Raman Spectroscopy in Archaeology and Art History* (eds H.G.M. Edwards and J.M. Chalmers), Royal Society of Chemistry, Cambridge (2005), pp 17–40.

D. Bourgeois and S.P. Church, Studies of dyestuffs in fibres by Fourier transform Raman spectroscopy, *Spectrochimica Acta Part A*, **46** (1990), 295–301.

F.A. Bovey, Structure of chains by solution NMR spectroscopy, in *Comprehensive Polymer Science* Vol. 1 (eds C. Booth and C. Price), Pergamon Press, Oxford (1989).

D.I. Bower and W.F. Maddams, *The Vibrational Spectroscopy of Polymers*, Cambridge University Press, Cambridge (1989).

E. Breitmaier, *Structure Elucidation by NMR in Organic Chemistry*, John Wiley & Sons Ltd, Chichester (1993).

N. Brianese, U. Casellato, F. Fenzi, *et al.*, Medieval and Renaissance glass technology in Tuscany. Part 4: the 14th century sites of Santa Cristina (Gambassi–Firenze) and Poggio Imperiale (Siena), *Journal of Cultural Heritage*, **6** (2005), 213–225.

B. Brodsky-Doyle, E.G. Bendit and E.R. Blout, Infrared spectroscopy of collagen and collagen-like peptides, *Biopolymers*, **14** (1975), 937–945.

R.H. Brody, H.G.M. Edwards and A.M. Pollard, Chemometrics methods applied to the differentiation of Fourier transform Raman spectra of ivories, *Analytica Chimica Acta*, **427** (2001a), 223–232.

R.H. Brody, H.G.M. Edwards and A.M. Pollard, A study of amber and copal samples using FT-Raman spectroscopy, *Spectrochimica Acta Part A*, **57** (2001b), 1325–1338.

R.H. Brody, H.G.M. Edwards and A.M. Pollard, Fourier transform Raman spectroscopic study of natural resins of archaeological interest, *Biospectroscopy*, **67** (2002), 129–141.

S.D. Brown, Chemometrics, in *Encyclopedia of Analytical Chemistry*, Vol. 1 (ed. R.A. Meyers), John Wiley & Sons Ltd, Chichester (2000), pp 9671–9678.

K.L. Brown and R.J.H. Clark, Analysis of pigmentary materials on the Vinland Map and Tartar relation by Raman microprobe spectroscopy, *Analytical Chemistry*, **74** (2002), 3658–3661.

L.D. Brown, A.S. Ray and P.S. Thomas, ^{29}Si and ^{27}Al NMR study of amorphous and paracrystalline opals from Australia, *Journal of Non-Crystalline Solids*, **332** (2003), 242–248.

S. Bruni, F. Cariati, F. Casadio and L. Toniolo, Spectrochemical characterisation by micro-FTIR spectroscopy of blue pigments in different polychrome works of art, *Vibrational Spectroscopy*, **20** (1999), 15–25.

T. Buffeteau and M. Pezolet, Linear dichroism in infrared spectroscopy, in *Handbook of Vibrational Spectroscopy*, Vol. 1 (eds J.M. Chalmers and P.R. Griffiths), John Wiley & Sons Ltd, Chichester (2002) pp 693–710.

E. Bulska and B. Wagner, A study of ancient manuscripts exposed to iron gall ink corrosion, in *Non-Destructive Microanalysis of Cultural Heritage Materials* (eds S.K. Janssens and R. van Grieken), Elsevier, Amsterdam (2004), pp 755–788.

L. Burgio and R.J.H. Clark, Library of FT Raman spectra of pigments, minerals, pigment media and varnishes, and supplement to existing library of Raman spectra of pigments with visible excitation, *Spectrochimica Acta Part A*, **57** (2001), 1491–1521.

L. Burgio, Case study: Raman spectroscopy – a powerful tool for the analysis of museum objects, in *Raman Spectroscopy in Archaeology and Art History* (eds. H.G.M. Edwards and J.M. Chalmers), Royal Society of Chemistry, Cambridge (2005), pp 179–191.

G. Burrafato, M. Calabrese, A. Cosentino *et al.*, ColoRaman project: Raman and fluorescence spectroscopy of oil, tempera and fresco paint pigments, *Journal of Raman Spectroscopy*, **35** (2004), 879–886.

T. Calligaro, S. Colinart, J.P. Poirot and C. Sudres, Combined external beam PIXE and μ-Raman characterisation of garnets used in Merovingian jewellery, *Nuclear Instruments and Methods in Physics Research B*, **189** (2002), 320–327.

F. Cappitelli, THM-GCMS and FTIR for the study of binding media in Yellow Islands by Jackson Pollock and Break Point by Fiona Banner, *Journal of Analytical and Applied Pyrolysis*, **71** (2004), 405–415.

F. Cappitelli and F. Koussiaki, THM-GCMS and FTIR for the investigation of paints in Picasso's Still Life, Weeping Woman and Nude Woman in a Red Armchair from the Tate Collection, London, *Journal of Analytical and Applied Pyrolysis*, **75** (2006), 200–204.

S. Carbonin, D. Ajo, I. Rizzo and F. de Zuane, Identification of synthetic spinels by means of photoluminescence spectroscopy, *Journal of Gemmology*, **27** (2001), 30–31.

J.M. Cardamone, Nondestructive evaluation of ageing in cotton textiles by Fourier transform reflection-absorption infrared spectroscopy, in *Historic Textile and Paper Materials II* (eds S.H. Zeronian and H.L. Needles), American Chemical Society, Washington (1989), pp 239–251.

P. Cardiano, R.C. Ponterio, S. Sergi, *et al.*, Epoxy-silica polymers as stone conservation materials, *Polymer*, **46** (2005), 1857–1864.

F. Cariati and S. Baini, Raman Spectroscopy in *Modern Analytical Methods in Art and Archaeology* (eds E. Ciliberto and G. Spoto), John Wiley & Sons Inc., New York (2000), pp 255–277.

F. Casadio and L. Toniolo, The analysis of polychrome works of art: 40 years of infrared spectroscopic investigations, *Journal of Cultural Heritage*, **2** (2001), 71–78.

U. Casellato, P.A. Vigato, U. Russo and M. Matteini, A Mössbauer approach to the physico-chemical characterisation of iron-containing pigments for historical wall paintings, *Journal of Cultural Heritage*, **1** (2000), 217–232.

U. Cassellato, F. Fenzi, P. Guerriero, *et al.*, Medieval and renaissance glass technology in Valdelsa (Florence). Part 1: raw materials, sands and non-vitreous finds, *Journal of Cultural Heritage*, **4** (2003), 337–353.

C. Casieri, S. Bubici, I. Viola and F. De Luca, A low resolution non-invasive NMR characterisation of ancient paper, *Solid State Nuclear Magnetic Resonance*, **26** (2004), 65–73.

K. Castro, M. Perez, M.D. Rodriquez-Laso and J.M. Madariaga, FTIR spectra database of inorganic art materials, *Analytical Chemistry* (2003), 214A–221A.

K. Castro, M. Perez-Alonso, M.D. Rodriquez-Laso, *et al.*, On-line FT–Raman and dispersive Raman spectra database of artists' materials (e-VISART database), *Analytical and Bioanalytical Chemistry*, **382** (2005), 248–258.

J. Cervantes, G. Mendoza-Diaz, D.E. Alvarez-Gasca and A. Martinez-Richa, Application of ^{29}Si and ^{27}Al magic angle spinning nuclear magnetic resonance to studies of the building materials of historical monuments, Solid State Nuclear Magnetic Resonance, 13 (1999), 263–269.

J.M. Chalmers, Infrared spectroscopy in analysis of polymers and rubbers, in Encyclopedia of Analytical Chemistry, Vol. 9 (eds R.A. Meyers), John Wiley & Sons Ltd, Chichester (2000), pp 7702–7759.

J.M. Chalmers, R.W. Hannah and D.W. Mayo, Spectra-structure correlations: Polymer spectra, in Handbook of Vibrational Spectroscopy, Vol. 3 (eds J.M. Chalmers and P.R. Griffiths), John Wiley & Sons Ltd, Chichester (2002), pp 1893–1918.

T.D. Chaplin, R.J.H. Clark, D. Jacobs, et al., The Gutenberg Bibles: analysis of the illuminations and inks using Raman spectroscopy, Analytical Chemistry, 77 (2005), 3611–3622.

R. Chen and K. Jakes, FTIR study of dyed and undyed cotton fibres recovered from a marine environment, in Historic Textiles, Papers and Polymers in Museums, (eds J.M. Cardamone and M.T. Baker), American Chemical Society, Washington (2001), pp 38–54.

T.H. Chen, T. Calligaro, S. Pages-Camagna and M. Menu, Investigation of Chinese archaic jade by PIXE and μRaman spectrometry, Applied Physics A, 79 (2004), 177–180.

O. Chiantore, D. Scalarone and T. Learner, Characterisation of artists' acrylic emulsion paints, International Journal of Polymer Analysis and Characterisation, 8 (2003), 67–82.

M.C. Christensen, Analysis of mineral salts from monuments by infrared spectroscopy, in IRUG2 Meeting Postprints, (ed B. Pretzel), Victoria and Albert Museum, London (1998), pp 93–100.

R. Chujo, A. Shimaoka, K. Nagaoka, A. Kurata and M. Inoue, Primary structure of archaeological silk and ancient climate, Polymer, 37 (1996), 3693–3696.

R.J.H. Clark, Pigment identification on medieval manuscripts by Raman microscopy, Journal of Molecular Structure, 347 (1995a), 417–428.

R.J.H. Clark, Raman microscopy: application to the identification of pigments on medieval manuscripts, Chemical Society Reviews, 24 (1995b), 187–196.

R.J.H. Clark and P.J. Gibbs, Identification of lead (II) sulfide and pararealgar on a 13th century manuscript by Raman microscopy, Chemical Communications, 11 (1997a), 1003–1004.

R.J.H. Clark and P.J. Gibbs, Non-destructive in situstudy of ancient Egyptian faience by Raman microscopy, Journal of Raman Spectroscopy, 28 (1997b), 99–103.

R.J.H. Clark, M.L. Curri and C. Laganara, Raman spectroscopy: the identification of lapis lazuli on medieval pottery fragments from the south of Italy, Spectrochimica Acta Part A, 53 (1997a), 597–603.

R.J.H. Clark, L. Curri, G.S. Henshaw and C. Laganara, Characterisation of brown-black and blue pigments in glazed pottery fragments from Castel Fiorentino (Foggia, Italy) by Raman microscopy, X-ray powder diffractometry and X-ray photoelectron spectroscopy, Journal of Raman Spectroscopy, 28 (1997b), 105–109.

R.J.H. Clark and P.J. Gibbs, Raman microscopy of a 13th century illuminated text, Analytical Chemistry, 70 (1998), 99A–105A.

R.J.H. Clark, Raman microscopy: sensitive probe of pigments on manuscripts, paintings and other artifacts, Journal of Molecular Structure, 480–481 (1999), 15–20.

R.J.H. Clark, Pigment identification by spectroscopic means: and arts/science interface, *C.R. Chimie*, **5** (2002), 7–20.

R.J.H. Clark, Raman microscopy in the identification of pigments on manuscripts and other artwork, in *Scientific Examination of Art: Modern Techniques in Conservation and Analysis*, National Academies Press, Washington DC (2005), pp 162–185.

C. Clementi, C. Miliani, A. Romani and G. Favaro, In situfluorimetry: a powerful non-invasive diagnostic technique for natural dyes used in artifacts. Part I. Spectral characterisation of orecein in solution, on silk and wool laboratory standards and a fragment of renaissance tapestry, *Spectrochimica Acta A*, **64** (2006), 906–912.

D.J. Cole-Hamilton, J.A. Chudek, G. Hunter and C.J.M. Martin, NMR imaging of water in wood, including water-logged archaeological artifacts, *Journal of the Institute of Wood Science*, **12** (1990), 111–113.

D.J. Cole-Hamilton, B. Kaye, J.A. Chudek and G. Hunter, Nuclear magnetic resonance imaging of waterlogged wood, *Studies in Conservation*, **40** (1995), 41–50.

P. Colomban, S. Jullian, M. Parker and P. Monge-Cadet, Identification of the high temperature impact/friction of aeroengine blades and cases by micro Raman spectroscopy, *Aerospace Science and Technology*, **3** (1999), 447–459.

P. Colomban and F. Treppoz, Identification and differentiation of ancient and modern European porcelains by Raman macro- and micro-spectroscopy, *Journal of Raman Spectroscopy*, **32** (2001), 93–102.

P. Colomban, G. Sagon and X. Faurel, Differentiation of antique ceramics from the Raman spectra of their coloured glazes and paintings, *Journal of Raman Spectroscopy*, **32** (2001), 351–360.

P. Colomban, Raman spectrometry, a unique tool to analyse and classify ancient ceramics and glasses, *Applied Physics A*, **79** (2004), 167–170.

P. Colomban, Case study: glasses, glazes and ceramics – recognition of ancient technology from the Raman spectra, in *Raman Spectroscopy in Archaeology and Art History* (eds H.G.M. Edwards and J.M. Chalmers), Royal Society of Chemistry, Cambridge (2005), pp 192–206.

D.C. Cook, Application of Mössbauer spectroscopy to the study of corrosion, *Hyperfine Interactions*, **153** (2004), 61–82.

D. Cordischi, D. Monna and A.L. Segre, ESR analysis of marble samples from Mediterranean quarries of archaeological interest, *Archaeometry*, **25** (1983), 68–76.

A. Cordy and K. Yeh, Blue dye identification on celllulosic fibres: indigo, logwood and Prussian blue, *Journal of the American Institute of Conservation*, **24** (1984), 33–39.

A. Corradi, C. Leonelli, P. Veronesi, *et al.*, Ancient glass deterioration in mosaics of Pompeii, *Surface Engineering*, **21** (2005), 402–405.

J. Corset, P. Dhamelincourt and J. Barbillat, Raman microscopy, *Chemistry in Britain*, **6** (1989), 612–616.

C. Coupry, G. Sagon and P. Gorguet-Ballesteros, Raman spectroscopic investigation of blue contemporary textiles, *Journal of Raman Spectroscopy*, **28** (1997), 85–89.

A. Cunningham, I.D. Gay, A.C. Oehlschlager and J.H. Langenheim, ^{13}C NMR and IR analyses of structure, ageing and botanical origin of Dominican and Mexican ambers, *Phytochemistry*, **22** (1983), 965–968.

S. Danillia, D. Bikiaris, L. Burgio, *et al.*, An extensive non-destructive and micro-spectroscopic study of two post-Byzantine overpainted icons of the 16th century, *Journal of Raman Spectroscopy*, **33** (2002), 807–814.

F. De Barquin and J.M. Dereppe, Drying of a white porous limestone monitored by NMR imaging, *Magnetic Resonance Imaging*, **14** (1996), 941–943.

G.E. De Benedetto, R. Laviano, L. Sabbatini and P.G. Zambonin, Infrared spectroscopy in the mineralogical characterisation of ancient pottery, *Journal of Cultural Heritage*, 3 (2002), 177–186.

G.E. De Benedetto, B. Fabbri, S. Gualtieri, L. Sabbatini and P.G. Zambonin, FTIR-chemometric tools as aids for data reduction and classification of pre-Roman ceramics, *Journal of Cultural Heritage*, 6 (2005), 205–211.

P. Deitemann, M. Kalin, S. Zumbuhl, *et al.*, A mass spectrometry and electron paramagnetic resonance study of photochemical and thermal ageing study of triterpenoid varnishes, *Analytical Chemistry*, 73 (2001), 2087–2096.

E.R. de la Rie, Fluorescence of paint and varnish layers (part I), *Studies in Conservation*, 27 (1982a), 1–7.

E.R. de la Rie, Fluorescence of paint and varnish layers (part II), *Studies in Conservation*, 27 (1982b), 65–69.

E.R. de la Rie, Fluorescence of paint and varnish layers (part III), *Studies in Conservation*, 27 (1982c), 102–108.

F. De Weerdt and J. Van Royen, Defects in coloured natural diamonds, *Diamonds and Related Materials*, 10 (2001), 474–479.

S.K. Dedushenko, I.B. Makhina, A.A. Marin, *et al.*, What oxidation state of iron determines the amethyst colour?, *Hyperfine Interactions*, 156–157 (2004), 417–422.

M.L. Dele-Dubois, P. Dhamelincourt, J.P. Poirot and H.J. Schubnel, Differentiation between natural gems and synthetic minerals by laser Raman microspectrometry, *Journal of Molecular Structure*, 143 (1986), 135–138.

M.L. Dele, P. Dhamelincourt, J.P. Poirot, *et al.*, Use of spectroscopic techniques for the study of natural and synthetic gems: application to rubies, *Journal of Raman Spectroscopy*, 28 (1997), 673–676.

R.C. Denney and R. Sinclair, *Visible and Ultraviolet Spectroscopy*, John Wiley and Sons Ltd, Chichester (1991).

M. Derrick, Fourier transform infrared spectral analysis of natural resins used in furniture finishes, *Journal of the American Institute for Conservation*, 28 (1989), 43–56.

M.R. Derrick, Infrared microspectroscopy in the analysis of cultural artefacts in *Practical Guide to Infrared Microspectroscopy* (ed H.J. Humecki), Marcel Dekker, New York (1995).

M.R. Derrick, D. Stulik and J.M. Landry, *Infrared Spectroscopy in Conservation Science*, The Getty Conservation Institute, Los Angeles (1999).

A.E. Derome, *Modern NMR Techniques for Chemistry Research*, Pergamon Press, Oxford (1987).

V. Desnica, K. Func, B. Hochleitner and M. Mantler, A comparative analysis of five chrome green pigments based on different spectroscopic techniques, *Spectrochimica Acta B*, 58 (2003), 681–687.

D.P.E. Dickson and F.J. Berry (eds), *Mössbauer Spectroscopy*, Cambridge University Press, Cambridge (1986).

M.T. Domenech-Carbo, F. Bosch Reig, J.V. Gimeno Adelantado and V. Penz Martinez, Fourier transform infrared spectroscopy and the analytical study of works of art for purposes of diagnosis and conservation, *Analytical Chimica Acta*, 330 (1996), 207–215.

M.T. Domenech-Carbo, V. Penz Martinez, J.V. Cumeno Adelantado, *et al.*, Fourier transform infrared spectroscopy and the analytical study of sculptures and wall decoration, *Journal of Molecular Structure*, 410–411 (1997), 559–563.

M.T. Domenech-Carbo, M.J. Casas-Catalan, A. Domenech-Carbo, *et al.*, Analytical study of canvas painting collection from the Basilica de la Virgen de los Desamparados using SEM/EDX, FT-IR, GC and electrochemical techniques, *Fresenius Journal of Analytical Chemistry*, **369** (2001a), 571–575.

M.T. Domenech-Carbo, A. Domenech-Carbo, J.V. Gimeno-Adelantado and F. Bosch-Reig, Identification of synthetic resins used in works of art by Fourier transform infrared spectroscopy, *Applied Spectroscopy*, **55** (2001b), 1590–1602.

L. D'Orazio, G. Gentile, C. Mancarella, E. Martuscelli and V. Massa, Water-dispersed polymers for the conservation and restoration of cultural heritage: a molecular, thermal, structural and mechanical characterisation, *Polymer Testing*, **20** (2001), 227–240.

G. Dupuis, M. Elias and L. Simonet, Pigment identification by fibre optics diffuse reflectance spectroscopy, *Applied Spectroscopy*, **56** (2002), 1329–1336.

M. Duttine, G. Villeneuve, G. Poupeau, *et al.*, Electron spin resonance of Fe^{3+} ion in obsidians from Mediterranean islands. Application to provenance studies, *Journal of Non-Crystalline Solids*, **323** (2003), 193–199.

H.G.M. Edwards, A.F. Johnson, I.R. Lewis and P. Turner, Raman spectroscopic studies of 'Pedigree Doll disease', *Polymer Degradation and Stability*, **41** (1993), 257–264.

H.G.M. Edwards and D.W. Farwell, Ivory and simulated ivory artefacts: Fourier transform Raman diagnostic study, *Spectrochimica Acta Part A*, **51** (1995), 2073–2081.

H.G.M. Edwards, D.W. Farwell, T. Seddon and J.K.F. Tait, Scrimshaw – real or fake – a Fourier transform Raman diagnostic study, *Journal of Raman Spectroscopy*, **26** (1995), 623–628.

H.G.M. Edwards and D.W. Farrell, Fourier transform Raman spectroscopy of amber, *Spectrochimica Acta Part A*, **52** (1996a), 1119–1125.

H.G.M. Edwards, D.W. Farrell and L. Daffner, Fourier transform Raman spectroscopic study of natural waxes and resins I, *Spectrochimica Acta Part A*, **52** (1996b), 1639–1648.

H.G.M. Edwards, E. Ellis, D.W. Farwell and R.C. Janaway, Preliminary study of the application of Fourier transform Raman spectroscopy to the analysis of degraded archaeological linen textiles, *Journal of Raman Spectroscopy*, **27** (1996), 663–669.

H.G.M. Edwards and M.J. Falk, Fourier transform Raman spectroscopic study of frankincense and myrrh, *Spectrochimica Acta Part A*, **53** (1997a), 2393–2401.

H.G.M. Edwards and M.J. Falk, Investigation of the degradation products of archaeological linens by Raman spectroscopy, *Applied Spectroscopy*, **51** (1997b), 1134–1138.

H.G.M. Edwards, D.W. Farwell, J.M. Holder and E.E. Lawson, Fourier transform Raman spectroscopy of ivory III: identification of mammalian specimens, *Spectrochimica Acta Part A*, **53** (1997a), 2403–2409.

H.G.M. Edwards, D.W. Farwell and D. Webster, FT Raman microscopy of untreated natural plant fibres, *Spectrochimica Acta Part A*, **53** (1997b), 2383–2392.

H.G.M. Edwards, D.W. Farwell, J.M. Holder and E.E. Lawson, Fourier transform Raman spectroscopy of ivory: II spectroscopic analysis and assignments, *Journal of Molecular Structure*, **435** (1997c), 49–58.

H.G.M. Edwards, D.W. Farwell, J.M. Holder and E.E. Lawson, Fourier transform Raman spectroscopy of ivory: a non-destructive diagnostic technique, *Studies in Conservation*, **43** (1998a), 9–16.

H.G.M. Edwards, M.J. Falk, M.G. Sibley, J. Alvarez-Benedi and F. Rull, FT-Raman spectroscopy of gums of technological significance, *Spectrochimica Acta Part A*, **54** (1998b), 903–920.

H.G.M. Edwards and J.K.F. Tait, FT-Raman spectroscopic study of decorated stained glass, *Applied Spectroscopy*, 52 (1998), 679–682.

H.G.M. Edwards, Art works studied using IR and Raman spectroscopy, in *Encyclopedia of Spectroscopy and Spectrometry*, Vol. 1 (eds G. Tranter, J. Holmes and J. Lindon), Elsevier, Amsterdam (2000), pp 2–17.

H.G.M. Edwards, E.M. Newton and J. Russ, Raman spectroscopic analysis of pigments and substrata in prehistoric rock art, *Journal of Molecular Structure*, 550–551 (2000), 245–256.

H.G.M. Edwards, D.W. Farwell, E.M. Newton, F. Rull Perez and S. Jorge Villar, Application of FT-Raman spectroscopy to the characterisation of parchment and vellum, I; novel information for paleographic and historiated manuscript studies, *Spectrochimica Acta Part A*, 57 (2001a), 1223–1234.

H.G.M. Edwards, D.W. Farwell, F. Rull Perez and J. Medina Garcia, Medieval cantorals in the Valladolid Biblioteca: FT-Raman spectroscopic study, *Analyst*, 126 (2001b), 383–388.

H.G.M. Edwards, Raman microscopy in art and archaeology: illumination of historical mysteries in rock art and frescoes, *Spectroscopy*, 17 (2002), 16–40.

H.G.M. Edwards, M. Gniadecka, S. Peterson, NIR-FT Raman spectroscopy as a diagnostic probe for mummified skin and nails, *Vibrational Spectroscopy*, 28 (2002), 3–15.

H.G.M. Edwards and F. Rull Perez, Application of Fourier transform Raman spectroscopy to the characterisation of parchment and vellum. II – Effect of biodeterioration and chemical deterioration on spectral interpretation, *Journal of Raman Spectroscopy*, 35 (2004), 754–760.

H.G.M. Edwards and D.L.A. de Faria, Infrared, Raman microscopy and fibre-optic Raman spectroscopy (FORS), in *Non-Destructive Microanalysis of Cultural Heritage Materials* (eds S.K. Janssens and R. van Grieken), Elsevier, Amsterdam (2004), pp 359–395.

H.G.M. Edwards and J.M. Chalmers, Practical Raman spectroscopy, in *Raman Spectroscopy in Archaeology and Art History* (eds H.G.M. Edwards and J.M. Chalmers), Royal Society of Chemistry, Cambridge (2005), pp 41–67.

H.G.M. Edwards, Case study; prehistoric art, in *Raman Spectroscopy in Archaeology and Art History* (eds H.G.M. Edwards and J.M. Chalmers), Royal Society of Chemistry, Cambridge (2005a), pp 84–96.

H.G.M. Edwards, Overview: biological materials and degradation, in *Raman Spectroscopy in Archaeology and Art History* (eds H.G.M. Edwards and J.M. Chalmers), Royal Society of Chemistry, Cambridge (2005b), pp 231–279.

H.G.M. Edwards and P. Wyeth, Case study: ancient textile fibres, in *Raman Spectroscopy in Archaeology and Art History* (eds H.G.M. Edwards and J.M. Chalmers), Royal Society of Chemistry, Cambridge (2005), pp 304–324.

H.G.M. Edwards and T. Munshi, Diagnostic Raman spectroscopy for the forensic detection of biomaterials and the preservation of cultural heritage, *Analytical and Bioanalytical Chemistry*, 382 (2005), 1398–1406.

H. Fabian, Fourier transform infrared spectroscopy in peptide and protein analysis, in *Encyclopedia of Analytical Chemistry*, Vol. 7 (ed R.A. Meyers), John Wiley & Sons Ltd, Chichester (2000), pp 5779–5803.

H. Fabian and W. Mantele, Infrared spectroscopy of proteins, in *Handbook of Vibrational Spectroscopy*, Vol. 5 (eds J.M. Chalmers and P.R. Griffiths), John Wiley & Sons Ltd, Chichester (2002), pp 3399–3425.

M. Fabbri, M. Picollo, S. Porcinai and M. Bacci, Mid-infrared fibre optics reflectance spectroscopy: a noninvasive technique for remote analysis of painted layers. Part I: Technical setup, *Applied Spectroscopy*, 55 (2001), 420–427.

M. Fabbri, M. Picollo, S. Porcinai and M. Bacci, Mid-infrared fibre optics reflectance spectroscopy: a noninvasive technique for remote analysis of painted layers. Part II: Statistical analysis of spectra, *Applied Spectroscopy*, 55 (2001), 428–433.

V.C. Farmer (ed), *The Infrared Spectra of Minerals*, Mineralogical Society, London (1974).

S. Felder-Casagrande and M. Odlyha, Development of standard paint films based on artists' materials, *Journal of Thermal Analysis*, 49 (1997), 1585–1591.

R.L. Feller, Dammar and mastic infrared analysis, *Science*, 120 (1954), 1069–1070.

J.M. Fernandez and A. la Iglesia, Estudio de la Coloracion Roja y Amarilla de Vidrios de la Catedral de Toldeo, *Boletin de la Sociedad Ceramica y Vidrio*, 33 (1999), 864–868.

J.R. Ferraro (ed), *Infrared Spectra Handbook of Minerals and Clays*, Sadtler Research Laboratories, Philadelphia (1982).

J.R. Ferraro, K. Nakamoto and C.W. Brown, *Introductory Raman Spectroscopy*, 2nd edn, Academic Press, San Diego (2003).

N. Ferrer and M.C. Sistach, Characterisation by FTIR spectroscopy of ink components in ancient manuscripts, *Restaurator*, 26 (2005), 105–117.

N. Ferrer and A. Vila, Fourier transform infrared spectroscopy applied to ink characterisation on one-penny postage stamps printed 1841-1880, *Analytica Chimica Acta*, 555 (2006), 161–166.

L.D. Field, S. Sternhell and J.R. Kalman, *Organic Structures from Spectra*, John Wiley & Sons Ltd, Chichester (2002).

K.H. Friolo, Chemical characterisation of weathered and unweathered stones in Sydney's heritage buildings, Honours Thesis, University of Technology, Sydney (2001).

J.A. Gadsen, *Infrared Spectra of Minerals and Related Inorganic Compounds*, Butterworth, London (1975).

M. Garcia-Heras, N. Carmona, C. Gil and M. Angeles Villegas, Neorenaissance/neobaroque stained glass windows from Madrid: a characterisation study of some panels signed by the Maimejean Fréres company, *Journal of Cultural Heritage*, 6 (2005), 91–98.

D.J. Gardiner and P.R. Graves (eds), *Practical Raman Spectroscopy*, Springer-Verlag, Berlin (1989).

P. Garside and P. Wyeth, Monitoring the deterioration of historic textiles: developing appropriate micromethodology, in *Historic Textiles, Papers and Polymers in Museums* (eds J.M. Cardamone and M.T. Baker), American Chemical Society, Washington (2001), pp 171–176.

P. Garside, S. Lahlil and P. Wyeth, Characterisation of historic silk by polarised attenuated total reflectance Fourier transform infrared spectroscopy for informed conservation, *Applied Spectroscopy*, 59 (2005), 1242–1247.

C. Genestar and C. Pons, Ancient covering plaster mortars from several convents and Islamic and Gothic palaces in Palma de Mallorca (Spain). Analytical characterisation, *Journal of Cultural Heritage*, 4 (2003), 291–298.

C. Genestar and C. Pons, Earth pigments in painting: characterisation and differentiation by means of FTIR spectroscopy and SEM-EDS microanalysis, *Analytical and Bioanalytical Chemistry*, 382 (2005), 269–274.

C. Genestar, C. Pons and A. Mas, Analytical characterisation of ancient mortars from the archaeological Roman city of Pollentia (Balearic Islands, Spain), *Analytica Chimica Acta*, 557 (2005), 373–379.

E.L. Ghisalberti and I.M. Godfrey, Application of nuclear magnetic resonance spectroscopy to the analysis of organic archaeological materials, *Studies in Conservation*, 43 (1998), 215–230.

L.T. Gibson, B.G. Cooksey, D. Littlejohn and N.H. Tennent, Investigation of the composition of a unique efflorescence on calcareous museum artefacts, *Analytica Chimica Acta*, 337 (1997), 253–264.

R.D. Gillard, S.M. Hardman and D.E. Watkinson, Recent advances in textile studies using FTIR microscopy, in *Conservation Science in the UK* (ed N.H. Tennent), James and James, London (1993), pp 71–76.

R.D. Gillard, S.M. Hardman, R.G. Thomas and D.E. Watkinson, The detection of dyes by FTIR microscopy, *Studies in Conservation*, 39 (1994), 187–192.

D. Goltz, J. McClelland, A. Schellenberg, *et al.*, Spectroscopic studies on the darkening of lead white, *Applied Spectroscopy*, 57 (2003), 1393–1398.

L.R. Green and F.A. Hart, Colour and chemical composition in ancient glass: an examination of some Roman and Wealdon glass by means of ultraviolet-visible-infrared spectrometry and electron microprobe analysis, *Journal of Archaeological Science*, 14 (1987), 271–282.

P.R. Griffiths and J.A. de Haseth, *Fourier Transform Infrared Spectrometry*, John Wiley & Sons Inc., New York (1986).

R. Grun, Electron spin resonance dating, in *Modern Analytical Methods in Art and Archaeology* (eds E. Ciliberto and G. Spoto), John Wiley & Sons Inc., New York (2000), pp 641–679.

Q. Guangyong, L. Shi, P. Xianjia and L. Guodong, Mössbauer spectroscopy study of terracotta warriors and horses of Qin Dynasty (221 BC), *Chinese Science Bulletin*, 34 (1989), 1777–1782.

B. Guineau, Non-destructive analysis of organic pigments and dyes using Raman microprobe, microfluorometer or absorption microspectrophotometer, *Studies in Conservation*, 34 (1989), 38–44.

H. Gunter, *NMR Spectroscopy*, 2nd edn, John Wiley & Sons Ltd, Chichester (1995).

D. Haberman, A. Banerjee, J. Meijer and A. Stephen, Investigation of manganese in salt- and freshwater pearls, *Nuclear Instruments and Methods in Physics Research B*, 181 (2001), 739–743.

T. Hainschwang, D. Simic, E. Fritsch, *et al.*, A gemological study of a collection of chameleon diamonds, *Gems and Gemmology*, 41 (2005), 20–34.

R.K. Harris, *Nuclear Magnetic Resonance Spectroscopy*, Longman, Harlow (1986).

L.M. Harwood and T.D.W. Claridge, *Introduction to Organic Spectroscopy*, Oxford University Press, Oxford (1997).

W. Hausler, Firing of clays studied by X-ray diffraction and Mössbauer spectroscopy, *Hyperfine Interactions*, 154 (2004), 121–141.

B. Havermans, H.A. Aziz and N. Penders, NIR as a tool for the identification of paper and inks in conservation research, *Restaurator*, 26 (2005), 172–180.

F. Hayashida, W. Hausler, J. Riederer and U. Wagner, Technology and organisation of Inka pottery production in the Leche Valley. Part II: Study of fired vessels, *Hyperfine Interactions*, 150 (2003), 153–163.

V. Hayez, J. Gillaume, A. Hubin and H. Terryn, Micro-Raman spectroscopy for the study of corrosion products on copper alloys: setting up s reference database and studying works of art, *Journal of Raman Spectroscopy*, **35** (2004), 732–738.

P.J. Hendra, W.F. Maddams, I.A.M. Royaud, *et al.*, The application of Fourier transform Raman spectroscopy to the identification and characterisation of polyamides I. Single number nylons, *Spectrochimica Acta Part A*, **46** (1990), 747–753.

P. Hendra, C. Jones and Warnes, *Fourier Transform Raman Spectroscopy: Introduction and Chemical Applications*, Ellis Horwood, New York (1991).

Y. Hsia and H. Huang, Mössbauer studies in Chinese archaeology: A review, *Hyperfine Interactions*, **150** (2003), 33–50.

H.J. Humecki (edn.), *Practical Guide to Infrared Microspectroscopy*, Marcel Dekker, New York (1999).

D.O. Hummel and F. Scholl, *Atlas of Polymer and Plastics Analysis*, Verlag, New York (1981).

S. Hvilsted, Analysis of emulsion paints, *Progress in Organic Coatings*, **13** (1985), 253–271.

M. Ikeya, *New Applications of Electron Spin Resonance*, World Scientific, London (1993).

R. Janes, M. Edge, J. Rigby, *et al.*, The effect of sample treatment and composition on the photoluminescence of anatase pigments, *Dyes and Pigments*, **48** (2001), 29–34.

M. Jones, Concepts and methods of ESR dating, *Radiation Measurements*, **27** (1997), 943–973.

A. Jurado-Lopez and M.D. Luque de Castro, Use of near infrared spectroscopy in a study of binding media used in painting, *Analytical and Bioanalytical Chemistry*, **380** (2004), 706–711.

J.E. Katon and A.J. Sommers, IR microspectroscopy: routine IR sampling methods extended to the microscopic domain, *Analytical Chemistry*, **64** (1992), 931A–940A.

J.E. Katon, Infrared microspectroscopy: a review of fundamentals and applications, *Micron*, **27** (1996), 3030–314.

I.P. Keen, G.W. White and P.M. Fredericks, Characterisation of fibres by Raman microprobe spectroscopy, *Journal of Forensic Sciences*, **43** (1998), 82–89.

B. Keisch, Analysis of works of art, in *Applications of Mössbauer Spectroscopy*, Vol. 1 (eds R.L. Cohen), Academic Press (1976), pp 263–286.

B. Keneghan, *Assessing plastic collections in museums by FTIR spectroscopy*, Infrared Users Group Postprints (1998), pp 21–24.

L.H. Kidder, A.S. Haka and E.N. Lewis, Instrumentation for FT-IR imaging, in *Handbook of Vibrational Spectroscopy*, Vol. 2 (eds J.M. Chalmers and P.R. Griffiths), John Wiley & Sons Ltd, Chichester (2002), pp 1386–1404.

L. Kiefert, H.A. Hanni and T. Ostertug, Raman spectroscopic applications to gemology, in *Handbook of Raman Spectroscopy: From Research Laboratory to the Process line* (eds I.R. Lewis and H.G.M. Edwards), Marcel Dekker, New York (2001), pp 469–489.

L. Kiefert, J.P. Chalain and S. Haberti, Case study: diamonds, gemstones and pearls – from the past to the present, in *Raman Spectroscopy in Archaeology and Art History* (eds H.G.M. Edwards and J.M. Chalmers), Royal Society of Chemistry, Cambridge (2005), pp 379–402.

M. Kihara, M. Takayama, H. Wariishi and H. Tahaka, Determination of the carbonyl groups in native lignin utilising Fourier transform Raman spectroscopy, *Spectrochimica Acta Part A*, **58** (2002) 2213–2221.

J.M. King, J.E. Shigley, T.H. Gells, et al., Characterisation of grading of natural colour yellow diamonds, Gems and Gemmology, 41 (2005), 88–115.

J. Koenig, Spectroscopy of Polymers, 2nd edn, Elsevier, Amsterdam (1999).

J. Kolar, M. Strlic, D. Muller-Hess, et al., Laser cleaning of paper using Nd:YAG laser running at 532 nm, Journal of Cultural Heritage, 4 (2003), 185s–187s.

K. Krishnan, J.R. Powell and S.L. Hill, Infrared microimaging, in Practical Guide to Infrared Microspectroscopy (ed H.J. Humecki), Marcel Dekker, New York (1995).

S. Kuckova, I. Nemec, R. Hynek, J. Hradilova and T. Grygar, Analysis of organic colouring and binding components in colour layer of art works, Analytical and Bioanalytical Chemistry, 382 (2005), 275–282.

A.V.R. Kumar and R. Balasubramaniam, Corrosion product analysis of corrosion resistant ancient Indian iron, Corrosion Science, 40 (1998), 1169–1178.

A.H. Kupstev and G.N. Zhizhin, Handbook of Fourier Transform Raman and Infrared Spectra of Polymers, Elsevier, Amsterdam (1998).

A. Kyek, F.E. Wagner, G. Lehrberger, et al., Celtic gold coins in the light of Mössbauer spectroscopy, electron microprobe analysis and X-ray diffraction, Hyperfine Interactions, 126 (2000), 235–240.

L. Laguardia, E. Vassallo, F. Cappitelli, et al., Investigation of the effects of plasma treatments on biodeteriorated ancient paper, Applied Surface Science, 252 (2005), 1159–1166.

J.B. Lambert, J.S. Frye and G.O. Poinar, Amber from the Dominican Republic: analysis by nuclear magnetic resonance spectroscopy, Archaeometry, 27 (1985), 43–51.

J.B. Lambert, C.W. Beck and J.S. Frye, Analysis of European amber by carbon-13 nuclear magnetic resonance spectroscopy, Archaeometry, 30 (1988), 248–263.

J.B. Lambert, J.S. Frye, T.A. Lee, et al., Analysis of Mexican amber by carbon-13 NMR spectroscopy, in Archaeological Chemistry IV (ed R.O. Allen), American Chemical Society, Washington (1989), pp 381–388.

J.B. Lambert, J.S. Frye and G.O. Poinar Amber from the Dominican Republic: analysis by nuclear magnetic resonance spectroscopy, Archaeometry, 27 (1985), 43–51.

J.B. Lambert, C.W. Beck and J.S. Frye, Analysis of European amber by carbon-13 nuclear magnetic resonance spectroscopy, Archaeometry, 30 (1988), 248–263.

J.B. Lambert, J.S. Frye, T.A. Lee, et al., Analysis of Mexican amber by carbon-13 NMR spectroscopy, in Archaeological Chemistry IV (ed R.O. Allen), American Chemical Society, Washington (1989), pp 381–388.

J.B. Lambert, J.S. Frye and G.O. Poinar, Analysis of North American amber by carbon-13 NMR spectroscopy, Geoarchaeology, 5 (1990), 43–52.

J.B. Lambert, J.S. Frye and G.W. Carriveau, The structure of oriental lacquer by solid state nuclear magnetic resonance spectroscopy, Archaeometry 33 (1991) 87–93.

J.B. Lambert, J.S. Frye and A. Jurkiewicz, The provenance and coal rank of jet by carbon-13 nuclear magnetic resonance spectroscopy, Archaeometry, 34 (1992), 121–128.

J.B. Lambert, S.C. Johnson and G.O. Poinar, Nuclear magnetic resonance characterisation of Cretaceous amber, Archaeometry 38 (1996) 325–335.

J.B. Lambert, Traces of the Past: Unravelling the Secrets of Archaeology Through Chemistry, Perseus Publishing, Cambridge (1997).

P.L. Lang, J. Cook, B. Fuller Morris, S. Cullison, S. Telles and T. Barrett, Characterisation of historic papers using attenuated total reflection infrared spectroscopy, Applied Spectroscopy, 52 (1998), 713–716.

J.H. Langenheim and C.W. Beck, Infrared spectra as a means of determining the botanical sources of amber, Science, 149 (1965), 52–55.

L.J. Larson, K.S.K. Shin and J.I. Zink, Photoluminescence spectroscopy of natural resins and organic binding media of paintings, *Journal of the American Institute for Conservation*, **30** (1991), 89–104.

L. Lazzarini and F. Antonelli, Petrographic and isotopic characterisation of the marble of the island of Tinos (Greece), *Archaeometry*, **45** (2003), 541–552.

T. Learner, The use of FTIR in the conservation of twentieth century paintings, *Spectroscopy Europe*, **8** (1996), 14–19.

T. Learner, The use of a diamond cell for the FTIR characterisation of paints and varnishes available to twentieth century artists, in *IRUG2 Meeting Postprints* (ed B. Pretzel), Victoria and Albert Museum, London (1998), pp 7–20.

T.J.S. Learner, *Analysis of Modern Paints*, Getty Conservation Institute, Los Angeles (2004).

T. Learner, Modern paints, in *Scientific Examination of Art: Modern Techniques in Conservation and Analysis*, National Academies Press, Washington DC (2005), pp 137–151.

N.Q. Liem, N.T. Thanh and P. Colomban, Reliability of Raman micro-spectroscopy in analyzing ancient ceramics: the case of ancient Vietnamese porcelain and celadon glazes, *Journal of Raman Spectroscopy*, **33** (2002), 287–294.

J. Lindblom, J. Holsa, H. Papunen, *et al.*, Differentiation of natural and synthetic gem-quality diamonds by luminescence properties, *Optical Materials*, **24** (2003), 243–251.

D.A. Long, Introduction to Raman Spectroscopy, in *Raman Spectroscopy in Archaeology and Art History* (eds H.G.M. Edwards and J.M. Chalmers), Royal Society of Chemistry, Cambridge (2005), pp 17–40.

G. Longworth, N.H. Tennent, M.J. Tricker and P.P. Vaishnava, Iron-57 Mössbauer spectral studies of medieval stained glass, *Journal of Archaeological Science*, **9** (1982), 261–273.

G. Longworth, Studies of ceramics and archaeological materials, in *Mössbauer Spectroscopy Applied to Inorganic Chemistry* (ed G.J. Long), Plenum, New York (1984).

P. Lopez-Arce, J. Garcia-Guinea, M. Gracia and J. Obis, Bricks in historical buildings of Toledo City: characterisation and restoration, *Materials Characterisation*, **50** (2003), 59–68.

E. Lopez-Ballester, M.T. Domenech-Carbo, J.V. Gimeno-Adelantado and F. Bosch-Reig, Study by FTIR spectroscopy of ageing of adhesives used in restoration of archaeological glass objects, *Journal of Molecular Structure*, **482–483** (1999), 525–531.

M.J.D. Low and N.S. Baer, Application of infrared Fourier transform spectroscopy to problems in conservation I: General principles, *Studies in Conservation*, **22** (1977), 116–128.

T. Lu, Y. Liu, J. Shigley, T. Moses and I.M. Reinitz, Characterisation of a notable historic gem diamond showing the alexandrite effect, *Journal of Crystal Growth*, **193** (1998), 577–584.

W.F. Maddams, A review of Fourier transform Raman spectroscopic studies of polymers, *Spectrochimica Acta Part A*, **50** (1994), 1967–1986.

A. Maddock, *Mössbauer Spectroscopy: Principles and Applications*, Horwood, Chichester (1997).

J.R. Mansfield, M. Attas, C. Majzels, *et al.*, Near infrared spectroscopic reflectance imaging: a new tool in art conservation, *Vibrational Spectroscopy*, **28** (2002), 59–66.

P. Maravelaki-Kalaitzaki, V. Zafiropulos and C. Fotakis, Excimer laser cleaning of encrustation on Pentelic marble: procedure and evaluation of the effects, *Applied Surface Science*, **148** (1999), 92–104.

P. Maravelaki-Kalaitzaki, Black crusts and patinas on Pentelic marble from the Parthenon and Erechtheum (Acropolis, Athens): characterisation and origin, *Analytica Chimica Acta*, **532** (2005), 187–198.

E. Marengo, E. Robetti and M.C. Liparota and M.C. Gennaro, Monitoring pigmented and wooden surfaces in accelerated ageing processes by FT–Raman spectroscopy and multivariate control charts, *Talanta*, **63** (2004) 987–1002.

E. Marengo, M.C. Liparota, E. Robotti and M. Bobba, Multivariate calibration applied to the field of cultural heritage: analysis of the pigments on the surface of a painting, *Analytica Chimica Acta*, **553** (2005), 111–122.

L. Maritan, L. Nodari, C. Mazzoli, *et al.*, Influence of firing conditions on ceramic products: experimental study on clay rich in organic matter, *Applied Clay Science*, **31** (2006), 1–15.

A. Martinez-Richa, R. Vera-Graziano, A. Rivera and P. Joseph-Nathan, A solid state ^{13}C NMR analysis of ambers, *Polymer*, **41** (2000), 743–750.

A. Matuszewska and M. Czaja, Aromatic compounds in molecular phase of Baltic amber – synchrotron luminescence analysis, *Talanta*, **56** (2002), 1049–1059.

V.J. McBrierty, NMR Spectroscopy of Polymers in the Solid State, in *Comprehensive Polymer Science*, Vol. 1 (eds C. Booth and C. Price), Pergamon Press, Oxford (1989).

R.J. Meilunas, J.G. Bentsen and A. Steinberg, Analysis of aged paint binders by FTIR spectroscopy, *Studies in Conservation*, **35** (1990), 33–51.

R.G. Messerschmidt and M.A. Harthcock (eds.), *Infrared Microspectroscopy: Theory and Applications*, Marcel Dekker, New York (1998).

A. Middleton and J. Ambers, Case study: analysis of nephrite jade using Raman microscopy and X-ray fluoresce spectroscopy, in *Raman Spectroscopy in Archaeology and Art History* (eds H.G.M. Edwards and J.M. Chalmers), Royal Society of Chemistry, Cambridge (2005), pp 403–411.

C. Miliani, A. Romani and G. Favaro, A spectrophotometric and fluorimetric study of some anthraquinoid and indigoid colourants used in artistic paintings, *Spectrochimica Acta Part A*, **54** (1998), 581–588.

P.A. Mirau, *NMR Characterisation of Polymers* in *Polymer Characterisation* (eds B.J. Hunt and M.I. James), Blackie, London (1993).

T. Miyoshi, Laser-induced fluorescence of oil colours and its applications to the identification of pigments in oil paintings, *Japanese Journal of Applied Physics*, **21** (1982), 1032–1036.

T. Miyoshi, Fluorescence from oil colours, linseed oil and poppy oil using N_2 laser excitation, *Japanese Journal of Applied Physics*, **24** (1985), 371–372.

I. Moroz, M. Roth, M. Boudeulle and G. Panczer, Raman microspectroscopy and fluorescence of emeralds from various deposits, *Journal of Raman Spectroscopy*, **31** (2000), 485–490.

V. Mosini and S. Munziante Cesaro, Comparison of Baltic amber and an aged Pinus Halepensis resin by means of infrared spectroscopy, *Phytochemistry*, **25** (1996), 244–245.

M.C.M. Moya Moreno, Fourier transform infrared spectroscopy and the analytical study of sculptures and wall decoration, *Journal of Molecular Structure*, **410–411** (1997), 559–563.

E. Murad, The characterisation of soils, clays and clay firing products, *Hyperfine Interactions*, **111** (1998), 251–259.

N. Na, Q. Ouyang, H. Ma, J. Ouyang and Y. Li, Non-destructive and in situ identification of rice paper, seals and pigments by FTIR and XRF spectroscopy, *Talanta*, **64** (2004), 1000–1008.

P. Nair and R.A. Lodder, Near-ir identification of woods for restoration of historic buildings and furniture, *Applied Spectroscopy*, **47** (1993), 287–291.

M. Nasraoui, J.C. Waerenborgh, M.I. Prudencio and E. Bilal, Typology of the granitic stones of the cathedral of Évora (Portugal): a combined contribution of geochemistry and ^{57}Fe Mössbauer spectroscopy, *Journal of Cultural Heritage*, **3** (2002), 127–132.

R. Newman, Some applications of infrared spectroscopy in the examination of painting materials, *Journal of the American Institute for Conservation*, **19** (1979), 42–62.

R. Newman, Tempera and other nondrying-oil media, in *Painted Wood: History and Conservation* (eds V. Dorge and F.C. Howlett), Getty Conservation Institute, Los Angeles (1998), pp 33–63.

K. Ochocinska, A. Kaminska and G. Sliwinski, Experimental investigations of stained paper documents cleaned by the Nd: YAG laser pulses, *Journal of Cultural Heritage* **4** (2003) 188s–193s.

M. Odlyha, Investigation of the binding media of paintings by thermoanalytical and spectroscopic techniques, *Thermochimica Acta*, **269/270** (1995), 705–727.

A. Orlando, M. Picollo, B. Radicati, *et al.*, Principal component analysis of near-infrared and visible spectra: an application to a XIIth century Italian work of art, *Applied Spectroscopy*, **49** (1995), 459–465.

M. Ortega-Aviles, P. Vandenabeele, D. Tenorio, *et al.*, Spectroscopic investigation of a 'Virgin of Sorrows' canvas painting: a multi-method approach, *Analytica Chimica Acta*, **550** (2005), 164–172.

V. Otieno-Alego, Raman microscopy: a useful tool for the archaeometric analysis of pigments, in *Radiation in Art and Archaeometry* (eds D.C. Creagh and D.A. Bradley), Elsevier, Amsterdam (2000), pp 76–100.

C. Paris, S. Lecomte and C. Coupry, ATR–FTIR spectroscopy as a way to identify natural protein-based materials, tortoiseshell and horn, from their protein-based imitation, galalith, *Spectrochimica Acta Part A*, **62** (2005), 532–538.

C. Paris and C. Coupry, Fourier transform Raman spectroscopic study of the first cellulose-based artificial materials in heritage, *Journal of Raman Spectroscopy*, **36** (2005), 77–82.

R.V. Parish, Gold-197 Mössbauer spectroscopy in the characterization of gold compounds, in *Mössbauer Spectroscopy Applied to Inorganic Chemistry*, Vol. 1 (ed G.J. Long), Plenum Press, New York (1984), pp 577–617.

A. Perardi, A. Zoppi and E. Castellucci, Micro-Raman spectroscopy for standard and in situcharacterisation of painting materials, *Journal of Cultural Heritage*, **1** (2000), 5269–5272.

M. Perez-Alonso, K. Castro, I. Martine-Arkarazo, *et al.*, Analysis of bulk and inorganic degradation products of stones, mortars and wall paintings by portable Raman microprobe spectroscopy, *Analytical and Bioanalytical Chemistry*, **379** (2004), 42–50.

M. Pinet, D. Smith and B. Lasnier, Utilité de la microsonde Raman pour l'identification non destructive des gemmes, in *La Microsonde Raman en Gemmologie*, Revue Française de Gemmologie No. Hors Série (1992), pp 11–61.

K. Polikreti and Y. Maniatis, A new methodology for the provenance of marble based on EPR spectroscopy, *Archaeometry*, **44** (2002), 1–21.

K. Polokreti, Y. Maniatis, Y. Bassiakos, et al., Provenance of archaeological limestone with EPR spectroscopy: the case of the Cypriote-type statuettes, Journal of Archaeological Science, 31 (2004), 1015–1028.

S.S. Potgieter-Vermaak, R.H.M. Godoi, R. Van Grieken, et al., Micro-structural characterisation of black crust and laser cleaning of building stones by micro-Raman and SEM techniques, Spectrochimica Acta Part A, 61 (2005), 2460–2467.

C.J. Pouchert (ed.), The Aldrich Library of FTIR Spectra, 3rd edn, Aldrich Chemical Company, Milwaukee (1995).

G. Pozza, D. Ajo, G. Ciari, F. De Zuane and M. Favaro, Photoluminescence of the inorganic pigments Egyptian blue, Han blue and Han purple, Journal of Cultural Heritage, 1 (2000), 393–398.

E. Pretsch, T. Clerc, J. Seibl and W. Simon, Tables of Spectral Data for the Structural Determination of Organic Compounds, Springer Verlag, Berlin (1989).

B. Price, J. Carlson and R. Newman, An infrared spectral library of naturally occurring minerals, in IRUG2 Meeting Postprints (ed B. Pretzel), Victoria and Albert Museum, London (1998), pp 103–126.

B. Price and B. Pretzel, Infrared and Raman Users Group Spectral Database, The Infrared and Raman Users Group, Philadelphia (2000).

E. Princi, S. Vicini, E. Pedemonte, et al., New polymeric materials for paper and textile conservation. I. Synthesis and characterisation of acrylic copolymers, Journal of Applied Polymer Science, 98 (2005), 1157–1164.

N. Proietti, D. Capitani, E. Pedemonte, et al., Monitoring degradation in paper: non-invasive analysis by unilateral NMR. Pat II, Journal of Magnetic Resonance, 170 (2004), 113–120.

N. Proietti, D. Capitani, R. Lamanna, et al., Fresco paintings studied by unilateral NMR, Journal of Magnetic Resonance, 177 (2005), 111–117.

G.Y. Qin and X.J. Pan, Mössbauer firing study of terracotta warriors and horses of the Qin Dynasty (221 BC), Archaeometry, 31 (1989), 3–12.

G.Y. Qin and Z. Lin, Mössbauer study of fired Lishan clay and terracotta warriors and horses of Qin Dynasty (221 BC), Hyperfine Interactions, 70 (1992), 1045–1048.

J.F. Rabek, Experimental Methods in Polymer Chemistry, John Wiley & Sons Ltd, Chichester (1980).

L. Rampazzi, A. Andreotti, I. Bonduce,, M.P. Colombini, C. Colombo and L. Toniolo, Analytical investigation of calcium oxalate films on marble monuments, Talanta, 63 (2004), 967–977.

B. Ranby and J.F. Rabek, ESR Spectroscopy in Polymer Research, Springer-Verlag, Berlin (1977).

I. Reiche, S. Pages-Campagna and L. Lambacher, In situ Raman spectroscopic investigations of the adorning gemstones on the reliquary Heinrich's Cross from the treasury of Basel Cathedral, Journal of Raman Spectroscopy, 35 (2004), 719–725.

D. Rendell and D. Mowthorpe, Fluorescence and Phosphorescence Spectroscopy, John Wiley & Sons Ltd, Chichester (1987).

L.A. Rijniers, L. Pel, H.P. Huinink and K. Kopinga, Salt crystallisation as damage mechanism in porous building materials – a nuclear magnetic resonance study, Magnetic Resonance Imaging, 23 (2005), 273–276.

L. Robinet and D. Thickett, Case study: application to Raman spectroscopy to corrosion products, in Raman Spectroscopy in Archaeology and Art History (eds H.G.M. Edwards and J.M. Chalmers), Royal Society of Chemistry, Cambridge (2005), pp 325–334.

N. Robinson, R.P. Evershed, J. Higgs, *et al.*, Proof of a pine wood origin for pitch from Tudor (Mary Rose) and Etruscan shipwrecks: application of analytical organic chemistry in archaeology, *Analyst*, **112** (1987), 637–644.

A. Romani, C. Miliani, A. Morresi, *et al.*, Surface morphology and composition of some 'lustro' decorated fragments of ancient ceramics from Derita (Central Italy), *Applied Surface Science*, **157** (2000), 112–122.

A. Romani, C. Zuccaccia and C. Clementi, An NMR and UV-visible spectroscopic study of the principal coloured component of Stil de grain lake, *Dyes and Pigments*, **71** (2006), 224–229.

V. Rouchon-Quillet, C. Remazeilles, J. Bernard, *et al.*, The impact of gallic acid on iron gall ink corrosion, *Applied Physics A*, **79** (2004), 389–392.

F. Rull Perez, Applications of IR and Raman spectroscopy to the study of medieval pigments, in *Handbook of Raman Spectroscopy: From Research Laboratory to the Process Line* (eds I.R. Lewis and H.G.M. Edwards), Marcel Dekker, New York (2001), pp 835–862.

D.A. Rusak, L.M. Brown and S.D. Martin, Classification of vegetable oils by principal component analysis of FTIR spectra, *Journal of Chemical Education*, **80** (2003), 541–543.

A.O. Salnick and W. Faubel, Photoacoustic FTIR spectroscopy of natural copper patina, *Applied Spectroscopy*, **49** (1995), 1516–1524.

M. Saltzman, Identifying dyes in textiles, *American Scientist*, **9** (1992), 474–481.

N. Salvado, S. Buti, M.J. Tobin, *et al.*, Advantages of the use of SR-FTIR microspectroscopy: applications to cultural heritage, *Analytical Chemistry*, **77** (2005), 3444–3451.

S.R. Sandler, W. Karo, J. Bonesteel and E.M. Pearce, *Polymer Synthesis and Characterisation: A Laboratory Manual*, Academic Press, San Diego (1998).

B. Schrader, H. Schulz, G.N. Adreev, *et al.*, Non-destructive NIR-FT-Raman spectroscopy of plant and animal tissues, of food and works of art, *Talanta*, **53** (2000), 35–45.

M.R. Schreiner, I. Prohaska, J. Rendl and C. Weigel, Leaching studies of potash-lime-silica glass with medieval glass composition, in *Conservation of Glass and Ceramics: Research, Practice and Training* (ed. N. Tennent), James and James, London (1999).

J.W.H. Schreurs and R.H. Brill, Iron and sulfur related colours in ancient glass, *Archaeometry*, **26** (1984), 199–209.

R.B. Scrozelli, S. Petrick, A.M. Rossi, *et al.*, Obsidian archaeological artefacts provenance studies in the Western Mediterranean basin: an approach by Mössbauer spectroscopy and electron paramagnetic resonance, *Earth and Planetary Sciences*, **332** (2001), 769–776.

I.T. Shadi, B.Z. Chowdhry, M.J. Snowden and R. Withnall, Semi-quantitative analysis of indigo carmine, using silver colloids, by surface enhances resonance Raman spectroscopy (SERRS) *Spectrochimica Acta Part A*, **59** (2003), 2201–2206.

S. Sharma, F. Casanova, W. Wache, *et al.*, Analysis of historical porous building materials by the NMR-MOUSE, *Magnetic Resonance Imaging*, **21** (2003), 249–255.

J.C. Shearer, D.C. Peters, G. Hoepfner and T. Newton, FTIR in the service of art conservation, *Analytical Chemistry*, **55** (1983), 874A–880A.

M. Shimoyama, T. Ninomiya and Y. Ozaki, Nondestructive discrimination of ivories and prediction of their specific gravity by Fourier transform Raman spectroscopy and chemometrics, *Analyst*, **128** (2003), 950–953.

D.A. Skoog and J.J. Leary, *Principles of Instrumental Analysis*, Harcourt Brace, Fort Worth (1992).

H.W. Siesler and K. Holland-Moritz, *Infrared and Raman Spectroscopy of Polymers*, Marcel Dekker, New York (1980).

D.A. Silva, H.R. Wenk and P.J.M. Monteiro, Comparative investigation of mortars from Roman Colosseum and cistern, *Thermochimica Acta*, **438** (2005), 35–40.

R.M. Silverstein, F.X. Webster and D. Kiemle, *Spectrometric Identification of Organic Compounds*, John Wiley & Sons Ltd, Chichester (2005).

M.C. Sistach, N. Ferrer and M.T. Romero, Fourier transform infrared spectroscopy applied to the analysis of ancient manuscripts, *Restaurator*, **19** (1998), 173–186.

A.G. Smallwood, P.S. Thomas and A.S. Ray, Characterisation of sedimentary opals by Fourier transform Raman spectroscopy, *Spectrochimica Acta Part A*, **53** (1997), 2341–2345.

R.C. Smith, *Understanding Magnetic Resonance Imaging*, CRC Press, Boca Raton (1997).

D.C. Smith, Overview: jewellery and precious stones, in *Raman Spectroscopy in Archaeology and Art History* (eds H.G.M. Edwards and J.M. Chalmers), Royal Society of Chemistry, Cambridge (2005a), pp 335–378.

D.C. Smith, Case study: meso-american jade, in *Raman Spectroscopy in Archaeology and Art History* (eds H.G.M. Edwards and J.M. Chalmers), Royal Society of Chemistry, Cambridge (2005b), pp 412–426

G.D. Smith and R.J.H. Clark, Raman microscopy in archaeological science, *Journal of Archaeological Science*, **31** (2004), 1137–1160.

A. Sodo, M. Nardone, D. Ajo, G. Pozza and M. Bicchieri, Optical and structural properties of gemological materials used in works of art and handicrafts, *Journal of Cultural Heritage*, **4** (2003), 317s–320s.

A.J. Sommer, Mid-infrared transmission microspectroscopy, in *Handbook of Vibrational Spectroscopy*, Vol. 2 (eds J.M. Chalmers and P.R. Griffiths), John Wiley & Sons Ltd, Chichester (2002), pp 1369–1385.

A. Spyros and D. Anglos, Study of ageing in oil paintings by 1D and 2D NMR spectroscopy, *Analytical Chemistry*, **76** (2004), 4929–4936.

S.J. Stewart, G. Cernicchiaro, R.B. Scorzelli, *et al.*, Magnetic properties and ^{57}Fe Mössbauer spectroscopy of Mediterranean prehistoric obsidians for provenance studies, *Journal of Non-Crystalline Solids*, **323** (2003), 188–192.

L. Stievano, M. Bertelle and S. Calogero, Application of ^{57}Fe Mössbauer spectroscopy for the characterisation of materials of archaeological interest: the work performed in Italy, *Hyperfine Interactions*, **150** (2003), 13–31.

D.P. Stronmen and K. Nakamoto, *Laboratory Raman Spectroscopy*, John Wiley & Sons Inc., New York (1984).

B.H. Stuart, *Infrared Spectroscopy: Fundamentals and Applications*, John Wiley & Sons Ltd, Chichester (2004).

M. Takayama, T. Johjima, T. Yamanaka, *et al.*, Fourier transform Raman assignment of guaiacyl and syringyl marker bands for lignin determination, *Spectrochimica Acta Part A*, **53** (1997), 1621–1628.

M. Takeda, M. Mabuchi and T. Tominaga, A tin-119 Mössbauer study of Chinese bronze coins, *Radiochemical and Radioanalytical Letters* **29** (1977) 191–198.

M.N. Taran, A.N. Taraschchan, H. Rager, *et al.*, Optical spectroscopy study of variously coloured gem-quality topazes from Ouro Preto, Minas Gerais, Brazil, *Physics and Chemistry of Minerals*, **30** (2003), 546–555.

G.W. Taylor, Detection and identification of dyes on Anglo-Scandinavian textiles, *Studies in Conservation*, **28** (1983), 153–160.

D. Tenorio, A. Cabral, P. Bosch, *et al.*, Differences in coloured obsidians from Sierra de Pachuca, Mexico, *Journal of Archaeological Science*, **25** (1998), 229–234.

J.S. Tsang and R.H. Cunningham, Some improvements in the study of cross sections, *Journal of the American Institute of Conservation*, **30** (1991), 163–177.

H. Tschauner and U. Wagner, Pottery from a Chimú workshop studied by Mössbauer spectroscopy, *Hyperfine Interactions*, **150** (2003), 165–186.

M.W. Tungol, E.G. Bartick and A. Montaser, Analysis of single polymer fibres by Fourier transform infrared microscopy: the results of case studies, *Journal of Forensic Science*, **36** (1991), 1027–1043.

M.W. Tungol, E.G. Bartick and A. Montaser, Forensic examination of synthetic textile fibres by microscopic infrared spectrometry, in *Practical Guide to Infrared Microspectroscopy* (ed H. Humecki), Marcel Dekker, New York (1995), pp 245–285.

P. Vandenabeele, L. Moens, H. Edwards and R. Dams, Raman spectroscopic database of azo pigments and application to modern art studies, *Journal of Raman Spectroscopy*, **31** (2000a), 509–517.

P. Vandenabeele, B. Wehling, L. Moens, *et al.*, Analysis with micro-Raman spectroscopy of natural organic binding media and varnishes used in art, *Analytica Chimica Acta*, **407** (2000b), 261–274.

P. Vandenabeele, F. Verpoort and L. Moens, Non-destructive analysis of paintings using Fourier transform Raman spectroscopy with fibre optics, *Journal of Raman Spectroscopy*, **32** (2001a), 263–269.

P. Vandenabeele, A. Hardy, H.G.M. Edwards and L. Moens, Evaluation of a principal components-based searching algorithm for Raman spectroscopic identification of organic pigments in 20th century artwork, *Applied Spectroscopy*, **55** (2001b), 535–533.

P. Vandenabeele, D.M. Grimaldi, H.G.M. Edards and L. Moens, Raman spectroscopy of different types of Mexican copal resins, *Spectrochimica Acta Part A*, **59** (2003), 2221–2229.

P. Vandenabeele, T.L. Weis, E.R. Grant and L.J. Moens, A new instrument adapted to in situRaman analysis of objects of art, *Analytical and Bioanalytical Chemistry*, **379** (2004), 137–142.

P. Vandenabeele and L. Moens, Pigment identification in illuminated manuscripts, in *Non-Destructive Microanalysis of Cultural Heritage Materials* (eds K. Janssens and R. van Grieken), Elsevier, Amsterdam (2004), pp 635–662.

P. Vandenabeele and L. Moens, Overview: Raman spectroscopy of pigments and dyes, in *Raman Spectroscopy in Archaeology and Art History* (eds H.G.M. Edwards and J.M. Chalmers), Royal Society of Chemistry, Cambridge (2005), pp 71–83.

P. Vandenabeele and H.G.M. Edwards, Overview: Raman spectroscopy of artefacts, in *Raman Spectroscopy in Archaeology and Art History* (eds H.G.M. Edwards and J.M. Chalmers), Royal Society of Chemistry, Cambridge (2005), pp 169–178.

H.W. van der Marel and H. Beutelspacher, *Atlas of Infrared Spectroscopy of Clay Minerals and their Admixtures*, Elsevier, Amsterdam (1976).

J. van der Weerd, Microspectroscopic analysis of traditional oil paint, PhD thesis, Universiteit van Amsterdam (2002).

J. van der Weerd, H. Brammer, J.J. Boon and RM.A. Heeren Fourier transform infrared microscopic imaging of an embedded paint cross-section, *Applied Spectroscopy*, **56** (2002), 275–283.

J. van der Weerd, M.K. van Veen, R.M.A. Heeren and J.J. Boon, Identification of pigments in paint cross sections by reflection visible light imaging microspectroscopy, *Analytical Chemistry*, 75 (2003), 716–722.

A. van Loon and J.J. Boon, Characterisation of the deterioration of bone black in the 17th century Oranjezaal paintings using electron-microscopic and micro-spectroscopic imaging techniques, *Spectrochimica Acta Part B*, 59 (2004), 1601–1609.

K.S. Venkateswarlu and V. Ramshesh, Application of Mössbauer spectroscopy in archaeology, *Journal of Archaeological Chemistry*, 2 (1984), 41–49.

S. Viel, D. Capitani, N. Proietti, F. Ziarelli and A.L. Segre, NMR spectroscopy applied to cultural heritage: a preliminary study on ancient wood characterisation, *Applied Physics A*, 79 (2004), 357–361.

M. Vilarigues and R.C. Da Silva, Ion beam and infrared analysis of medieval stained glass, *Applied Physics A*, 79 (2004), 373–378.

I. Viola, S. Bubici, C. Casieri and F. De Luca, The Codex Major of the Cellectio Altaempsiana: a non-invasive NMR study of paper, *Journal of Cultural Heritage*, 5 (2004), 257–261.

S.T. Vohra, F. Bucholtz, G.M. Nau, *et al.*, Remote detection of trichloroethylene in soil by a fibre-optic infrared reflectance probe, *Applied Spectroscopy*, 50 (1996), 985–990.

M. Wadsak, T. Aastrup, I Odnevall Wallinder, *et al.*, Multianalytical in situinvestigation of the initial atmospheric corrosion of bronze, *Corrosion Science*, 44 (2002), 791–802.

U. Wagner, F.E. Wagner, W. Hausler and I. Shimada, The use of Mössbauer spectroscopy in studies of archaeological ceramics, in *Radiation in Art and Archaeometry* (eds D.C. Creagh and D.A. Bradley), Elsevier, Amsterdam (2000), pp 417–443.

B. Wagner, E. Bulska, B. Stohl, *et al.*, Analysis of Fe valence states in iron gall inks from 16th century manuscripts by ^{57}Fe Mössbauer spectroscopy, *Analytica Chimica Acta*, 527 (2004), 195–202.

F.E. Wagner and A. Kyek, Mössbauer spectroscopy in archaeology: Introduction and experimental conditions, *Hyperfine Interactions*, 154 (2004), 5–33.

F.E. Wagner and U. Wagner, Mössbauer spectra of clays and ceramics, *Hyperfine Interactions*, 154 (2004), 35–82.

A. Wallert and R. Boytner, Dyes from the Tumilaca and Chiribaya Cultures, South Coast of Peru, *Journal of Archaeological Science*, 23 (1996), 853–861.

A. Wang, J. Han, L. Guo,*et al.*, Database of standard Raman spectra of minerals and related inorganic crystals, *Applied Spectroscopy*, 48 (1994), 959–968.

D.H. Williams and I. Fleming, *Spectroscopic Methods in Organic Chemistry*, 5th edn, McGraw-Hill, London (1995).

M.A. Wilson, I.M. Godfrey, J.V. Hanna, *et al.*, The degradation of wood in old Indian Ocean shipwrecks, *Organic Geochemistry* 20 (1993) 599–610.

D. Wise and A. Wise, Application of Raman microspectroscopy to problems in the conservation, authentification and display of fragile works of art on paper, *Journal of Raman Spectroscopy*, 35 (2004), 710–718.

R. Withnall, I.T. Shadi and B.Z. Chowdhry, Case study: the analysis of dyes by SERRS, in *Raman Spectroscopy in Archaeology and Art History* (eds H.G.M. Edwards and J.M. Chalmers), Royal Society of Chemistry, Cambridge (2005), pp 152–166.

N.L. Yang, J. Liutkas and H. Haubenstock, An ESR study of initially formed intermediates in the photodegradation of poly(vinyl chloride), in *Polymer Characterisation of ESR and NMR* (eds A.E. Woodward and F.A. Bovey), ACS Symposium Series 142, American Chemical Society, Washington (1980), pp 35–48.

R. Zboril, M. Mashlan, L. Machala, *et al.*, Characterisation and thermal behaviour of garnets from almandine-pyrope series at 1200 °C, *Hyperfine Interactions*, **156–157** (2004), 403–410.

Y.F. Zhang and Y.F. Hsia, Studies of archaeological problems by Mössbauer spectroscopy, *Hyperfine Interactions*, **68** (1991), 131–142.

5

Atomic Spectroscopy

5.1 INTRODUCTION

In atomic spectroscopy a material is decomposed into atoms in a flame, furnace or plasma. The concentrations of the atoms in the vapour produced are measured by the absorption or emission of ultraviolet or visible radiation by the gaseous atoms. The techniques of atomic spectroscopy have the advantage that they are sensitive and individual elements can be identified in a complex sample.

There are three main types of atomic spectroscopy: absorption, emission and fluorescence. In atomic absorption, the atoms absorb part of the light from a source; the remainder of the light from the source reaches a detector. Atomic emission results from atoms that are in an excited state as a result of the high thermal energy of a flame. In atomic fluorescence atoms are excited by a lamp or laser and fluorescence is measured by the detector. Although atomic fluorescence is potentially a thousand times more sensitive than atomic absorption, the equipment for atomic fluorescence is not available for routine use. In this chapter, the atomic absorption and atomic emission techniques that are suitable for use for the study of cultural materials are outlined.

5.2 ATOMIC ABSORPTION SPECTROSCOPY

In atomic absorption spectroscopy (AAS) a sample in aqueous solution is aspirated in a flame at high temperature (2000–6000 K) and atomised by the process [Harris, 2003; Skoog and Leary, 1992; Young and Pollard, 2000]. Light of a suitable wavelength for a particular element is shone

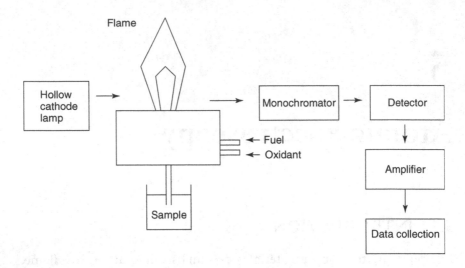

Figure 5.1   Layout of apparatus used for AAS

through the flame and some of the light is absorbed by the atoms
of the sample. The amount of light absorbed is proportional to the
concentration of the element in the solution. Measurements are made
separately for each element in turn to achieve a complete analysis, so the
technique can be relatively slow. However, it does have the advantage
that it is sensitive and small quantities of elements (of the order of ppm)
can be detected. The layout of an apparatus used for AAS is illustrated
in Figure 5.1. A hollow-cathode lamp is commonly used as a source.
A sample size of about 10 mg is required, so although this technique
is destructive, damage to an object can be minimal. The sample is
accurately weighed and usually dissolved in strong acids. The sample
solutions need to be diluted to provide concentrations that fall within
the detection range of the instrument.

AAS using a flame employs a premix burner in which the fuel, oxidant
and sample are mixed before being introduced into the flame. A nebuliser
creates an aerosol of the liquid sample by passing the liquid through a
capillary. The most common fuel-oxidiser combination is acetylene and
air, producing a flame temperature of 2400–2700 K. Although flames
have traditionally been used to atomise an analyte in AAS, flames are
being replaced by the graphite furnace (GFAAS). An electrically heated
graphite furnace provides greater sensitivity than flames and less sample
is required. Sample (1–100 μL as opposed to 1–2 ml for flame analysis)
is injected into the furnace through a hole and argon gas is passed over
the furnace to prevent oxidation. Solid samples may also be analysed.

Table 5.1 Resonance lines used in Atomic Absorption Spectroscopy (AAS)

Element	Wavelength (nm)
Mg	285.21
Al	396.15
S	180.73
Ca	422.67
V	318.54
Cr	425.44
Mn	279.48
Fe	248.33
Co	240.73
Ni	232.00
Cu	324.75, 327.40
Zn	213.86
As	228.80, 234.90
Rb	780.02
Sr	460.73
Pd	244.80
Ag	328.07, 338.29
Cd	228.80
Sn	235.48
Sb	217.58
Pt	265.95
Au	242.80, 267.60
Hg	253.65
Pb	217.00, 283.31

A graphite furnace confines the atomised sample in the optical path for several seconds providing higher sensitivity (1–10 ppb), compared to 1–10 ppm for flame AAS.

The most sensitive resonance lines used in AAS for common elements [Young and Pollard, 2000] are listed in Table 5.1. There are several problems that can occur during AAS analysis. Some species, such as nickel and antimony, have close lines which can be difficult to separate if both elements are present in a sample. Lanthanides are not analysed well by AAS and titanium, aluminium and vanadium can oxidise in a flame to form oxides that go unmeasured. Matrix effects also need to be considered, but the production of standards close to the composition of the unknown sample solutions can minimise these effects.

5.2.1 Ceramics

The elemental analysis of ceramics using AAS enables the provenance of such objects to be determined [al-Saad, 2002; Cariati *et al.*, 2003; Garcia et al., 2005; 2006; Kneisel et al., 1997; Pollard and Hatcher, 1986; Sanchez et al., 2002; Shingleton et al., 1994]. In one study of Chinese greenwares, AAS was used to determine the chemical composition which could be associated with production in different parts of China from different periods [Pollard and Hatcher, 1986]. Small cores were removed from 133 samples and 25 mg of powder were dissolved in acid solution. Quantitative analysis of aluminium oxide (Al_2O_3), calcium oxide (CaO), magnesium oxide (MgO), iron[III] oxide (Fe_2O_3), titanium dioxide (TiO_2), sodium oxide (Na_2O), manganese oxide (MnO) and potassium oxide (K_2O) was carried out and the amount of silicon dioxide was estimated by the difference of the sum of the measured oxides from 100 %. Cluster analysis was carried out on the data, which divided into two groups corresponding to the earlier (3rd–10th century) wares and the later (10th–14th century) wares.

5.2.2 Glass

AAS can be used to characterise historic glasses [Angelini et al., 2004; Longworth et al., 1982; Schreiner et al., 1999]. The composition of glass can be determined using this technique. AAS has also proved useful for characterising the leaching behaviour of model glasses with compositions comparable to medieval stained glass [Schreiner et al., 1999]. Leaching experiments were carried out on ground and polished glass specimens in aqueous solutions of hydrochloric, nitric, sulfuric and oxalic acids. The amount of potassium and calcium leached from the glasses at various concentration and times were measured using AAS. The leaching process was shown to be dependent upon the glass composition and the type and concentration of acid.

5.2.3 Stone

AAS can be utilised to characterise historic building materials [Leyson et al., 1990; Marinoni et al., 2002; Moropoulou et al., 1996; Perry and Duffy, 1997; Zamudio-Zamudio et al., 2003]. This approach can be used to examine degradation products and assist in an understanding of a deterioration process. For example, in a study of sandstone and limestone from Trinity College, Dublin, AAS was used to measure the calcium (Ca^{2+}) ion concentration [Perry and Duffy, 1997]. An enrichment of Ca^{2+} on the stones and a depletion of Ca^{2+} at the mortar surface indicated that Ca^{2+} from the mortar was washed into solution and deposited onto the surface. The Ca^{2+} was observed to precipitate mainly as calcite, which blocked the pores of the stone and was believed to be responsible for restricting water movement.

5.2.4 Written Material

Elemental analysis of paper and ink has been carried out using AAS [Budnar *et al.*, 2006; Bulska *et al.*, 2001; Tang, 1978; Simon *et al.*, 1977; Wagner *et al.*, 1999, 2001]. Both flame and graphite furnace AAS have been used to determine the metal content in ancient manuscripts [Budnar *et al.*, 2006, Wagnar *et al.*, 1999, 2001]. The elements iron and copper can indicate the presence of species responsible for the degradation of paper, so a measure of their concentrations can aid in the preservation process. Paper samples of milligram quantity are suitable for analysis.

5.2.5 Paintings

Graphite furnace AAS has been used to identify pigments [Goltz *et al.*, 2004]. Samples were gently removed from the surface of a painting with a cotton swab. Only 1–2 μg of pigment was removed during the process and the swab was cleaned with nitric acid. The identification of metals in the samples could be correlated with specific pigments. For example, GFAAS indicated the presence of copper and the use of Raman spectra confirmed the presence of phthalocyanine green. Also, a combination of these two techniques is useful for identifying zinc oxide as a white pigment.

5.2.6 Metals

The composition of metal objects can be readily determined using AAS methods [Johanssen *et al.*, 1986; Klein, 1999; Scott and Seeley, 1983]. Information regarding production methods of such objects can be gained from such analyses. One study used AAS methods to examine a pre-Hispanic gold–copper alloy chisel from Colombia [Scott and Seeley, 1983]. A small fragment was removed from the artefact for analysis and the results of the analysis were gold (48.34 %), copper (38.79 %), silver (10.71 %) and iron (0.04 %). The silver present is a result of using native Colombian gold which is known to contain some silver.

5.3 ATOMIC EMISSION SPECTROSCOPY

In atomic emission spectroscopy (AES), also known as optical emission spectroscopy (OES), transitions of an atom from an excited state to its ground state and the resulting emission of energy is studied [Harris, 2003; Skoog and Leary, 1992; Young and Pollard, 2000]. The radiation emitted is characteristic of the element studied and the emission intensity is proportional to the concentration of the element in the sample. A spectrum of emission intensity versus wavelength showing multiple element excitation lines is produced meaning that multiple elements can be detected in one experimental run. Alternatively, a monochromator can be fixed at one wavelength to analyse a single element at a particular emission line. Sources of atomisation and excitation for liquid samples include flames, furnaces and plasmas.

Where a flame is used as an excitation source in AES, the flame is responsible for desolvating and vaporising a sample into free atoms. For AES, the flame must excite the atoms to higher energy levels. Some common fuels include hydrogen and acetylene gas and oxidants may be air, oxygen or nitrous oxide (N_2O) to produce flames in the temperature range 2000–2800 K. A common burner can have the sample solution directly aspirated into the flame. Alternatively, the sample may be nebulised, premixing the sample with fuel and oxidant before it reaches the burner.

The use of plasmas, particularly inductively coupled plasma (ICP), has superceded the use of flames in atomic spectroscopy [Boumans, 1987; Harris 2003; Montaser and Golightly, 1992]. ICP–AES is now widely used for elemental analysis and has a sensitivity of the order of 0.5–100 ppb, compared to 100 ppm for flame AES. An ICP is produced by ionising a flowing gas such as argon with a Tesla coil. Free electrons are accelerated by a field that oscillates about the coil. The accelerated electrons collide with atoms and transfer their energy to the gas. The electrons absorb enough energy from the coil to produce a temperature of 6000–1000 K in the plasma. The concentration of the analyte required for a suitable signal is greatly reduced. An ultrasonic nebuliser is used to produce an aerosol of solid particles reaching the plasma. The layout of an ICP–AES spectrometer is illustrated in Figure 5.2.

5.3.1 Glass

ICP–AES has been employed for the analysis of trace elements in historic glasses [Azzoni *et al.*, 2002; Bellot-Gurlet *et al.*, 2005; Bianchin *et al.*, 2005; Bressy *et al.*, 2005; Corradi *et al.*, 2005; El Nady and Zimmer, 1985; Hartmann *et al.*, 1997; Kilikoglou *et al.*, 1997; Zhang *et al.*, 2005]. This technique is, of course, destructive with samples undergoing acid digestion prior to analysis. However, only small (milligram)

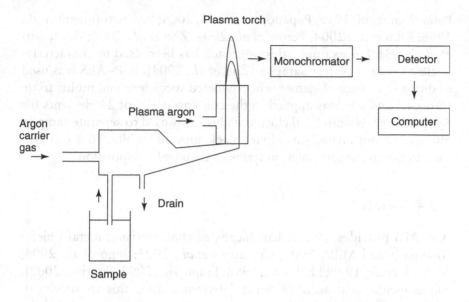

Figure 5.2 Layout of an ICP-AES spectrometer

quantities are required. This approach is generally used in conjunction with other analytical techniques for measuring elements such as PIXE (Chapter 6) and ICP–MS (see Chapter 7). ICP–AES is useful for detecting light elements that may not be detected using other techniques. For instance, ICP–AES has been used to measure the sodium oxide and magnesium oxide compositions of early Chinese glazes and aided in the classification of these glasses [Zhang et al., 2005].

5.3.2 Stone

ICP–AES can be employed to characterise building stone of historical importance [Friolo et al., 2003; Welton et al., 2003]. This technique has been used to analyse the elemental composition of heritage sandstones [Friolo et al., 2003]. In particular, a comparison of weathered and unweathered stones noted an increase in iron and a decrease in both silicon and aluminium in weathered clay samples. Such changes supported the supposition of Fe^{3+} substituting for Al^{3+} and/or Si^{4+} in the kaolinite structure.

5.3.3 Ceramics

Ceramics may be characterised using ICP–AES [Cariati et al., 2003; Cardiano et al., 2004; Cecil, 2004; Grave et al., 2005; Lapuente and

Perez-Arantegui, 1999; Papadopolou *et al.*, 2004; Perez-Arantegui *et al.*, 1996; Feliu *et al.*, 2004; Fermo *et al.*, 2004; Zhu *et al.*, 2004; Zucchiatti *et al.*, 2003]. For example, this approach has been used to characterise ancient Chinese pottery samples [Zhu *et al.*, 2004]. ICP–AES was used to quantify a range of elements in powdered specimens and multivariate statistical analysis was applied to the concentrations of 24 elements for 48 specimens. Hierarchical cluster analysis was used to separate samples into three groups based on colour–black, grey and white. PCA was also used to group the specimens in terms of chemical composition.

5.3.4 Metals

ICP–AES provides an excellent means of characterising metal objects [Bourgarit and Mille, 2003; Coustures *et al.*, 2003; Ingo *et al.*, 2004; Klemenc *et al.*, 1999; Hall *et al.*, 1998; Lonnqvist, 2003; Ponting, 2002]. The elemental analysis of elements determined using this approach can provide valuable information regarding the source of material and the manufacturing process. For instance, ICP–AES was used in a study of copper alloy Roman coins to determine the chemical composition [Lonnqvist, 2003]. The coins included those collected from the South Wall in Jerusalem (50 coins) and Masada at the Dead Sea (46 coins) and dated from 6 to 66 AD. The analysis showed that two notably different bronze alloys were used in the period 6–66 AD: a pure tin bronze (copper 87 %, tin 10 %, lead 0.2 %) and a leaded tin bronze (copper 78 %, tin 10 %, lead 11 %).

5.4 LASER INDUCED BREAKDOWN SPECTROSCOPY

Laser induced breakdown spectroscopy (LIBS), also known as laser induced plasma spectroscopy, has emerged in recent years as a new emission technique [Anglos, 2001]. LIBS uses a short pulsed laser (~ 7 ns) at the surface of a sample to remove a very small amount which is excited. The laser (usually 1064 nm) produces a microplasma with a temperature of $10\,000 - 20\,000$ K. This approach uses fibre optics to collect the plasma emission so the analysis of an object can be recorded in situ. This technique has the considerable advantage of eliminating the need for a sample solution. As such, time is saved on sample preparation, and the process is less destructive. A sample of about 20–200 ng is typically consumed in an experiment and changes to a sample surface

Table 5.2 Emission lines for detection of certain elements using Laser Induced Breakdown Spectroscopy (LIBS)

Element	Wavelength (nm)
Ag	272.2, 328.1, 338.3
Al	257.5, 265.3, 266.0, 308.2, 309.3, 394.4, 396.2
As	274.5, 278.0, 286.0, 289.9
Au	264.2, 267.6, 274.8, 288.3, 290.7–291.4, 293.2, 302.9, 312.3, 320.5, 323.1
Ca	315.9, 317.9–318.1, 393.4, 396.8, 422.7
Cr	357.9, 359.3, 360.5, 375.8, 427.5, 429.0
Cu	261.8, 276.6, 282.4, 296.1, 301.1, 303.6, 310.0, 310.9, 312.6, 324.8, 327.4, 329.1, 330.8
Fe	258.5, 259.8–259.9, 260.7, 261.2, 261.4, 261.8, 262.6, 262.8, 263.1, 271.9–272.1, 274.9–275.0, 296.7, 298.4, 299.4, 302.1
Mg	279.6, 280.3, 285.2
Mn	370.6, 371.9, 380.7, 382.4, 383.4, 384.1, 403.1–403.5, 404.1, 407.9–408.4, 413.1–413.5, 423.5–423.5, 446.2
Pb	257.7, 261.4, 266.3, 280.2, 282.3, 283.3, 287.3, 357.3, 364.0, 368.4, 374.0, 402.0, 405.8
Si	263.1, 288.2, 390.6,
Sn	266.1, 270.7, 278.0, 281.4, 284.0, 286.3, 291.4, 300.9, 303.4, 317.5, 326.2, 333.1
Ti	323.5, 334.2, 334.9, 336.1, 337.0–337.3, 338.4

are practically invisible to the naked eye. The spatial resolution is of the order of 100 μm. LIBS can also be operated at atmospheric pressure and spectra are recorded very rapidly. LIBS has a depth profiling capacity if successive laser pulses are delivered at the same point and spectra are recorded separately.

Common emission lines observed in LIBS for certain elements are listed in Table 5.2 [Melessanaki *et al.*, 2002]. The choice of suitable emission lines depends on the sample type and the information required. A well separated line provides reliable analysis. For quantitative analysis using LIBS, matrices of well known composition are used and the relative intensity of the spectral line is compared with the concentration of the corresponding element in the matrix. Alternatively, calibration-free LIBS can be used – an approximation that eliminates the need for an internal standard or reference samples [Burgio *et al.*, 2000a].

5.4.1 Paintings

LIBS has proved effective for characterising paintings. The elemental composition of inorganic pigments to be identified [Anglos, 2001; Anglos *et al.*, 1997; Bicchieri *et al.*, 2001; Burgio *et al.*, 2001, 2000a,

Table 5.3 Elements identified in inorganic pigments by Laser Induced Breakdown Spectroscopy (LIBS)

Pigment	Element
Lead white	Pb
Titanium white	Ti
Zinc white	Zn
Lithopone	Ba, Zn, Ca
Chalk	Ca
Barium sulfate, Barytes	Ba
Gypsum	Ca
Cadmium yellow	Cd, Zn, Ba
Cobalt yellow	Co
Orpiment	As
Naples yellow	Pb, Sb
Lead tin yellow	Pb, Sn
Strontium yellow	Sr, Cr
Barium yellow	Ba, Cr
Yellow ochre	Fe, Si, Al
Cadmium red	Cd
Cinnabar/Vermillion	Hg
Red ochre	Fe, Al
Realgar	As
Mars red/Haematite	Fe
Red lead/Minium	Pb
Lapis lazuli/Ultramarine	Al, Si, Na
Egyptian blue	Cu, Si, Ca
Cobalt blue	Co, Al, Na
Cerulean blue	Co, Sn
Prussian blue	Fe, Ca
Azurite	Cu, Si
Malachite	Cu, Si
Viridian green	Cr
Emerald green	Cu, As
Verdigris	Cu
Ivory black/Bone black	Ca, P
Manganese black	Mn
Magnetite/Mars black	Fe

2000b; Castillejo *et al.*, 2000; Melessanaki *et al.*, 2001]. A number of the inorganic pigments that may be analysed using LIBS and the elements identified are listed in Table 5.3 [Anglos, 2001]. Depending on the elements present in a pigment, the LIBS spectra can appear simple with the emission lines clearly resolved, or complex where there is overlap between lines. Overlapping lines observed in low resolution spectra can be resolved by carrying out the experiment at high resolution. However, such a configuration may reduce the number of elements that can be

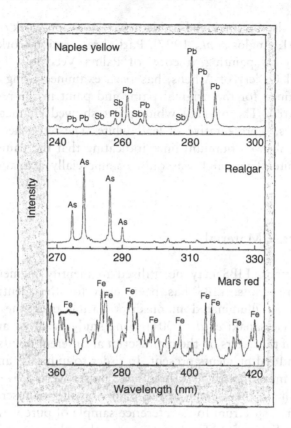

Figure 5.3 LIBS spectra of Naples yellow, realgar and Mars red pigments [Anglos, 2001]

detected. The LIBS spectra of some pigments, Naples yellow, realgar and Mars red, are illustrated in Figure 5.3 [Anglos, 2001]. Despite the complex nature of the Mars red spectrum, the spectral features are quite characteristic of the presence of iron.

The ability to carry out depth profiling using LIBS means that the surface layers of a painting may be characterised [Anglos, 2001; Anglos *et al.*, 1997; Burgio *et al.*, 2000a, 2000b]. As each pulse ablates a thin layer of material and exposes a fresh surface to the next pulse, delivery of a certain number of laser pulses at the same point enables the spectra of successive layers to be recorded. For paintings the ablation depth per pulse is 0.5–2 μm. Such an approach has been applied to the characterisation of a 19th century Byzantine icon [Burgio *et al.*, 2000b]. This study was able to produce distinctive spectra for the paint, ground layers and silver foil that made up the surface of the icon by an ablation procedure.

LIBS may also be employed to identify restoration work on paintings [Anglos 2001; Anglos *et al.*, 1997]. Partial restoration work on a later 18th century oil painting, a copy of Palma Vecchio's 'La Bella' at the National Gallery of Athens, has been examined using LIBS. LIBS spectra obtained for the original paint and paint in the restored area were compared. The original white paint showed characteristic lead emission lines, representative of lead white. The retouched regions of the painting showed titanium lines indicating that the paint contained titanium white. This paint was only commercially available after the early 1900s.

5.4.2 Written Material

As for paintings, LIBS may be utilised to identify pigments used in historic manuscripts. LIBS has been used for the identification of pigments in an illuminated manuscript dating from the 12th–13th centuries [Melessanaki *et al.*, 2001]. One study involved an examination of one of the letters on the manuscript at different points to identify the white and red pigments present. In order to minimise any potential damage, no more that two or three laser pulses were delivered at each point examined. The spectra of the paints on the manuscript, as well as showing the spectrum for a reference sample of pure vermillion, are illustrated in Figure 5.4. The spectrum for the white paint shows lines due to lead, which enables the pigment to be identified as lead white. This pigment was widely used at the time the manuscript was produced and is the only white pigment to contain lead. The spectrum obtained for the red paint shows magnesium lines, indicating that the paint is vermillion. This is also confirmed by comparison with the spectrum of the vermillion standard. The spectra for both paint samples also show lines due to magnesium, silicon, aluminium and calcium. These are believed to be due to impurities in the pigments, from the binding medium or from environmental deposits.

LIBS lends itself to the analysis of paper and has been used to study pigments and binders in contemporary papers [Hakkanen *et al.*, 2001; Hakkanen and Korppi-Tommola, 1998]. As the most commonly used pigments in modern papers are calcium carbonate, kaolinite and talc ($3MgO.4SiO_2.H_2O$), the elements that are monitored using LIBS are aluminium, calcium, carbon, magnesium and silicon. The useful emission lines for paper analysis are observed in the 200–450 nm region.

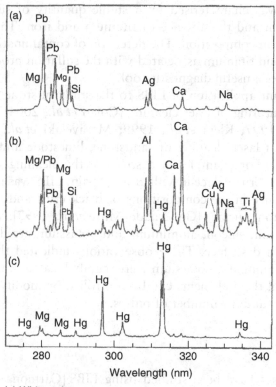

(a) White paint
(b) Red paint
(c) Pure vermillion (reference)

Figure 5.4 LIBS spectra of paints on a manuscript. Reprinted from Spectrochimica Acta B, Vol 56, Melessanaki *et al.*, 2337–2346, 2001, with permission from Elsevier

5.4.3 Stone

LIBS is an excellent technique for characterising the encrustation that can form on the surface of stone objects. This approach has been used to study the nature of the encrustation formed on Pentelic marble from the Forum Romano in Athens [Maravelaki-Kalaitzaki *et al.*, 2001]. Depth profiling could be carried out before cleaning the marble. Separate dendritic black and thin black encrustations resulting from the interaction of atmospheric pollutants with the marble were identified. The external layer, ranging in thickness from 100 to 200 μm, was found to consist of variable concentrations of alumino-silicates, gypsum and particulates based on the gradients of sulfur, iron, silicon, aluminium and titanium intensities. The LIBS spectra of the internal layer of about 100 μm in thickness was characterised by a stable quantity of sulfur, silicon and aluminium and the absence of titanium and iron. This layer was homogeneous in composition. The detection of contamination elements such as iron and titanium associated with the sulfation process by LIBS in situ provides a useful diagnostic tool.

An important application of LIBS to the study of stone surfaces has been the monitoring of laser cleaning [Colao *et al.*, 2004; Gobernado-Mitre *et al.*, 1997; Klein *et al.*, 1999; Maravelaki *et al.*, 1997]. The effectiveness of laser cleaning of sandstone, limestone and marble has been reported. For example, in a study of the cleaning of limestone from the Santa Cruz Palace in Valladolid, Spain, LIBS was used to show a clear difference in the compositions of black crust and laser-cleaned areas of the stone surface [Gobernado-Mitre *et al.*, 1997]. The emission lines due to iron, silicon, aluminium and potassium are decreased for the laser cleaned surface. These observations indicated that the iron, silicon, aluminium and potassium were mainly located in the crust and the removal of these elements can be controlled by modifying the laser energy density and the number of pulses.

5.4.4 Glass

Historic glasses have been studied using LIBS [Carmona *et al.*, 2005; Muller and Stege, 2003]. The presence of all the elemental components of glass samples, silicon, aluminium, sodium, potassium and magnesium, can be detected by LIBS. LIBS also allows for the analyses of glass chromophores such as cobalt, copper, iron, manganese and chromium at a ppm level. A wider range of chromophores than UV-visible spectroscopy may be detected. Quantitative and semi-quantitative approaches have been attempted and it is possible to establish calibration curves of the elemental constituents of glass.

This technique also shows promise as a technique for the diagnosis of corroded glasses [Carmona *et al.*, 2005]. The depth profiling ability of LIBS allows for the detection of different corrosion layers resulting from weathering. Differences in the corrosion crust, the gel layer and the bulk layer can be detected. The corrosion layer, mainly consisting of salts containing calcium, potassium and sulfur, produces emission lines of calcium, potassium and sulfur. The gel layer below the corrosion layer shows emission lines due to silicon, the main component of this layer, as well as some residual calcium, potassium and sulfur lines from the crust. With further ablation the sulfur line disappears and the silicon, calcium and potassium lines increase in intensity as the bulk glass is reached.

LIBS also has been used as a means of monitoring the laser cleaning of glass using a similar approach to that used to study the laser cleaning of stone surfaces [Klein *et al.*, 1999].

5.4.5 Ceramics

The pigment, lustre and glaze composition of ceramics may be characterised using LIBS [Colao *et al.*, 2002; Lazic *et al.*, 2003; Melessanaki *et al.*, 2002; Yoon *et al.*, 2001]. For example, LIBS was used to analyse the glaze of ancient Korean pottery samples [Yoon *et al.*, 2001]. The flux and colouring materials were identified using this approach. Quantitative analysis of the calcium oxide, magnesium oxide and iron[III] oxide components was carried out on a series of samples using intensity ratios of lines with the silicon emission line as a standard. Of course, because of potential interference from the layers beneath the surface coating, care must be taken when attempting quantitative analysis.

5.4.6 Metals

LIBS can be used to study the elemental composition of metal objects and to identify surface deposits on a metal [Acquaviva *et al.*, 2004; Anglos, 2001; Colao *et al.*, 2002; Melessanaki *et al.*, 2002]. The technique has the advantage that both minor as well as major elements may be identified. For example, a comparative study of coins of different origin using LIBS was able to demonstrate clear differences in their composition [Anglos, 2001]. A 16th century Byzantine coin was analysed and produced an emission spectrum with copper and silver emission lines. The relative intensities of the lines demonstrated that the coin was composed mainly of copper with small amounts of silver present. The spectrum of a 1926 Greek coin showed that it was predominantly copper and zinc. A 1952 Russian coin was also studied and shown to be brass as the major emission lines in the LIBS spectrum were copper and zinc. Small amounts of nickel and manganese were also found in this coin.

Of particular use to conservators is the characterisation of a metal surface after treatment. For instance, LIBS has been employed to study a historic gun found on the Adriatic sea bed near Italy [Acquaviva *et al.*, 2004]. The gun showed an encrustation on its surface which was believed responsible for deterioration. LIBS was used to study the untreated surface of the gun and the surface after cleaning. Calcium emission lines were observed in the LIBS spectra of the gun before treatment and were attributed to the encrustation. The LIBS spectra obtained after cleaning, which involved the mechanical removal of the encrustation as well as treatment by chemical agents, showed that the calcium lines were absent. Thus, LIBS was able to be used to demonstrate the efficacy of the cleaning process.

REFERENCES

S. Acquaviva, M.L. DeGiorgi, C. Marini and R. Poso, Elemental analysis by laser induced breakdown spectroscopy as restoration test on a piece of ordinance, *Journal of Cultural Heritage*, 5 (2004), 365–369.

Z. al-Saad, Chemical composition and manufacturing technology of a collection of various types of Islamic glazes excavated from Jordan, *Journal of Archaeological Science*, 29 (2002), 803–810.

I. Angelini, G. Artioloi, P. Bellintani, *et al.*, Chemical analyses of Bronze Age glasses from Frattesina di Rovigo, Northern Italy, *Journal of Archaeological Science*, 31 (2004), 1175–1184.

D. Anglos, S. Couris and C. Fotakis, Laser diagnostics of painted artworks: laser-induced breakdown spectroscopy in pigment identification, *Applied Spectroscopy*, 51 (1997), 1025–1030.

D. Anglos, Laser-induced breakdown spectroscopy in art and archaeology, *Applied Spectroscopy*, 55 (2001), 186A–205A.

C.B. Azzoni, D. Di Martino, C. Chiavari, *et al.*, Electron paramagnetic resonance of mosaic glasses from the Mediterranean area, *Archaeometry*, 44 (2002), 543–554.

L. Bellot-Gurlet, G. Poupeau, J. Salomon, *et al.*,Obsidian provenance studies in archaeology: a comparison between PIXE, ICP–AES and ICP–MS, *Nuclear Instruments and Methods in Physics Research B*, 240 (2005), 583–588.

S. Bianchin, N. Brianese, U. Casellato, *et al.*, Medieval and renaissance glass technology in Valdelsa (Florence). Part 2: vitreous finds and sands, *Journal of Cultural Heritage*, 6 (2005), 39–54.

M. Biccheri, M. Nardone, P.A. Russo, *et al.*, Characterisation of azurite and lazurite based pigments by laser induced breakdown spectroscopy and micro-Raman spectroscopy, *Spectrochimica Acta Part B*, 56 (2001), 915–922.

P.W.J.M. Boumans, *Inductively Coupled Plasma Emission Spectroscopy. Part 1. Methodology, Instrumentation and Performance*, John Wiley & Sons Inc., New York (1987).

D. Bourgarit and B. Mille, The elemental analysis of ancient copper-based artefacts by inductively coupled plasma atomic emission spectrometry: an optimised methodology reveals some secrets of the Vix crater, *Measurement Science and Technology*, 14 (2003), 1538–1555.

C. Bressy, G. Poupeau and K.A. Yener, Cultural interactions during the Ubaid and Halaf periods: Tell Kurdu (Amuq Valley, Turkey) obsidian sourcing, *Journal of Archaeological Science*, 32 (2005), 1560–1565.

M. Budnar, M. Ursic, J. Simcic, *et al.*, Analysis of iron gall inks by PIXE, *Nuclear Instruments and Methods in Physics Research B*, 407 (2006), 407–416.

E. Bulska, B. Wagner and M.G. Sawicki, Investigation of complexation and solid-liquid extraction of iron from paper by UV–VIS and atomic absorption spectrometry, *Mikrochimica Acta*, **136** (2001), 61–66.

L. Burgio, M. Corsi, R. Fantoni, *et al.*, Self-calibrated quantitative elemental analysis by laser induced plasma spectroscopy: application to pigment analysis, *Journal of Cultural Heritage*, **1** (2000a), 281–286.

L. Burgio, R.J.H. Clark, T. Stratoudaki, *et al.*, Pigment identification in painted artworks: a dual analytical approach employing laser-induced breakdown spectroscopy and Raman spectroscopy, *Applied Spectroscopy*, **54** (2000b), 463–469.

L. Burgio, K. Melessanaki, M. Doulgeridis, *et al.*, Pigment identification in paintings employing laser induced breakdown spectroscopy and Raman microscopy, *Spectrochimica Acta Part B*, **56** (2001), 905–913.

P. Cardiano, S. Ioppolo, C. De Stefano, *et al.*, Study and characterisation of the ancient bricks of monastery of 'San Filippo di Fragala' in Frazzano (Sicily), *Analytica Chimica Acta*, **519** (2004), 103–111.

F. Cariati, P. Fermo, S. Gilardoni, *et al.*, A new approach for archaeological ceramics analysis using total reflection X-ray fluorescence spectrometry, *Spectrochimica Acta Part B*, **58** (2003), 177–184.

N. Carmona, M. Oujja, E. Rebollar, *et al.*, Analysis of corroded glasses by laser induced breakdown spectroscopy, *Spectrochimica Acta Part B*, **60** (2005), 1155–1162.

M. Castillejo, M. Martin, D. Silva, *et al.*, Analysis of pigments in polychromes by use of laser-induced breakdown spectroscopy and Raman microscopy, *Journal of Molecular Structure*, **550–551** (2000), 191–198.

L.G. Cecil, Inductively coupled plasma emission spectroscopy and postclassic Petén slipped pottery: an examination of pottery wares, social identity and trade, *Archaeometry*, **46** (2004), 385–404.

F. Colao, R. Fantoni, V. Lazic and V. Spizzichino, Laser induced breakdown spectroscopy for semi-quantitative and quantitative analyses of artworks–application on multi-layered ceramics and copper based alloys, *Spectrochimica Acta Part B*, **57** (2002), 1219–1234.

F. Colao, R. Fantoni, V. Lazic, *et al.*, LIBS used as a diagnostic tool during the laser cleaning of ancient marble from Mediterranean areas, *Applied Physics A*, **79** (2004), 213–219.

A. Corradi, C. Leonelli, P. Veronesi, *et al.*, Ancient glass deterioration in mosaics of Pompeii, *Surface Engineering*, **21** (2005), 402–405.

M.P. Coustures, D. Beziat, F. Tollon, *et al.*, The use of trace element analysis of entrapped slag inclusions to establish ore-bar iron links: example of two Gallo–Roman iron-making sites in France (Leo Lertys, Montagne Noire and Les Ferrys, Loiret), *Archaeometry*, **45** (2003), 599–613.

A.B.M. El Nady and K. Zimmer, Spectrochemical analysis of medieval glasses by means of inductively coupled plasma and glow discharge sources, *Spectrochimica Acta Part B*, **40** (1985), 999–1003.

M.J. Feliu, M.C. Edreira and J. Martin, Application of physical–chemical analytical techniques in the study of ancient ceramics, *Analytica Chimica Acta*, **502** (2004), 241–250.

P. Fermo, F. Cariati, D. Ballabio, *et al.*, Classification of ancient Etruscan ceramics using statistical multivariate analysis of data, *Applied Physics A*, **79** (2004), 299–307.

K.H. Friolo, B.H. Stuart and A. Ray, Characterisation of weathering of Sydney sandstones in heritage buildings, *Journal of Cultural Heritage*, **4** (2003), 211–220.

R. Garcia Gimenez, R. Vigil de la Villa, P. Recio de la Rosa, *et al.*, Analytical and multivariate study of Roman age architectural terracotta from northeast of Spain, *Talanta*, **65** (2005), 861–868.

R. Garcia Gimenez, R. Vigil de la Villa, M.D. Petit Dominguez and M.I. Rucandio, Application of chemical, physical and chemometric analytical techniques to the study of ancient ceramic oil lamps, *Talanta* , **68** (2006), 1236–1246.

I. Gobernado-Mitre, A.C. Prieto, V. Zafiropulos, *et al.*, On-line monitoring of laser cleaning of limestone by laser induced breakdown spectroscopy and laser induced fluorescence, *Applied Spectroscopy*, **51** (1997), 1125–1129.

D.M. Goltz, J. Coombs, C. Marion, *et al.*, Pigment identification in artwork using graphite furnace atomic absorption spectrometry, *Talanta*, **63** (2004), 609–616.

P. Grave, L. Lisle and M. Maccheroni, Multivariate comparison of ICP–OES and PIXE/PIGE analysis of east Asian storage jars, *Journal of Archaeological Science*, **32** (2005), 885–896.

H. Hakkanen, J. Houni, S. Kaski and J.E.I. Korppi-Tommola, Analysis of paper by laser induced plasma spectroscopy, *Spectrochimica Acta Part B*, **56** (2001), 737–742.

H.J. Hakkanen and J.E.I. Korppi-Tommola, Laser induced plasma emission spectrometric study of pigments and binders in paper coatings: matrix effects, *Analytical Chemistry*, **70** (1998), 4724–4729.

M.E. Hall, S.P. Brimmer, F.H. Li and Y. Yablonsky, ICP–MS and ICP–OES studies of gold from a later Sarmatian burial, *Journal of Archaeological Science*, **25** (1998), 545–552.

D.C. Harris, *Quantitative Chemical Analysis*, 6th edn, W.H. Freeman, New York (2003).

G. Hartmann, I. Kappel, K. Groto and B. Arndt, Chemistry and technology of prehistoric glass from Lower Saxony and Hesse, *Journal of Archaeological Science*, **24** (1997), 547–559.

G.M. Ingo, E. Angelini, T. De Caro, *et al.*, Combined use of GDOES, SEM–EDS, XRD and OM for the microchemical study of the corrosion products on archaeological bronzes, *Applied Physics A*, **79** (2004), 199–203.

E.M. Johansson, S.A.E. Johansson, K.G. Malmqvist and I.M.B. Wiman, The feasibility of the PIXE technique in the analysis of stamps and art objects, *Nuclear Instruments and Methods in Physics Research B*, **14** (1986), 45–49.

V. Kilikoglou, Y. Bassiakos, R.C. Doonan and J. Stratis, NAA and ICP analysis of obsidian from Central Europe and the Aegean: Source characterisation and provenance determination, *Journal of Radioanalytical and Nuclear Chemistry*, **216** (1997), 87–93.

S. Klein, Iron age leaded tin bronzes from Khirbet Edh-Dharih, Jordan, *Journal of Archaeological Science*, **26** (1999), 1075–1082.

S. Klein, T. Stratoudaki, V. Zafiropolous, *et al.*, Laser induced breakdown spectroscopy for on-line control of laser cleaning of sandstone and stained glass, *Applied Physics A*, **69** (1999), 441–444.

S. Klemenc, B. Budic and J. Zupan, Statistical evaluation of data obtained by inductively coupled plasma atomic emission spectrometry (ICP–AES) for archaeological copper ingots, *Analytica Chimica Acta*, **389** (1999), 141–150.

E.A. Kneisel, N.A. Ciszkowski, W.J. Bowyer, *et al.*, Identifying clay sources of prehistoric pottery using atomic spectroscopy, *Microchemical Journal*, **56** (1997), 40–46.

P. Lapuente and J. Perez-Arantegui, Characterisation and technology from studies of clay bodies of local Islamic production in Zaragoza (Spain), *Journal of European Ceramic Society*, **19** (1999), 1835–1846.

V. Lazic, F. Colao, R. Fantoni, *et al.*, Characterisation of lustre and pigment composition in ancient pottery by laser induced fluorescence and breakdown spectroscopy, *Journal of Cultural Heritage*, **4** (2003), 303s–308s.

G. Longworth, N.H. Tennent, M.J. Tricker and P.P. Vaishnava, Iron-57 Mössbauer spectral studies of medieval stained glass, *Journal of Archaeological Science*, **9** (1982), 261–273.

L.A. Leyson, E.J. Roekens, R.E. Van Grieken and G. De Geyter, Characterisation of the weathering crust of various historical buildings in Belgium, *The Science of the Total Environment*, **90** (1990), 117–147.

K.K.A. Lonnqvist, A second investigation into the chemical composition of the Roman provincial (procuratorial) coinage of Judaea, AD 6–66, *Archaeometry*, **45** (2003), 45–60.

P.V. Maravelaki, V. Zafiropulos, V. Kilikogou, *et al.*, Laser induced breakdown spectroscopy as a diagnostic technique for the laser cleaning of marble, *Spectrochimica Acta Part B* **52** (1997) 41–53.

P. Maravelaki-Kalaitzaki, D. Anglos, V. Kilikoglou and V. Zafiropulus, Compositional characterisation of encrustation on marble with laser induced breakdown spectroscopy, *Spectrochimica Acta Part B*, **56** (2001), 887–903.

N. Marinoni, A. Pavese, R. Bugini and G. Di Silvestro, Black limestone used in Lombard architecture, *Journal of Cultural Heritage*, **3** (2002), 241–249.

K. Melessanaki, V. Papadakis, C. Balas and D. Anglos, Laser induced breakdown spectroscopy and hyper-spectral imaging analysis of pigments on an illuminated manuscript, *Spectrochimica Acta Part B* **56** (2001) 2337–2346.

K. Melessanaki, M. Mateo, S.C. Ferrence, *et al.*, The application of LIBS for the analysis of archaeological ceramic and metal artefacts, *Applied Surface Science*, **197–198** (2002), 156–163.

A. Montaser and D.W. Golightly, *Inductively Coupled Plasma in Analytical Atomic Spectrometry*, 2nd edn, VCH, New York (1992).

A. Moropoulou, T. Tsiourva, K. Bisbikou, *et al.*, Hot lime technology imparting high strength to historic mortars, *Construction and Building Materials*, **10** (1996), 151–159.

K. Muller and H. Stege, Evaluation of the analytical potential of laser induced breakdown spectroscopy (LIBS) for the analysis of historical glasses, *Archaeometry*, **45** (2003), 421–433.

D.N. Papadopoulou, G.A. Zachariadis, A.N. Anthemidis, *et al.*, Comparison of a portable micro-X-ray fluorescence spectrometry with inductively coupled plasma atomic emission spectrometry for the ancient ceramics analysis, *Spectrochimica Acta Part B*, **59** (2004), 1877–1884.

J. Perez-Arantegui, M.I. Urunuela and J.R. Castillo, Roman glazed ceramics in the Western Mediterranean: chemical characterisation by inductively coupled plasma atomic emission spectrometry of ceramic bodies, *Journal of Archaeological Science*, **23** (1996), 903–914.

S.H. Perry and A.P. Duffy, The short-term effects of mortar joints on salt movement in stone, *Atmospheric Environment*, **31** (1997), 1297–1305.

A.M. Pollard and H. Hatcher, The chemical analysis of oriental ceramic body compositions: Part 2–greenwares, *Journal of Archaeological Science*, **13** (1986), 261–287.

M.J. Ponting, Roman military copper-alloy artefacts from Israel: questions of organisation and ethnicity, *Archaeometry*, **44** (2002), 555–571.

S. Sanchez Ramos, F. Bosch Reig, J.V. Gimeno Adelantado, *et al.*, Study and dating of medieval ceramic tiles by analysis of enamels with atomic absorption spectroscopy, X-ray fluorescence and electron probe microanalysis, *Spectrochimica Acta B*, 57 (2002), 689–700.

M.R. Schreiner, I. Prohaska, J. Rendl and C. Weigel, Leaching studies of potash–lime–silica glass with medieval glass composition, in *Conservation of Glass and Ceramics: Research, Practice and Training* (ed N.H. Tennent), James and James, London (1999), pp 72–83.

D.A. Scott and N.J. Seeley, The examination of a pre-Hispanic gold chisel from Colombia, *Journal of Archaeological Science*, 10 (1983), 153–163.

K.L. Shingleton, G.H. Odell and T.M. Harris, Atomic absorption spectrophotometry analysis of ceramic artefacts from a prehistoric site in Oklahoma, *Journal of Archaeological Science*, 21 (1994), 343–358.

P.J. Simon, B.C. Glesson and T.R. Copeland, Categorisation of papers by trace metal content using atomic absorption spectrometric and pattern recognition techniques, *Analytical Chemistry*, 49 (1977), 2285–2288.

D.A. Skoog and J.J. Leary, *Principles of Instrumental Analysis*, Harcourt Brace, Fort Worth (1992).

L.C. Tang, Determination of iron and copper in 18th century and 19th century books by flameless atomic absorption spectroscopy, *Journal of the American Institute for Conservation*, 17 (1978), 19–32.

B. Wagner, S. Carbos, E. Bulska and A. Hulanicki, Determination of iron and copper in old manuscripts by slurry sampling graphite furnace atomic absorption spectrometry and laser ablation inductively coupled plasma mass spectrometry, *Spectrochimica Acta Part B*, 54 (1999), 797–804.

B. Wagner, E. Bulska, T. Meisel and W. Wegscheider, Use of atomic spectrometry for the investigation of ancient manuscripts, *Journal of Analytical Atomic Spectrometry*, 16 (2001), 417–420.

R.G. Welton, S.J. Cuthbert, R. McLean, *et al.*, A preliminary study of the phycological degradation of natural stone masonry, *Environmental Geochemistry and Health*, 25 (2003), 139–145.

Y. Yoon, T. Kim, M. Yang, *et al.*, Quantitative analysis of pottery glaze by laser induced breakdown spectroscopy, *Microchemical Journal*, 68 (2001), 251–256.

S.M.M. Young and A.M. Pollard, Atomic Spectroscopy and Spectrometry, in *Modern Analytical Methods in Art and Archaeology*, (eds E. Ciliberto and G. Spoto), John Wiley & Sons Inc, New York (2000), pp 21–53.

T.J. Zamudio-Zamudio, A. Garrido-Alfonseca, D. Tenorio and M. Jimenez-Reyes, Characterisation of 16th and 18th century building materials from Veracruz City, Mexico, *Microchemical Journal*, 74 (2003) ,83–91.

B. Zhang, H.S. Cheng, B. Ma, *et al.*, PIXE and ICP-AES analysis of early glass unearthed from Xinjiang (China), *Nuclear Instruments and Methods in Physics Research B*, 240 (2005), 559–564.

J. Zhu, J. Shan, P. Qiu, *et al.*, The multivariate statistical analysis and XRD analysis of pottery at Xigongqiao site, *Journal of Archaeological Science*, 31 (2004), 1685–1691.

A. Zucchiatti, A. Bouquillon, J. Castaing and J.R. Gaborit, Elemental analyses of a group of glazed terracotta angels from the Italian renaissance, as a tool for the reconstruction of a complex conservation history, *Archaeometry*, 3 (2003), 391–404.

6

X-ray Techniques

6.1 INTRODUCTION

X-rays are an electromagnetic radiation with wavelengths of the order of 10^{-10} m. They are typically generated via the bombardment of a metal with high energy electrons. The manner in which X-ray radiation interacts with or is emitted by matter has been exploited in a number of experimental techniques that are widely employed for the structural and compositional analysis of materials of cultural importance. X-ray diffraction (XRD) exploits the fact that the wavelengths of X-rays are comparable to the spacing between regularly spaced atoms. This enables diffraction patterns to be formed and the compounds present to be identified. In X-ray fluorescence (XRF), a beam of X-rays are used to ionise atoms in a material of interest. The characteristic radiation that is emitted provides information about the type and amount of the elements present in a sample. Electron microprobe analysis can be used for analysis of elements as well. The emission of X-rays induced by the use of high energy ion beams can also be examined in techniques such as proton induced X-ray emission (PIXE). As with XRF, qualitative and quantitative analysis can be carried out. The energy of X-rays is also exploited in X-ray photoelectron spectroscopy (XPS) and Auger electron spectroscopy (AES). In XPS, the energy of the incident radiation causes electrons to be ejected from the inner cores of atoms. In AES, the emission of a second electron occurs after high energy radiation has expelled another. As the energies released are characteristic of the elements present, these techniques can be used to identify elements in a material of interest.

Analytical Techniques in Materials Conservation   Barbara H. Stuart
© 2007 John Wiley & Sons, Ltd

6.2   X-RAY DIFFRACTION

X-ray diffraction (XRD) is a technique used to determine the arrangement of atoms in solids [Janssens, 2004]. As the wavelengths of some X-rays are about equal to the distance between the planes of atoms in crystalline solids, reinforced diffraction peaks of radiation with varying intensity are produced when a beam of X-rays strikes a crystalline solid. The diffraction of X-rays by a crystal structure is illustrated in Figure 6.1. Two waves reach the crystal at an angle θ and are diffracted at the same angle by adjacent layers. When the first wave strikes the top layer and the second wave strikes the next layer the waves are in phase. Constructive interference of the rays occurs at an angle θ if the path length difference is equal to a whole number of wavelengths, $n\lambda$, where n is an integer and λ is the wavelength. The fundamental relationship in XRD is the Bragg equation:

$$n\lambda = 2d \sin \theta \qquad (6.1)$$

where d is the distance between the layers of atoms.

There are a few different types of X-ray sources used in XRD. Sealed X-ray tubes, radioactive sources and rotating anode tubes can be used. The anode materials used are chromium, iron, copper or molybdenum. Alternatively, a synchrotron source, in which electrons are accelerated in a circle by electromagnetic fields, may be employed if access is available. Data can be collected at a faster rate when using synchrotron source, which can be useful when a large number of samples need to be examined. Conventionally, data collection can take several hours.

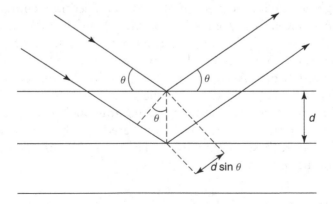

Figure 6.1   X-ray diffraction

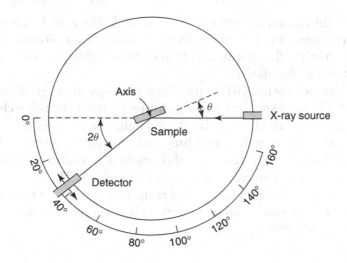

Figure 6.2 Schematic diagram of a XRD diffractometer

The most commonly used XRD technique is a powder method [Janssens, 2004]. In the powder diffraction technique a monochromatic X-ray beam is directed at a powdered sample spread on a support. The layout of a typical diffractometer is illustrated in Figure 6.2. The sample holder can be rotated. The intensities of the diffracted beams are recorded by a detector mounted on a movable carriage that can also be rotated. The angular position is measured in terms of 2θ. As the detector moves at a constant velocity, a computer plots the diffracted beam intensity as a function of 2θ, the diffraction angle. The pattern obtained is characteristic of the material under study and comparison with a database of XRD patterns allows the material to be identified. A drawback of powder diffraction is that the patterns of the constituent phases of a mixture strongly overlap, making quantitative analyses more complex. However, the Rietveld refinement, a method that has been developed, separates the individual constituent of a XRD pattern. Sample properties such as particle size distribution, orientation and lattice constants and instrumental effects can be taken into account and the experimental data fit to a simple model. It is important that homogeneous samples are carefully prepared for this technique to provide satisfactory results.

The single crystal diffraction technique may also be applied to very small samples. A tiny crystal of about 0.1 mm a side is required for this approach. The sample is placed in the centre of a four-circle diffractometer, a device for rotating the sample and the detector so that the entire diffraction pattern may be recorded. The raw data consist of the X-ray intensities at all angles in the diffractometer. Fourier analysis is used to analyse the measurements and convert then into the locations of the atoms in the material.

There are two useful XRD scattering methods, wide angle X-ray scattering (WAXS) and small angle X-ray scattering (SAXS). Such techniques can be employed to characterise the crystalline structures that appear in materials such as polymers, fibres and biological materials. WAXS uses angles from 5 to 120° and this method is useful for obtaining information about structures of the order of 1–50 Å. The size of crystals can be determined. SAXS uses angles from 1 to 5° and can be employed to obtain information about structures with larger interatomic distances in the range 50–700 Å.

6.2.1 Metals

XRD has been widely used to investigate metal objects. It can be employed to identify metal types to gain insight into the manufacturing process and, in particular, to characterise corrosion products [Chiavari et al., 2006; Creagh et al., 2004; De Ryck et al., 2003; Garcia-Heras et al., 2004; Leyssens et al., 2005; Linke et al., 2004a, 2004b; Pantos et al., 2005; Rojas-Rodriquez et al., 2004; Scott, 2004; Uda et al., 2005]. For instance, XRD has been used to investigate the corrosion products on lead elements from Spanish medieval stained glass windows [Garcia-Heras et al., 2004]. The XRD data for the lead showed the presence of anglesite ($PbSO_4$), lanarkite ($Pb_2O.SO_4$) and litharge (PbO). The presence of such compounds is indicative of atmospheric corrosion of lead – lead is initially oxidised to form lead oxides which eventually form lead sulfates in the presence of sulfur dioxide.

6.2.2 Paintings

XRD has been widely used to identify inorganic pigments used in paintings [Ajo et al., 2004; Casellato et al., 2000; Creagh, 2005; Hochleitner et al., 2003; Mazzocchin et al., 2003a, 2004; O'Neill et al., 2004; Schreiner et al., 2004]. This technique is particularly useful for differentiating pigments of the same chemical structure but with different crystalline phases. Synchrotron radiation XRD studies have been carried out on white pigments used in Australian aboriginal bark paintings [Creagh, 2005; O'Neill et al., 2004]. The quantitative analyses of data obtained for white pigments used by artists from different locations demonstrated that the mineral species used was diverse. Such information aids in establishing the authenticity of such works of art, as analysis of the pigment can potentially verify the clan area from which a painting originates.

6.2.3 Written Material

The structural properties of materials used in historic written works, including parchment and paper, can be characterised using XRD [Na et al., 2004; Wess et al., 2001]. For example, a synchrotron radiation XRD study of historic parchment from the middle ages was used to examine the collagen molecules that made up the material [Wess et al., 2001]. The XRD of the parchment shows important structural changes to the collagen fibrils compared to modern samples and shows potential as a technique for monitoring ageing.

6.2.4 Ceramics

XRD is one of the most direct techniques for the characterisation of the mineral content of ceramic materials and can be used to investigate inclusions, as well as the matrix. Information regarding the raw materials and the firing technology can be obtained from XRD data [Barone et al., 2003; Broekmans et al., 2004; Drebushchek et al., 2005; Maritan et al., 2006; Ruvalcaba-Sil et al., 1999; Zhu et al., 2004]. For example, in a study of cooking pottery from Tell Beydar in Syria, XRD was used to identify mineral phases [Broekmans et al., 2004]. The absence of kaolinite and the presence of montmorillonite and illite detected by XRD provided information about the firing temperature. Kaolinite is lost by 600 °C and montmorillonite and illite are stable up to 850 °C, indicating a firing temperature in the range 600–850 °C. XRD is also useful for the identification of pigments used on ceramics [Clark et al., 1997; Mazzocchin et al., 2003b; Uda et al., 1999].

Statistical analysis has been applied to the XRD data obtained for a series of Chinese pottery samples [Zhu et al., 2004]. Cluster analysis was used to group samples into one of three groups based on the concentration of 24 elements.

6.2.5 Stone

Many studies have employed XRD to characterise the materials used in historic buildings, especially mortars [Biscontin et al., 2002; Genestar and Pons, 2003; Genestar *et al.*, 2005; Ioannou *et al.*, 2005; Lopez-Arce *et al.*, 2003; Moropoulou *et al.*, 1995; Riccardi *et al.*, 1998; Silva *et al.*, 2005]. Mortar from the Colosseum in Rome has been characterised using XRD [Silva *et al.*, 2005]. The study was carried out determine whether the cement used in the construction of the Colosseum in 80 AD was lime (mainly lime and quartz) or pozzolanic cement (sand, lime and volcanic earth), a water resistant cement developed by the Romans. The XRD data displayed sharp peaks characteristic of calcite, with no other crystalline phases detected. This data, together with FTIR spectra and thermal analysis, confirmed that the mortar was lime that has converted to calcite.

XRD can also be utilised to characterise degraded material on stone statues and buildings [Maravelaki-Kalaitzaki, 2005; Maravelaki-Kalaitzaki *et al.*, 1999; Riotino *et al.*, 1998; Valls del Barrio *et al.*, 2002]. For example, XRD was used to characterise a black crust about 200μm thick on the surface of Pentelic marble monuments from the Acropolis in Athens [Maravelaki-Kalaitzaki, 2005]. XRD showed the presence of gypsum, calcite and kaolinite in the crust. Gypsum originates from the transformation of calcite in the presence of sulfur oxides in the atmosphere. Quartz and kaolinite were detected due to the presence of wind-blown particles embedded in the crust.

6.3 X-RAY FLUORESCENCE SPECTROSCOPY

X-ray fluorescence (XRF) spectroscopy is a widely used non-destructive technique for the measurements of the elemental composition of materials [Ferretti, 2000; Janssens *et al.*, 2000; Milazzo, 2004; Moens *et al.*, 2000; Skoog and Leary, 1992]. In this technique a sample is placed in a beam of high energy photons produced by an X-ray tube. The X-ray may be absorbed by an atom and transfers all of its energy to an innermost electron. If the primary X-ray has sufficient energy, electrons are ejected from the inner shells and create vacancies. Such vacancies produce an unstable atom and as the atom returns to its stable condition, electrons from the outer shells transfer to the inner shells. During this process a characteristic X-ray is emitted with an energy of the difference between the two binding energies of the corresponding atomic shells. As each element contains a unique set of energy levels, each element produces X-rays with a unique set of energies.

In most cases the innermost K and L shells are involved in XRF. The shells observed in atoms are illustrated in Figure 6.3. In the production of K lines in XRF, an electron from the L or M shell falls to fill the

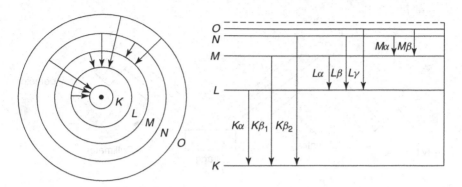

Figure 6.3 Electron transitions for the shells involved in emission of X-rays [Moens *et al.*, 2000]

vacancy. In the process, it emits a characteristic X-ray and, in turn, produces a vacancy in the L or M shells. In the production of L lines, when a vacancy is created in the L shell by either the primary X-ray or by a previous event, an electron from the M or N shell moves to occupy the vacancy. In this process, it emits a characteristic X-ray and, in turn, produces a vacancy in the M or N shell. The characteristic X-rays produced in XRF are labelled K, L, M or N, denoting the shells from which they originate. Where X-rays originate from the transitions of electrons from higher shells, an α, β or γ designation is used to label the X-rays. For example, a Kα X-ray is produced from a transition of an electron from the L to the K shell, while a Kβ X-ray is produced from a transition of an electron from the M to a K shell, and so on. A further designation is made as α_1, α_2, β_1, β_2, etc., to represent the transition of electrons from the orbits of higher and lower binding energy electrons within the shells into the same lower shell.

The two main types of XRF spectrometers are wavelength-dispersive and energy-dispersive [Ferretti, 2000; Janssens, 2004; Moens *et al.*, 2000; Skoog and Leary, 1992]. In wavelength-dispersive XRF (WDXRF) spectrometry, the sample is irradiated in a vacuum chamber by the primary X-rays of an X-ray source. The layout of a WDXRF instrument is shown in Figure 6.4. The secondary X-ray radiation resulting from the sample consists of different wavelengths originating from different elements present in the sample. The secondary radiation is guided through a collimator that allows only X-rays with parallel propagation to pass. The X-rays reach a Bragg crystal where they are reflected according to Bragg's law. In WDXRF, the rotation of the crystal makes it possible to detect X-rays of different wavelength. The radiation selected

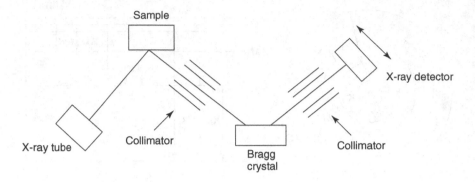

Figure 6.4 Layout of a WDXRF spectrometer

by the Bragg reflector will pass a second collimator before it reaches a detector. The signal may be converted into a spectrum, or if only some elements are of interest, the intensities of the appropriate wavelengths are recorded.

Energy-dispersive XRF (EDXRF) spectrometers have developed and allowed for the production of portable spectrometers. While WDXRF measures wavelength, in EDXRF the energy of the fluorescent radiation is measured. The development of semiconductor detectors has allowed instruments to be developed without the need of previous dispersion using a Bragg reflector: the X-ray photons entering the detector (such as lithium–silicon) produce a detectable signal. The signal produced in this method is weak and must be amplified. While the spectral resolution obtained in EDXRF is poor in comparison to WDXRF, EDXRF does have the advantage that the instrumentation is simpler and less expensive. The simplification of the instrument means that portable instruments have been developed [Ferretti, 2000; Janssens, 2004; Vittiglio *et al.*, 2004]. While in standard instruments, an X-ray tube is used as a source, portable instruments can use radioisotope sources to allow the rapid examination of samples. Microscopic XRF spectrometers have also been developed, allowing for the examination of a very small area of a sample surface [Janssens *et al.*, 2000; Vittiglio *et al.*, 2004]. This permits the study of the distribution of elements in the sample of interest [Longoni *et al.*, 1998]. Such instruments may also be portable and produce a resolution of 70–100 μm. The recent development of confocal μ-XRF allows depth profiling to be carried out [Kanngiesser *et al.* 2003; Smit *et al.*, 2004].

A wide range of samples can be investigated using XRF. Liquids and solids in many forms and sizes may be studied. Usually the sample chamber is evacuated to a pressure of about 10 Pa, but may be operated at normal atmospheric pressure or with helium gas' depending on the nature of the sample. Although the technique is non-destructive, as such, samples must often be crushed and pressed in pellets or fused in glass beads. However, it is possible to carry out experiments on bulk objects if they are a suitable size for the instrument sample compartment. Portable spectrometers are designed for in situ investigations and a sample does not have to be removed from the object of interest [Moioli and Seccaroni, 2000]. The portable spectrometers have the added advantage of allowing the mapping of an object's surface.

6.3.1 Metals

The provenance and manufacturing technology of metallic objects, especially coins, have been widely studied using XRF [Bichlmeier *et al.*, 2002; Ferretti *et al.*, 1997; Ferretti, 2000; Grolimund *et al.*, 2004; Guerra 2000; Janssens, 2004; Janssens *et al.*, 2000; Karydas *et al.*, 2004; Klockenkamper, 1978; Linke *et al.*, 2004a; Mantler and Schreiner, 2000; Moens *et al.*, 2000]. For example, a WDXRF study of a collection of Roman coins determined the concentration of silver at different depths [Klockenkamper, 1978; Moens *et al.*, 2000]. This study demonstrated a higher concentration of silver at the surface of the coins, fitting into the observation of a change of values in the Roman Empire. From about 200 AD, less silver was used for coin alloys in the Roman Empire, but the surface of coins was enriched.

XRF has also been used to discern genuine and counterfeit coins [Klockenkamper *et al.*, 1990; Reiff *et al.*, 2001]. A XRF examination of gold coins dated 1872 to 1914 determined that genuine coins showed a gold and silver content of about 90 % and 0.4 %, respectively [Klockenkamper *et al.*, 1990]. Counterfeit coins were shown to contain more gold and either more or less silver. In addition, the study demonstrated that some counterfeits show gold enrichment at the surface.

The portable non-destructive XRF approach is useful for examining large metal objects such as statues [Creagh *et al.*, 2004; Creagh, 2005; Ferretti *et al.*, 1997]. A portable XRF spectrometer with a γ-radiation source has been used to examine the armour of Joe Byrne, a bushranger (an outlaw) who was a member of the Kelly Gang active in Australia in the 1880s, held in the National Museum of Australia [Creagh *et al.*, 2004; Creagh, 2005]. The XRF study was used to provide information about the fabrication of the armour, a process that is debated. The components of the steel used to manufacture the armour were manganese, iron, arsénic, tin and lead and the compositions determined for the various sections of the armour indicated that the armour was produced from scrap steel available on a farm. The XRF data also showed the presence of tungsten and additional lead on a mark on the breastplate, evidence that the armour had been used in a gunfight. However, if the mark was made by a bullet, it could not have been during the Glenrowan siege, when the Kelly Gang was killed or captured, as tungsten was not used in ball ammunition until the end of World War I.

6.3.2 Glass

XRF provides a very effective technique for determining the composition of historic glass and so provides information about the raw materials and the technology used in production [Falcone *et al.*, 2002; Ferreti, 2000; Garcia-Heras *et al.*, 2005; Janssens *et al.*, 2000; Janssens *et al.*, 2003; Moens *et al.*, 2000; Schreiner *et al.*, 2004; Smit, 2005a; Wobrauschek *et al.*, 2000]. The ability of this technique to provide a trace analysis enables glasses to be classified. For example, a μ-XRF study of Roman glass samples from Israel quantified both the major and trace elements of these glasses and enabled samples to be categorised [Janssens *et al.*, 2000]. While glass from the Roman period has a composition of typically 66–72 % silicon dioxide, 16–18 % sodium oxide and 7–8 % calcium oxide, and the major elemental composition can be used to confirm that a glass is Roman, the identification of the trace elements provides more specific information. In this study, cluster analysis was performed for the XRF data of the Roman glass samples and the samples could be categorised into one of two groups. There were significant differences in the groups in the average concentrations of copper, tin, lead and antimony. These elements are associated with the colour of the glass and copper, tin, and lead probably originate from bronze chips added to the glass to provide a green colour with antimony added to provide transparency. It is also possible to study glass corrosion using XRF techniques [Janssens *et al.*, 2000]. An elemental analysis of a corrosion layer can provide a better understanding of the mechanism of corrosion by quantifying the migration of elements.

6.3.3 Ceramics

XRF has been used in many studies of ceramic materials [Barone *et al.*, 2003; Ferretti, 2000; Klein *et al.*, 2004; LaBrecque *et al.*, 1998; Leung and Lao, 2000; Mirti, 2000; Moens *et al.*, 2000; Pillay *et al.*, 2000; Yu, 2000; Puyandeera *et al.*, 1997; Yap and Vijayakumar, 1990]. As ceramics can be linked by the characteristics of the clay or temper, XRF can be used to determine the characteristic elements in ceramics of known and unknown provenance. Multivariate statistical methods such as PCA are generally used to interpret the data and link samples. Small quantities of the sample can be taken for a more precise destructive determination using WDXRF. Alternatively, a non-destructive approach using EDXRF can be applied.

EDXRF has been used to study Chinese blue and white porcelains of the Ming dynasty and a comparison made with modern imitation porcelain [Yu and Miao, 1996; Yu, 2000]. The concentration of the elements titanium, manganese, iron, cobalt, nickel, copper, zinc, gallium, lead, rubidium, strontium, yttrium and zirconium were determined for the surfaces of the porcelain samples. EDXRF spectra of a Ming porcelain and that of a contemporary porcelain are shown in Figure 6.5. The similarity in appearance of the spectra demonstrates that it can be quite difficult to date a porcelain from just a qualitative study of the data. This is particularly the case when an imitation has been deliberately modified to produce an EDXRF spectrum similar to that of a genuine porcelain. However, quantitative analysis using PCA demonstrated that the contemporary porcelain could not be incorporated into any clusters of Ming porcelains. Thus, it is more reliable to consider all the elements together when studying the provenance of ceramics.

6.3.4 Stone

EDXRF provides a rapid and sensitive non-destructive technique for the characterisation of gems [Joseph *et al.*, 2000]. For example, a study of natural and synthetic rubies showed that the natural form contains potassium, titanium, chromium, iron, copper, zinc, strontium and barium while the synthetic form contains chromium, nickel, copper and zinc. Iron was observed for all the naturally occurring rubies from different countries studied, and the synthetic rubies contained no iron.

Portable XRF provides potential as a non-destructive technique for the characterisation of gemstones [Pappalardo *et al.*, 2005]. A study of red stones in Hellenistic jewellery was carried out using XRF, as well as PIXE, in order to identify the stones and to provide a geographical provenance for the stones. The stones were recognised as garnets and the determination of trace elements, such as chromium and yttrium, could be used to classify the red garnets into different types.

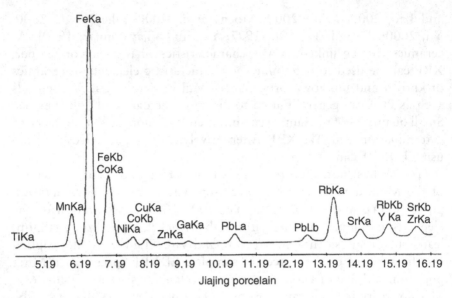

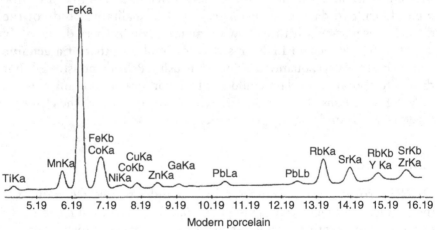

Figure 6.5 XRF spectra of a Jiajing porcelain and modern porcelain. X-Ray Spectrometry, Yu & Miao, 1996, John Wiley & Sons Limited. Reproduced with permission

6.3.5 Paintings

XRF can aid in the identification of inorganic pigments of paints as many pigments can be characterised by the presence of one or two detectable elements [Ferretti, 2000; Janssens, 2004; Klockenkamper *et al.*, 2000, 1993; Mantler and Schreiner, 2000; Mazzocchin *et al.*, 2003a, 2003b; Moens *et al.*, 2000; Scott, 2001; Szokefalvi-Novy *et al.*, 2004]. With normal XRF methods, sampling of paint is only possible when the pigment layer is not covered by varnish, so for older paintings the sample should be obtained during restoration. However, the more recent development of μ-XRF methods allows for the non-destructive analysis of paint layers [Smit *et al.*, 2004]. Some of the pigments that can be identified using XRF are listed in Table 6.1 [Ferretti, 2000; Janssens, 2004; Klockenkamper *et al.*, 1993, 2000]. Of course, it can be seen that a XRF spectrum does not always provide unambiguous information regarding pigments because many pigments share the key elements. XRF does have the advantage that quantitative analysis of minor, as well as major elements, can be carried out, enabling the original pigment ratios to be studied.

A XRF study of a purported Modigliani painting provided significant information about the pigment composition [Moens *et al.*, 2000]. The study of this painting revealed the presence of cerulean blue and emerald green. These results were compared with an earlier study of 15 authentic Modigliani paintings, none of which used cerulean blue or emerald green. This information, together with a radiographic examination, enabled the painting to be labelled a fake.

6.3.6 Written Material

Pigments used in illuminated manuscripts can be studied using the same approach as that already described for paintings [Janssens, 2004; Mantler and Schreiner, 2000]. Useful information regarding inks, especially gall ink, used in manuscripts can also be obtained from the application of XRF [Janssens, 2004; Janssens *et al.*, 2000; Kanngiesser *et al.*, 2003; Moens *et al.*, 2000]. XRF has been employed to examine Gutenberg Bibles [Janssens *et al.*, 2000; Janssens, 2004; Mommsen *et al.*, 1996]. A XRF study of the ink and paper of 22 different 15th century works from different locations in Germany, Italy and Switzerland was carried out [Mommsen *et al.*, 1996]. The recipes for the inks used by the different printers were kept a secret so XRF was able to provide information regarding the ink composition. A number of the leaves examined showed that the printed areas could not be distinguished from the blank areas using XRF, suggesting that the inks were mainly carbon-based, such as lamp black or soot. For other manuscripts nickel, copper and lead were detected in the ink, and for others potassium, calcium and iron were additionally detected. However, it was shown from a study of works of different periods by specific printers that the composition determined by XRF was different in these works, indicating that the printers changed the composition of their inks fairly frequently.

Table 6.1 Elements identified in inorganic pigments using XRF

Colour	Pigment	Elements
White	Antimony white	Sb
	Chalk	Ca
	Gypsum	Ca
	Lithopone	Zn, Ba
	Permanent white	Ba
	Titanium white	Ti
	White lead	Pb
	Zirconium oxide	Zr
	Zinc oxide	Zn
Yellow	Auripigmentum/Orpiment	As
	Cadmium yellow	Cd
	Chrome yellow	Cr
	Cobalt yellow	K, Co
	Lead-tin yellow	Sn
	Massicot	Pb
	Naples yellow	Pb, Sb
	Strontium yellow	Sr, Cr
	Titanium yellow	Ni, Sb, Ti
	Yellow ochre/Limonite	Fe
	Zinc yellow	Zn, Cr
Red	Cadmium red	Cd, Se
	Cadmium vermillion	Cd, Hg
	Chrome red	Pb, Cr
	Molybdate red	Pb, Cr, Mo
	Realgar	As
	Red lead	Pb
	Red ochre/Red earth	Fe
	Vermillion/Cinnabar	S, Hg
Blue	Azurite	Cu
	Cerulean blue	Co, Sn
	Cobalt blue	Co, Al
	Cobalt violet	Co
	Egyptian blue	Ca, Cu, Si
	Manganese blue	Ba, Mn
	Prussian blue	Fe
	Smalt	Si, K, Co
	Ultramarine	Si, Al, Na, S
Green	Basic copper sulfate	Cu
	Chromium oxide	Cr
	Chrysocolla	Cu
	Cobalt green	Co, Zn
	Emerald green	Cu, As
	Guignet's green	Cr
	Malachite	Cu
	Verdigris	Cu

Table 6.1 (*continued*)

Colour	Pigment	Elements
Black	Antimony black	Sb
	Black iron oxide/Iron black	Fe
	Cobalt black	Co
	Ivory black/Bone black	P, Ca
	Manganese oxide	Mn

6.4 ELECTRON MICROPROBE ANALYSIS

Electron microprobe analysis (EMPA) is a non-destructive technique that may be used to determine the chemical composition of small samples [Pollard and Heron, 1996; Reed, 2005]. The technique, which is related to XRF, uses an electron microscope with an X-ray detector. It utilises a focussed beam of high energy electrons to generate characteristic X-rays of the elements present in a material. The composition of a sample is determined by comparing the intensities of the X-rays generated with those of a standard. EMPA has the advantage that volumes as small as several μm^3 can be examined at a ppm level. EMPA has the ability to focus on a particular region of a sample. It is also possible to scan areas to produce elemental maps of the sample surface by rastering the electron beam across the surface. However, EMPA does require that the sample is electrically conducting, so samples such as glass or ceramics need to be coated with a layer of gold or carbon as described in Chapter 3. The technique is not as common as XRF and has not been as widely employed in the field of materials conservation.

6.4.1 Ceramics

EMPA has been used in several studies to examine ceramic glazes [Roque et al., 2005, 2006; Whitney, 1980]. In one study, the elemental composition of lustre development was monitored on ceramic sample exposed to different thermal treatments [Roque et al., 2006].

6.4.2 Glass

EMPA has been employed to compare ancient glass samples excavated from sites in England [Green and Hart, 1987]. The technique was used to compare Roman glass and glass from the 16th century and significant compositional differences were observed.

6.4.3. Stone

The degradation of building stone has been examined using EMPA [Leyson *et al.*, 1987]. A weathering crust of the walls of a Belgian 13–15th century limestone cathedral has been studies and gypsum crystals were shown to be predominant.

6.5   PROTON INDUCED X-RAY EMISSION

Proton induced X-ray emission (PIXE) is an ion beam technique that allows the concentration of elements in a material to be determined by examining the emission of characteristic X-rays [Calligaro *et al.*, 2004; Dran *et al.*, 2000]. PIXE has the advantage that it is non-destructive and can be performed in an air or helium atmosphere on even large or fragile objects. The technique involves the emission of X-rays induced by the interaction of energetic light ions, usually protons of a few MeV, with the atoms in the material of interest. The process involves expelling an electron from an inner shell of the atom (the K shell). The electron vacancy is filled by an electron originating from an external shell (the L shell). The excited state that results releases the excess electron binding energy ($E_K - E_L$) by the emission of an X-ray of this energy, which is characteristic of the atom.

PIXE was developed during the 1980s, utilising the highly focused particle beams produced by van de Graaf accelerators. These beams can be used externally from the accelerator ('tapped off') and focused onto a sample outside the accelerator. External beams also have the advantage that the sampling can be operated at atmospheric pressure. Sometimes the atmosphere is helium to reduce beam degradation. The beam produced by an accelerator has a diameter of several micrometres. Typically, protons of energy 2–3 MeV are utilised and the induced X-rays are collected by a solid-state detector. The lowest energy detectable by the detector is about 1 keV so all the elements with an atomic number greater than 11 can be detected simultaneously by their K or L lines. The detection limit is in the μg g^{-1} range, making PIXE suitable for trace element analysis.

Apart from PIXE, there are several other methods in ion beam analysis (IBA) which can be used in cultural heritage applications [Calligaro *et al.*, 2004]. Rutherford back scattering spectrometry (RBS) and nuclear reaction analysis (NRA), comprising proton-induced γ-induced emission (PIGE), are other ion beam techniques that may be encountered. RBS is a purely elastic process based on the electrostatic repulsion between a positively-charged projectile such as a helium atom. NRA occurs when the projectile and the target nuclei come close enough to undergo a nuclear reaction with the emission of characteristic photons or charged particles. Such techniques can be coupled with PIXE [Calligaro et al., 2004; Duerden et al., 1986]. For example, PIXE and RBS can be used in combination so that medium and heavy element composition is measured by PIXE and the light element composition determined from the RBS spectra [Ruvalcaba-Sil et al., 1999]. These techniques are not used to study cultural materials as frequently as PIXE as they have to be used under vacuum, which may risk damage to a sample.

6.5.1 Written Material

PIXE is often a more suitable technique than XRF for the elemental analysis of pigment composition in historical manuscripts. The paint layers in manuscripts are usually much thinner than in other forms so the XRF is much weaker [MacArthur et al., 1990]. Additionally, PIXE analysis with an external ion beam is very well suited to the study of such pigments as the reduced beam size allows analysis of a pigment without interference from adjacent areas of different colour. Qualitative analysis of pigments by PIXE is straightforward, but for quantitative information consideration must be given to the fact that the thickness of the paint layer is unknown. However, the ratio of elements can be used to quantitatively compare pigments.

PIXE has provided valuable information regarding historical documents and can be used to characterise inks, pigments, paper, papyrus and parchment [Budnar et al., 2001, 2005; Cahill et al., 1987; Calligaro et al., 2003; Cambria et al., 1993; Del Carmine et al., 1996; Demortier, 1988; Dran et al., 2000; Giuntini et al., 1995; Kolar et al., 2006; Kusko et al., 1984; Kusko and Schwab, 1987; Lovestrom et al., 1995; MacArthur et al., 1990; Mando et al., 2005; Neelmeijer and Mader, 2002; Olsson et al., 2001; Remazeilles et al., 2001; Ursic et al., 2006; Vodopivec et al., 2005]. With the development of PIXE methods that can be carried out without a vacuum, fragile documents can be examined without the risk of damage.

Information regarding the printing techniques and the authenticity of significant documents including the Gutenberg Bible [Demortier, 1988; Kusko et al., 1984; Kusko and Schwab, 1987], the Vinland Map [Cahill et al., 1987] and manuscripts by Galileo [Del Carmine et al., 1996; Giuntini et al., 1995] have all been examined using PIXE techniques. In studies of the Gutenberg Bible, analysis of the ink showed that copper/lead ratios can be used to determine the chronology of the written work [DeMortier, 1988; Kusko and Schwab, 1987]. During the long process of producing the Bible, different batches of ink with different copper/lead ratios were used. By identifying this ratio by PIXE for different sections, the organisation of the printing can be understood. The blank paper in the Bible was also analysed using this technique. Iron/calcium and manganese/calcium concentration ratios were used to identify the type of paper.

Documents produced using papyrus, parchment and vellum have been investigated using PIXE [Kusko and Schwab, 1987; Lovestrom *et al.*, 1995; Olssen *et al.*, 2001]. Although these materials tend to show a greater variability in composition, it has been demonstrated that it is possible to distinguish variations in element analysis in parchments from different times and places. PIXE can also be used to trace missing or modified characters on such documents.

6.5.2 Paintings

PIXE can provide quantitative information about the major, minor and trace elements for the pigments found in paintings [Denker and Opitz-Coutureau, 2004; Kusko *et al.*, 1990; Neelmeijer *et al.*, 2000]. PIXE can be used for depth-profiling, so it can provide quantitative elemental analysis about different paint layers [Brissaud *et al.*, 1996, 1999; Lill *et al.*, 2002; Mando *et al.*, 2005; Neelmeijer *et al.*, 1996, 2000; Neelmeijer and Mader, 2002]. By using different particle energies the sequence of pigments in paint and the thickness of the layers can be determined.

6.5.3 Glass

Quantitative analysis of the major, minor and trace elements of glass objects by PIXE and PIGE can provide important information regarding the production process and new materials used for these objects [Duerden *et al.*, 1986; Fleming and Swann, 1999; Kuisma-Kursula, 2000; Mader *et al.*, 2005; Smit *et al.*, 2000, 2005a; Zhang *et al.*, 2005]. For example, a PIXE study of 16th century glasses excavated in Ljubljana, Slovenia was used to investigate the origins of the fragments [Smit *et al.*, 2000]. PCA was carried out on the PIXE data obtained for a group of 257 museum glass samples and late Roman and medieval glass samples. The 16th century Ljubljana glasses formed two different clusters which could be attributed to the different production centres and manufacture recipes. The rubidium/strontium ratio varied between these groups of samples.

PIXE and PIGE have also been used to evaluate glass corrosion. The study of model medieval glass demonstrated that it is feasible to distinguish thin altered layers from deep corrosion destruction using these techniques [Mader *et al.*, 1998; Mader and Neelmeijer, 2004].

RBS has also been used to characterise glasses [Jembrih *et al.*, 2001; Kossinides *et al.*, 2002; Lanford, 1986; Mader *et al.*, 2005; Mader and Neelmeijer, 2004]. The composition of a glass can be determined using this technique and so can be used for authentication and to investigate changes associated with deterioration.

6.5.4 Ceramics

Different aspects of ceramics, including glazes, pigments and the clay body, may be studied using PIXE analysis [Demortier 1988; Lin *et al.*, 1999; Pappalardo 1999; Robertson *et al.*, 2002; Ruvalcaba-Sil *et al.*, 1999; Uda *et al.*, 1999, 2002; Zucchiati *et al.*, 1998, 2003]. The clay elemental composition can be quantitatively determined using this approach and used as a 'fingerprint' for identification purposes. For instance, a PIXE study of pottery recovered from a renaissance ship-wreck in Italy determined the amount of 23 elements on a range of samples [Zucchiati *et al.*, 1998]. As iron was abundant in all the samples, it was used as the internal reference element to normalise the amounts of the other elements.

6.5.5 Stone

PIXE and PIGE have been applied to the characterisation of precious stones [Bellot-Gurlet *et al.*, 2005; Calligaro *et al.*, 1998, 2000, 2002, 2003; Chen *et al.*, 2004; Constantinescu *et al.*, 2002; Dran *et al.*, 2000; Kim *et al.*, 2003; Pappalardo *et al.*, 2005; Querre *et al.*, 1996]. For example, PIXE was used to examine red inlays of a Parthian statuette originating from 3rd century BC Babylon. The stones were readily identified as rubies as the spectra exhibit characteristic lines due to alumina and the other elements associated with the impurities that are present in rubies. This study was also able to determine the geographical origin of the rubies in the statuette. Rubies obtained from nine well-known sources around the world were examined using PIXE. Markedly different amounts of titanium, vanadium, chromium and iron were observed for the rubies coming from different locations, with a good agreement found for the amounts for rubies from the same location. Hierarchical clustering, PCA and discriminant analysis were used to compare the spectra of the stones from the statuette with the spectra obtained from the rubies of different origins. The analysis showed that the statuette stones were very close in composition to rubies from Burma, and to a lesser extent, to those of Sri Lanka.

6.5.6 Metals

PIXE is widely used for the study of metal objects, particularly coins [Demortier, 1988, 1992, 2004; Demortier and Ruvalcaba, 2005; Denker and Blaich, 2002; Denker *et al.*, 2005; Enguita *et al.*, 2002; Flament and Marchetti, 2004; Guerra, 2000; Guerra and Calligaro, 2003; Johanssen *et al.*, 1986; Linke *et al.*, 2004b; Smit *et al.*, 2005b]. This non-destructive technique provides better results than XRF for lighter elements and allows small details to be analysed with a small beam size. PIXE has been particularly successfully in demonstrating metal fabrication techniques for precious metal objects and for the detection of fakes and repairs. A PIXE investigation of a silver statuette found in Bonn in 1896 was used to examine potentially altered sections of the piece [Demortier, 1988]. The statuette was composed of the God Mercury, showing good moulding, with a ram at the base, showing a more course moulding. The PIXE analysis found that concentrations of silver and lead were significantly different in Mercury and the ram, and tin was an additional element present in the ram. The results indicate that the ram had been added to the statuette later.

An example of the use of external beam PIXE for the elucidation of metal fabrication procedures is a study of ancient gold jewellery kept at the Musée du Louvre [Demortier, 1986, 1988, 2004]. This study was able to demonstrate that four soldering procedures had been performed over a distance of less than 6 mm on the jewellery. The variation in concentrations of silver, copper and gold in the jewellery could be correlated with procedures carried out at different temperatures. As such, the results of this PIXE study indicate that the goldsmiths during the 4th century BC had a good knowledge of the temperatures at which different joining procedures could be performed.

PIXE and RBS can also be used to characterise the corrosion layers that may form on metal surfaces [Abraham *et al.*, 2001; Denker *et al.*, 2005]. For example, in a study of corroded archaeological bronze objects, changes in the concentrations of certain trace elements between the copper metal and the corrosion layer were quantified. Such changes were consistent with the migration of copper from the substrate to form an oxide and the migration of other elements from the substrate to the interface.

6.6   X-RAY PHOTOELECTRON SPECTROSCOPY AND AUGER SPECTROSCOPY

In electron spectroscopy the kinetic energy of emitted electrons is recorded. The resulting spectrum consists of a plot of the number of emitted electrons as a function of the energy of the emitted electrons. One of the more popular forms of electron spectroscopy is X-ray photoelectron spectroscopy (XPS) (also known as electron spectroscopy for chemical analysis, ESCA). In this technique a sample is irradiated with X-rays and the photoelectrons ejected are analysed [Briggs and Sean, 1990; Carlson, 1975, 1978; Hubin and Terryn, 2004; Hufner, 1995; Lambert and McLaughlin, 1976; Spoto and Ciliberto, 2000]. The principles of XPS, where the electron is emitted from core energy levels, are illustrated in Figure 6.6.

XPS experiments are conducted in a high vacuum chamber. When the sample is irradiated, the emitted photoelectrons are collected by a lens system and focused into an energy analyser. The analyser counts the number of electrons with a given kinetic energy (E_K). The binding energies (E_B) of the photoelectrons are obtained using the Einstein equation:

$$E_B = h\nu - E_K - \Phi \tag{6.2}$$

where $h\nu$ is the X-ray photon energy and Φ is the work function of the sample. Φ is a characteristic energy barrier for the ejection of an electron.

The core electron binding energies are characteristic of the atomic core level from which the photoelectron was emitted and thus enable

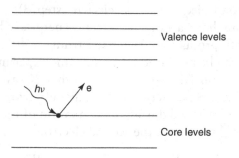

Figure 6.6   Emission of electrons in XPS

the surface element composition to be determined. All elements except hydrogen or helium may be detected. Core binding energies occur in the 50–1500 eV range. For certain elements, different types of core electrons may be emitted and multiple peaks are observed in the spectrum. For example, two peaks are observed for aluminium, at 120 eV (2s electron) and 75 eV (2p electron). Spin-orbit coupling, where the spins of p or d electrons interact with the orbital energies, can also result in further multiplicity.

XPS spectra have been obtained for every element. Ionisation of the core elements from each atomic orbital provides a collection of ionisation energies for each element [Briggs and Sean, 1990; Carlson, 1975, 1978; Hubin and Terryn, 2004; Hufner, 1995; Lambert and McLaughlin, 1976; Spoto and Ciliberto, 2000]. These energies are used for qualitative analysis. Electronic binding energies for some common elements are listed in Table 6.2. The core binding energies in an atom are influenced by the local electronic environment, and consequently, an atom in a molecule can exhibit a small range of binding energies known as chemical shifts.

As XPS involves the study of electrons ejected from the first several nanometres of a sample, the technique is a suitable method for surface analysis. As a consequence, it is important when carrying out an XPS experiment that the sample surface must be uniform and not contaminated.

The other commonly used form of electron spectroscopy is Auger spectroscopy (also known as Auger electron spectroscopy, AES) [Briggs and Sean, 1990; Carlson, 1975; Hubin and Terryn, 2004; Skoog and Leary, 1992; Spoto and Ciliberto, 2000]. The steps leading to the ejection of an Auger electron are shown in Figure 6.7. The incident X-ray radiation ionises a core electron (step 1) leaving a vacancy. An electron from a higher energy level fills the vacancy (step 2) and, in the process, releases energy. The energy may be reabsorbed by the atom and cause a second electron to be ejected (step 3). These secondary or Auger electrons have a range of kinetic energies which are plotted against intensity to produce an Auger spectrum.

Auger spectra can be produced using the same experimental apparatus as that used to obtain XPS spectra using either X-ray radiation or an electron beam. The kinetic energy of the Auger electrons is the difference between the energy released during the relaxation of the excited ion and the energy required to remove the second electron:

$$E_K = (E_B - E_B') - E_B' = E_B - 2E_B' \qquad (6.3)$$

Table 6.2 Binding energies for selected elements observed in XPS

Element	Core level binding energy (eV)						
	$K(1s)$	$L_1(2s)$	$L_2(2p)$	$L_3(2p)$	$M_1(3s)$	$M_2(3p)$	$M_3(3p)$
H	14						
He	25						
Li	55						
Be	111						
B	188						
C	284						
N	399						
O	532	24					
F	686	31					
Ne	867	45					
Na	1072	63		31			
Mg	1305	89		52			
Al	1560	118	74		73		
Si	1839	149	100		99		
P	2149	189	136		135		
S	2472	229	165		164		
Cl	2823	270	202		200	18	
Ar	3203	320	247		245	25	
K		377	297		294	34	18
Ca		438	350		347	44	26
Ti		564	461		455	59	34
V		628	520		513	66	38
Cr		695	584		575	74	43
Mn		769	652		641	84	49
Fe		846	723		710	95	56
Co		926	794		779	101	60
Ni		1008	872		855	112	68
Cu		1096	951		931	120	74
Zn		1194	1044		1021	137	87

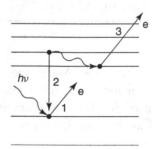

Figure 6.7 Steps leading to the ejection of an Auger electron

Table 6.3 Auger electron energies for selected elements

Element	Major energy peaks (eV)
C	270
N	380
O	510
F	650
Na	30, 990
Mg	47, 1180
Al	66, 1380
Si	91, 1610
P	116
S	150
Cl	180
K	250
Ca	290
Cr	35, 486, 527
Mn	40, 540, 588, 637
Fe	44, 594, 647, 700
Co	52, 651, 711, 771
Ni	60, 716, 781, 847
Cu	60, 772
Ta	26, 163, 173, 199, 209, 341
W	20, 164, 176, 205, 216, 347
Pt	13, 45, 68, 160, 171, 238, 844, 917
Zn	58, 830, 910, 990
Ge	24, 44, 48, 84, 1140, 1173
As	36, 42, 47, 91
Zr	20, 90, 115, 145
Nb	22, 102, 164, 194
Mo	27, 120, 185, 220
Ag	48, 56, 270, 308, 362
Sn	22, 64, 310, 360, 423, 254
Au	42, 71, 151, 165, 244, 261
Pb	46, 57, 92, 116, 246, 265
Bi	60, 102, 128, 258, 270

Each element has a characteristic Auger spectrum. The technique is very sensitive for higher elements (carbon, oxygen, nitrogen, sulfur) but although it can be used for heavier elements, the spectra become increasingly complicated as the atomic number increases. The notation used to assign peaks in Auger spectroscopy describes the types of orbital transitions involved in the production of the electron. For instance, the KLL transition involves an initial removal of a K electron followed by a transition of an L electron to the K orbital with the ejection of a second L electron. The K, L, M, ... shells refer to those with principal quantum numbers n = 1, 2, 3, . . . and so on. Auger electron energies for some common elements are listed in Table 6.3 [Riviere, 1990].

A useful aspect of AES is that Auger electron energies are independent of the energy of the incident radiation. This enables Auger peaks to be distinguished in an XPS spectrum. When the energy of the existing radiation is changed, the kinetic energies of the photoelectrons change, but those of the Auger electrons remain the same.

AES can be used to examine different regions of a surface. Scanning AES uses an electron beam which can be focussed. The degree of focussing can be varied to sample different areas of a surface. The electron analyser can be tuned to the energy of a particular Auger transition for a particular element, thus enabling the distribution of elements to be determined.

A wider range of conservation materials have been studied using XPS compared to AES. This is due to the problems that are encountered in AES when studying non-conducting materials. There is also an issue with AES of changes in composition of a range of inorganic and organic materials caused by the electron beam in this technique [Spoto and Ciliberto, 2000].

6.6.1 Glass

XPS and AES may be used to study ancient glasses [Angelini et al., 2004; Bertoncello et al., 2002; Dal Bianco et al., 2005; Dawson et al., 1978; Lambert et al., 1999; Lambert and McLaughlin, 1976, 1978; McNeil, 1984; Schreiner et al., 2004]. Although XRF is widely used to determine the composition of such materials, AES has the advantage of a high sensitivity for elements of low atomic number, such as sodium in glass. However, before using AES on glass, it is necessary to be aware of the effects of an electron beam on the surface composition [Dawson et al., 1978; Spoto and Ciliberto, 2000]. There is evidence that the beam can modify the composition of elements such as sodium and silicon, so it is particularly important to control the parameters used in an AES analysis of ancient glass.

XPS provides a useful means for examining the oxidation states of glass colorants and so may be used to identify the chemical species responsible for various colours. For example, XPS has proved useful for characterising Egyptian glass and was able to determine the species responsible for blue, green and red colours [Lambert et al., 1999; Lambert and McLaughlin, 1976, 1978]. Element analysis showed that blue glass did not contain a significant quantity of cobalt, even though the element is usually responsible for a blue colour in glass. Copper was present and, if copper is responsible for red and blue – green colours, it must be the result of different oxidation states. Relative binding energies and peak intensities were used to distinguish the Cu^{2+} oxidation state from the Cu^+ and Cu oxidation states. Red glass contained the lower oxidation states of copper and blue-green glasses the Cu^{2+} state.

6.6.2 Ceramics

The main advantage of the use of XPS for the analysis of ceramics is for the determination of the chemical state of the surface [Lambert et al., 1999; Spoto and Ciliberto, 2000]. For example, a XPS study of pottery samples from the Bahamas and Puerto Rico was able to distinguish iron oxidation states [Lambert et al., 1990]. Although XPS also has the advantages for ceramics of minimal sample preparation and light element detection, there are issues associated with the heterogeneous nature of the samples. While XPS provides information about the ceramic surface, it does not provide a characterisation of the bulk material.

6.6.3 Stone

XPS provides a useful tool for investigating the effect of pollutants on the surface of stones. For example, a XPS study of the effects of polluted atmospheres on Carrara marble was used to identify the species responsible for the weathering of the marble [Spoto and Ciliberto, 2000]. XPS has also been used to compare the structures of weathered and unweathered sandstone used in heritage buildings. XPS spectra of weathered and unweathered clay samples are shown in Figure 6.8 and indicate that both samples contain silicon, aluminium, oxygen and carbon. However, the presence of iron peaks at a binding energy of 720 eV was only observed in the weathered clay samples. To avoid interferences from iron impurities, the XPS spectrum was also obtained for weathered clay samples after the removal of non-structural iron and the results still indicate the presence of iron. This finding was used in combination with other techniques to support the suggestion that Fe^{3+} not only exists as an impurity, but also substitutes other metals, such as Al^{3+} and Si^{4+} in the aluminosilicate clay structure of sandstone.

6.6.4 Metals

As metals are highly conductive materials, they are readily studied using XPS and AES [Lambert et al., 1999; Spoto and Ciliberto, 2000]. XPS is

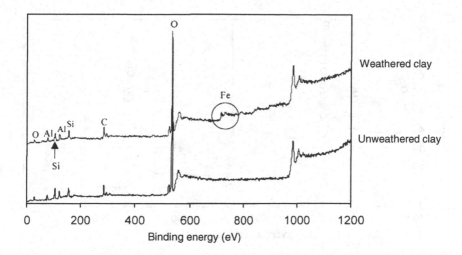

Figure 6.8 XPS spectra for weathered and unweathered clay from sandstone

particularly useful for identifying the oxidation states of metals and for examining corrosion and patinas. Ancient bronzes have been regularly studied using these techniques [Paparazzo *et al.*, 1995, 2001; Paparazzo and Moretto, 1995, 1999; Spoto *et al.*, 2000; Spoto and Ciliberto, 2000; Squarcialupi *et al.*, 2002]. XPS is used to distinguish the oxidation states of copper in bronze – Cu, Cu^+, Cu^{2+} show a shift in peak position and Cu^{2+} also shows shake-up satellites. Scanning AES is used to determine the distribution of elements and their chemical state on the bronze surface.

Lead objects, particularly Roman fistulae, have been studied using XPS and AES [Paparazzo, 1994; Paparazzo and Moretto, 1995, 1998]. XPS and AES can be used to provide information about the production methods. The presence of tin at a joint indicates that the pipes are soldered with a lead/tin solder rather than welded. Additionally, the detection of peaks at 137 eV due to lead and at 139 eV due to lead oxides at the pipe joint indicates the formation of protective oxides not present in the other parts of the pipe.

6.6.5 Paintings

XPS has been used to examine surface pigments in paintings [Clark *et al.*, 1997; Lambert *et al.*, 1999; Spoto and Ciliberto, 2000]. The colours of pigments are often associated with the oxidation states of an element which may be identified by XPS. XPS also has the advantage that the

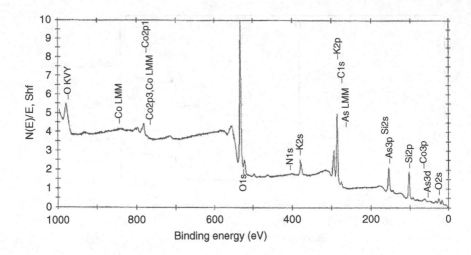

Figure 6.9 XPS spectrum of smalt. With kind permission of Springer Science and Business Media, Applied Physics A, Vol 79, 2004, 309–314, C Altavilla and E W Ciliberto, figure 1

chemical environments of light elements such as carbon and oxygen may be characterised, making it possible to study organic materials. The surface-sensitive nature of XPS can be exploited for the investigation of any deterioration processes on the surface of paintings [Lambert *et al.*, 1999; Spoto *et al.*, 2000].

XPS has been employed to study the pigment smalt and the results enable smalt to be discriminated from other blue pigments [Altavilla and Ciliberto, 2004]. The XPS spectrum of smalt is shown in Figure 6.9 and the binding energy peaks for silicon, cobalt, potassium, arsenic and oxygen and Auger lines for cobalt, arsenic and oxygen are observed. Traces of arsenic are derived from ores used during the preparation of the cobalt oxide component of smalt – the pigment is obtained by grinding blue coloured glass.

6.6.6 Written Material

XPS has proved to be a useful analytical technique for investigating the surface composition of paper and for monitoring degradation process [Istone, 1995; Laguardia *et al.*, 2005; Spoto and Ciliberto, 2000]. XPS has also been used to study the effects of deacidification of paper [Spoto *et al.*, 2000]. However, as the X-ray radiation used in XPS may cause damage to paper samples, a cold stage and brief exposure times should be used to minimise changes to the material surface. The applicability of this technique to paper studies lies with its sensitivity for carbon and oxygen. Curve-fitting analysis of the carbon 1s and oxygen 1s peaks of cellulose shows three and two bands, respectively, contributing to these peaks. The intensity ratios of these component peaks can be used to characterise paper samples. The ratio of the peaks due to C–O and O–C–O carbon atoms at 286.7 and 288.1 eV, respectively, and likewise the ratio of the C–O–H and O–C–O oxygen atom peaks at 532.9 and 533.5 eV, respectively, have been used to characterise cellulose. The overall oxygen/carbon ratio can also be used as the presence of other components such as lignin affects this ratio.

While AES tends not to be used for manuscript studies because of possible alteration caused by electron bombardment, scanning AES has proved applicable to the dating of manuscript inks [Spoto *et al.*, 2000]. AES can be used to study iron concentration on manuscripts, as iron from iron gall ink migrates over the substrate surface. The migration is dependent on age. It is also possible to scan a manuscript for ink removal and so determine whether or not a manuscript was written over a period of time.

REFERENCES

M.H. Abraham, G.W. Grime, M.A. Marsh and J.P. Northover, The study of thick corrosion layers on archaeological metals using controlled laser ablation in conjunction with an external beam microprobe, *Nuclear Instruments and Methods in Physics Research B*, **181** (2001), 688–692.

D. Ajo, U. Casellato, E. Fiorin and P.A. Vigato, Ciro Ferri's frescoes: a study of painting materials and technique by SEM–EDS microscopy, X-ray diffraction, micro FTIR and photoluminescence spectroscopy, *Journal of Cultural Heritage*, **5** (2004), 333–348.

C. Altavilla and EW. Ciliberto, Decay characterisation of glassy pigments: an XPS investigation of smalt paint layers, *Applied Physics A*, **79** (2004), 309–314.

I. Angelini, G. Artioli, P. Bellintani, *et al.*, Chemical analyses of Bronze Age glasses from Frattesina di Rovigo, Northern Italy, *Journal of Archaeological Science*, **31** (2004), 1175–1184.

G. Barone, V. Crupi, S. Galli, *et al.*, Spectroscopic investigation of Greek ceramic artefacts, *Journal of Molecular Structure*, **651–653** (2003), 449–458.

L. Bellot-Gurlet, G. Poupeau, J. Salomon, *et al.*, Obsidian provenance studies in archaeology: a comparison between PIXE, ICP-AES and ICP–MS, *Nuclear Instruments and Methods in Physics Research B*, **240** (2005), 583–588.

R. Bertoncello, L. Milanese, U. Russo, *et al.*, Chemistry of cultural glasses: the early medieval glasses of Montselice's Hill (Padova, Italy), *Journal of Non-Crystalline Solids*, **306** (2002), 249–262.

S. Bichlmeier, K. Janssens, J. Heckel, *et al.*, Comparative material characterisation of historical and industrial samples by using a compact micro-XRF spectrometer, *X-ray Spectrometry*, **31** (2002), 87–91.

G. Biscontin, M. Pellizon Birelli and E. Zendri, Characterisation of binders employed in the manufacture of Venetian historical mortars, *Journal of Cultural Heritage*, **3** (2002), 31–37.

D. Briggs and M.P. Sean (eds), *Practical Surface Analysis Vol. 1: Auger and X-ray Photoelectron Spectroscopy*, 2nd edn, John Wiley & Sons Inc., New York (1990).

I. Brissaud, G. Lagarde and P. Midy, Study of multilayers by PIXE technique. Application to paintings, *Nuclear Instruments and Methods in Physics Research B*, **117** (1996), 179–185.

I. Brissaud, A. Guillo, G. Lagarde, *et al.*, Determination of the sequence and thicknesses of multilayers in an easel painting, *Nuclear Instruments and Methods in Physics Research B*, **155** (1999), 447–451.

T. Broekmans, A. Adriaens and E. Pantos, Analytical investigations of cooking pottery from Tell Beydar (NE Syria), *Nuclear Instruments and Methods in Physics Research B*, **226** (2004), 92–97.

M. Budnar, J. Vodopevic, P.A. Mando, *et al.*, Distribution of chemical elements of iron gall ink writing studied by the PIXE method, *Restaurator*, **22** (2001), 228–241.

M. Budnar, M. Ursic, J. Simcic, *et al.*, Analysis of iron gall inks by PIXE, *Nuclear Instruments and Methods in Physics Research B*, **243** (2005), 407–416.

T.A. Cahill, R.N. Schwab, B.H. Kusko, *et al.*, The Vinland Map revisited: new compositional evidence on its inks and parchment, *Analytical Chemistry*, **59** (1987), 829–833.

T. Calligaro, A. Mossman, J.P. Poirot and C. Querre, Provenance study of rubies from a Parthian statuette by PIXE analysis, *Nuclear Instruments and Methods in Physics Research B*, **136–138** (1998), 846–850.

T. Calligaro, J.C. Dran, J.P. Poirot, *et al.*, PIXE/PIGE characterisation of emeralds using an external micro-beam, *Nuclear Instruments and Methods in Physics Research B*, **161–163** (2000), 769–774.

T. Calligaro, S. Colinart, J.P. Poirot and C. Sudres, Combined external-beam PIXE and μ – Raman characterisation of garnets used in Merovingian jewellery, *Nuclear Instruments and Methods in Physics Research B*, **189** (2002), 320–327.

T. Calligaro, J.C. Dran and M. Klein, Application of photo-detection to art and archaeology at the C2RMF, *Nuclear Instruments and Methods in Physics Research B*, **504** (2003), 213–221.

T. Calligaro, J.C. Dran and J. Saloman, Ion beam microanalysis, in *Non-Destructive Micro Analysis of Cultural Heritage Materials* (eds K. Jannsens and R. Van Grieken), Elsevier, Amsterdam (2004), pp 227–276.

R. Cambria, P. Del Carmine, M. Grange, *et al.*, A methodological test of external beam PIXE analysis on inks of ancient manuscripts, *Nuclear Instruments and Methods in Physics Research B*, **75** (1993), 488–492.

T.A. Carlson, *Photoelectron and Auger Spectroscopy*, Plenum Press, New York (1975).

T.A. Carlson, *X-ray Photoelectron Spectroscopy*, Dowden, Hutchinson and Ross, Stroudsburg (1978).

U. Casellato, P.A. Vigato, U. Russo and M. Matteini, Mössbauer approach to the physico-chemical characterisation of iron-containing pigments for historical wall paintings, *Journal of Cultural Heritage*, **1** (2000), 217–232.

T.H. Chen, T. Calligaro, S. Pages-Campagna and M. Menu, Investigation of Chinese archaic jade by PIXE and μ – Raman spectrometry, *Applied Physics A*, **79** (2004), 177–180.

C. Chiavari, A. Colledan, A. Frignani and G. Brunoro, Corrosion evaluation of traditional and new bronzes for artistic castings, *Materials Chemistry and Physics*, **95** (2006), 252–259.

R.J.H. Clark, L. Curri, G.S. Henshaw and C. Laganara, Characterisation of brown-black and blue pigments in glazed pottery fragments from Castel Fiorentino (Foggia, Italy) by Raman spectroscopy, X-ray powder diffractometry and X-ray photoelectron spectroscopy, *Journal of Raman Spectroscopy*, **28** (1997), 105–109.

B. Constantinescu, R. Bugoi and G. Sziki, Obsidian provenance studies of Transylvania's Neolithic tools using PIXE, micro-PIXE and XRF, *Nuclear Instruments and Methods in Physics Research B*, **189** (2002), 373–377.

D.C. Creagh, G. Thorogood, M. James and D.L. Hallam, Diffraction and fluorescence studies of bushranger armour, *Radiation Physics and Chemistry*, **71** (2004), 839–840.

D.C. Creagh, The characterisation of artefacts of cultural heritage significance using physical techniques, *Radiation Physics and Chemistry*, **74** (2005), 426–442.

B. Dal Bianco, R. Bertoncello, L. Milanese and S. Barison, Glass corrosion across the Alps: a surface study of chemical corrosion of glasses found in marine and ground environments, *Archaeometry*, **47** (2005), 351–360.

P.T. Dawson, O.S. Heavens and A.M. Pollard, Glass surface analysis by Auger electron spectroscopy, *Journal of Physics C: Solid State Physics*, **11** (1978), 2183–2193.

P. Del Carmine, L. Giuntini, W. Hooper, *et al.*, Further results from PIXE analysis of inks in Galileo's notes on motion, *Nuclear Instruments and Methods in Physics Research B*, **113** (1996), 354–358.

G. Demortier, LARN experience in non-destructive analysis of gold artefacts, *Nuclear Instruments and Methods in Physics Research B* **14** (1986), 152–155.

G. Demortier, Application of nuclear microprobes to material of archaeological interest, *Nuclear Instruments and Methods in Physics Research B*, **30** (1988), 434–443.

G. Demortier, Ion beam analysis of gold jewellery, *Nuclear Instruments and Methods in Physics Research B*, **64** (1992), 481–487.

G. Demortier, Precious metals and artefacts, in *Non-Destructive Micro Analysis of Cultural Heritage Materials* (eds K. Jannssens and R. Van Grieken), Elsevier, Amsterdam (2004), pp 493–564.

G. Demortier and J.L. Ruvalcaba-Sil, Quantitative ion beam analysis of complex gold-based artefacts, *Nuclear Instruments and Methods in Physics Research B*, **239** (2005), 1–15.

A. Denker and M.C. Blaich, PIXE analysis of middle ages objects using 68 MeV protons, *Nuclear Instruments and Methods in Physics Research B*, **189** (2002), 315–319.

A. Denker and J. Opitz-Coutureau, Paintings – high-energy protons detect pigments and paint layers, *Nuclear Instruments and Methods in Physics Research B*, **213** (2004), 677–682.

A. Denker, W. Bohne, J. Optitz-Coutureau, *et al.*, Influence of corrosion layers on quantitative analysis, *Nuclear Instruments and Methods in Physics Research B*, **239** (2005), 65–70.

L. De Ryck, A. Adriaens, E. Pantos and F. Adams, A comparison of microbeam techniques for the analysis of corroded ancient bronze objects, *Analyst*, **128** (2003), 1104–1109.

J.C. Dran, T. Calligaro and J. Saloman, Particle induced X-ray emission, in *Modern Analytical Methods in Art and Archaeology* (eds E. Ciliberto and G. Spoto), John Wiley & Sons Inc., New York (2000), pp 135–165.

V.A. Drebushchek, L.N. Mylnikova, T.N. Drebushchek and V.V. Boldyrev, The investigation of ancient pottery: application of thermal analysis, *Journal of Thermal Analysis and Calorimetry*, **82** (2005), 617–626.

P. Duerden, E. Clayton, J.R. Bird and D.D. Cohen, Recent ion beam analysis studies on archaeometry and art, *Nuclear Instruments and Methods in Physics Research B*, **14** (1986), 50–57.

O. Enguita, A. Climent-Font, G. Garcia, *et al.*, Characterisation of metal threads using differential PIXE analysis, *Nuclear Instruments and Methods in Physics Research B*, **189** (2002), 328–333.

R. Falcone, A. Renier and M. Verita, Wavelength-dispersive X-ray fluorescence analysis of ancient glasses, *Archaeometry*, **44** (2002), 531–542.

M. Ferretti, L. Miazzo and P. Poioli, The application of a non-destructive XRF method to identify different alloys in the bronze statue of the Capitoline Horse, *Studies in Conservation*, **42** (1997), 241–246.

M. Ferretti, X-ray fluorescence applications for the study and conservation of cultural heritage, in *Radiation in Art and Archaeometry* (eds D.C. Creagh and D.A. Bradley), Elsevier, Amsterdam (2000), pp 285–296.

C. Flament and P. Marchetti, Analysis of ancient silver coins, *Nuclear Instruments and Methods in Physics Research B*, **226** (2004), 179–184.

S.J. Fleming and C.P. Swann, Raman mosaic glass: a study of production processes using PIXE spectrometry, *Nuclear Instruments and Methods in Physics Research B*, **150** (1999), 622–627.

M. Garcia-Heras, M.A. Villegas, E. Cano, *et al.*, A conservation assessment on metallic elements from Spanish Medieval stained glass windows, *Journal of Cultural Heritage*, **5** (2004), 311–317.

M. Garcia-Heras, N. Carmona, C. Cul and M. Angeles Villegas, Neorenaissance/neobaroque stained glass windows from Madrid: a characterisation study of some panels signed by the Maimejean Fréres company, *Journal of Cultural Heritage*, **6** (2005), 91–98.

C. Genestar and C. Pons, Ancient covering plaster mortars from several convents and Islamic and Gothic palaces in Palma de Mallorca (Spain). Analytical characterisation, *Journal of Cultural Heritage*, **4** (2003), 291–298.

C. Genestar, C. Pons and A. Mas, Analytical characterisation of ancient mortars from the archaeological Roman city of Pollentia (Balearic Islands, Spain), *Analytica Chimica Acta*, **557** (2005), 373–379.

L. Giuntini, F. Lucarelli, P.A. Mando, *et al.*, Galileo's writings; chronology by PIXE, *Nuclear Instruments and Methods in Physics Research B*, **95** (1995), 389–392.

L.R. Green and F.A. Hart, Colour and chemical composition in ancient glass: an examination of some Roman and Wealdon glass by means of ultraviolet–visible–infrared spectrometry and electron microprobe analysis, *Journal of Archaeological Science*, **14** (1987), 271–282.

D. Grolimund, M. Senn, M. Trottmann, *et al.*, Shedding new light on historical metal samples using micro-focussed synchrotron X-ray fluorescence and spectroscopy, *Spectrochimica Acta*, **59** (2004), 1627–1635.

M.F. Guerra, The study of the characterisation and provenance of coins and other metal-work using XRF, PIXE and activation analysis, in *Radiation in Art and Archaeometry* (eds D.C. Creagh and D.A. Bradley), Elsevier, Amsterdam (2000), pp 379–416.

M.F. Guerra and T. Calligaro, Gold cultural heritage objects: a review of studies of provenance and manufacturing technologies, *Measurement Science and Technology*, **14** (2003), 1527–1537.

B. Hochleitner, V. Desnica, M. Mantler and M. Schreiner, Historical pigments: a collection analysed with X-ray diffraction analysis and X-ray fluorescence analysis in order to create a database, *Spectrochimica Acta Part B*, **58** (2003), 641–649.

A. Hubin and H. Terryn, X-ray photoelectron and Auger electron spectroscopy in *Non-Destructive Microanalysis of Cultural Heritage Materials* (eds K. Janssens and R. van Grieken), Elsevier, Amsterdam (2004), pp 277–312.

S. Hufner, *Photoelectron Spectroscopy*, Springer-Verlag, Berlin (1995).

I. Ioannou, C. Hall, W.D. Hoff, *et al.*, Synchrotron radiation energy dispersive X-ray diffraction analysis of salt distribution in Lepine limestone, *Analyst*, **130** (2005), 1006–1008.

W.K. Istone, X-ray photoelectron spectroscopy, in *Surface Analysis of Paper* (eds T.E. Connors and S. Banerjee), CRC Press, Boca Raton (1995), pp 235–268.

K. Janssens, G. Vittiglio, I. Deraedt, *et al.*, Use of microscopic XRF for non-destructive analysis in art and archaeometry, *X-ray Spectrometry*, **29** (2000), 73–91.

K. Janssens, X-ray based methods of analysis, in *Non-Destructive Microanalysis of Cultural Heritage Materials* (eds K. Jannsens and R. Van Grieken), Elsevier, Amsterdam (2004), pp 129–226.

D. Jembrih, C. Neelmeijer, *et al.*, M. Schreiner, Iridescent Art Nouveau glass – IBA and XPS for the characterisation of thin iridescent layers, *Nuclear Instruments and Methods in Physics Research B*, **181** (2001), 698–702.

E.M. Johansson, S.A.E. Johansson, K.G. Malmqvist and I.M.B. Wiman, The feasibility of the PIXE technique in the analysis of stamps and art objects, *Nuclear Instruments and Methods in Physics Research B*, **14** (1986), 45–49.

C. Jokubonis, P. Wobrauschek, S. Zamini, *et al.*, Results of quantitative analysis of Celtic glass artefacts by energy dispersive X-ray fluorescence spectrometry, *Spectrochimica Acta Part B*, **58** (2003), 627–633.

D. Joseph, M. Lal., P.S. Shinde and B.D. Padalla, Characterisation of gem stones (rubies and sapphires) by energy-dispersive X-ray fluorescence spectrometry, *X-ray Spectrometry*, **29** (2000), 147–150.

B. Kanngiesser, W. Malzer and I. Reiche, A new 3D micro X-ray fluorescence analysis set-up – first archaeometric applications, *Nuclear Instruments and Methods in Physics Research B*, **211** (2003), 259–264.

A.G. Karydas, D. Kotzamani, R. Bernard, J.N. Barrandon and C. Zarkadas, A compositional study of a museum jewellery collection (7th–1st BC) by means of a portable XRF spectrometer, *Nuclear Instruments and Methods in Physics Research B*, **226** (2004), 15–28.

J. Kim, A.W. Simon, V. Ripoche, *et al.*, Proton-induced x-ray emission analysis of turquoise artefacts from Salado Platform Mound sites in the Tonto Basin of central Arizona, *Measurement Science and Technology*, **14** (2003), 1579–1589.

M. Klein, F. Jesse, H.U. Kasper and A. Golden, Chemical characterisation of ancient pottery from Sudan by x-ray fluorescence spectrometry (XRF), electron microprobe analysis (EMPA) and inductively coupled plasma mass spectrometry (ICPMS), *Archaeometry*, **46** (2004), 339–356.

R. Klockenkamper, Detection of surface enrichment by x-ray spectrometry – gold enrichment on the surface of Roman gold coins, *Zeitschrift fur Analytische Chemie Fresenius*, **290** (1978), 212–216.

R. Klockenkamper, M. Becker and H. Otto, Röntgenspektralanalyse von echtren und gefälschten Reichsgoldmünzen, *Spectrochimica Acta Part B*, **45** (1990), 1043–1051.

R. Klockenkamper, A. von Bohlen, L. Moens and W. Devos, Analytical characterisation of artists' pigments used in old and modern paintings by total-reflection x-ray fluorescence, *Spectrochimica Acta Part B*, **48** (1993), 239–246.

R. Klockenkamper, A. von Bohlen and L. Moens, Analysis of pigments and inks in oil paintings and historical manuscripts using total reflection x-ray fluorescence spectrometry, *X-ray Spectrometry*, **29** (2000), 119–129.

J. Kolar, A. Stolfa, M. Strlic, *et al.*, Historical iron gall ink containing documents – properties affecting their condition, *Analytica Chimica Acta*, **555** (2006), 167–174.

S. Kossinides, M. Kokkoris, A.G. Karydas, *et al.*, Analysis of ancient glass using ion beams and related techniques, *Nuclear Instruments and Methods in Physics Research B*, **195** (2002), 408–413.

P. Kuisma-Kursula, Accuracy, precision and detection limits of SEM–WDS, SEM–EDS and PIXE in the multi-elemental analysis of medieval glass, *X-ray Spectrometry*, **29** (2000), 111–118.

B.H. Kusko, T.A. Cahill, R.A. Eldred and R.N. Schwab, Proton milliprobe analyses of the Gutenberg Bible, *Nuclear Instruments and Methods in Physics Research B*, **3** (1984), 689–694.

B.H. Kusko and R.N. Schwab, Historical analysis by PIXE, *Nuclear Instruments and Methods in Physics Research B*, **22** (1987), 401–406.

B.H. Kusko, M. Menu, T. Calligaro and J. Saloman, PIXE at the Louvre Museum, *Nuclear Instruments and Methods in Physics Research B*, **49** (1990), 288–292.

L. Laguardia, E. Vassallo, F. Cappitelli, *et al.*, Investigation of the effects of plasma treatments on biodeteriorated ancient paper, *Applied Surface Science*, **252** (2005), 1159–1166.

J.J. LaBreque, J.E. Vaz, J.M. Cruxent and P.A. Rosales, A simple radioisotope X-ray fluorescence method for provenance studies of archaeological ceramics employing principal component analysis, *Spectrochimica Acta Part B*, **53** (1998), 95–100.

J.B. Lambert and C.D. McLaughlin, X-ray photoelectron spectroscopy: a new analytical method for the examination of archaeological artefacts, *Archaeometry*, **18** (1976), 169–180.

J.B. Lambert and C.D. McLaughlin, in *Archaeological Chemistry II* (ed G. Carter), Advances in Chemistry 171, American Chemical Society, Washington DC (1978), pp 189–199.

J.B. Lambert, L. Xue, J.M. Weydert and J.H. Winter, Oxidation states of iron in Bahamian pottery by X-ray photoelectron spectroscopy, *Archaeometry*, **32** (1990), 47–54.

J.B. Lambert, C.D. McLaughlin, C.E. Shawl and L. Xue, X-ray photoelectron spectroscopy and archaeology, *Analytical Chemistry* (1999), 614A–620A.

W.A. Lanford, Ion beam analysis of glass surfaces: dating, authentication and conservation, *Nuclear Instruments and Methods in Physics Research B*, **14** (1986), 123–126.

P.L. Leung and H. Luo, A study of provenance and dating of ancient Chinese porcelain by X-ray fluorescence spectrometry, *X-ray Spectrometry*, **29** (2000), 34–38.

L. Leyson, E. Roekens, Z. Komy and R. Van Grieken, A study of the weathering of an historic building, *Analytica Chimica Acta*, **195** (1987), 247–255.

K. Leyssens, A. Adriaens, M.G. Dowsett, *et al.*, Simultaneous in situ time resolved SR–XRD and corrosion potential analysis to monitor the corrosion on copper, *Electrochemistry Communications*, 7 (2005), 1265–1270.

J.O. Lill, M. Strom, M. Brenner and A. Lindroos, Ion beam characterisation of paint layers made according to late 18th century techniques, *Nuclear Instruments and Methods in Physics Research B*, **189** (2002), 303–307.

E.K. Lin, Y.C. Yu, C.W. Wang, *et al.*, PIXE analysis of ancient Chinese Changsha porcelain, *Nuclear Instruments and Methods in Physics Research B*, **150** (1999), 581–584.

R. Linke, M. Schreiner, G. Demortier, *et al.*, The provenance of medieval silver coins: analysis with EDXRF, SEM/EDX and PIXE, in *Non-Destructive Micro Analysis of Cultural Heritage Materials* (eds K. Jannsens and R. Van Grieken), Elsevier, Amsterdam (2004a), pp 605–633.

R. Linke, M. Schreiner and G. Demortier, The application of photon, electron and proton induced X-ray analysis for the identification and characterisation of medieval silver coins, *Nuclear Instruments and Methods in Physics Research B*, **226** (2004b), 172–178.

A. Longoni, C. Fiorini, P. Leutenegger, *et al.*, A portable XRF spectrometer for non-destructive analysis in archaeometry, *Nuclear Instruments and Methods in Physics Research A*, **409** (1998), 395–400.

P. Lopez-Arce, J. Garcia-Guinea, M. Gracia and J. Obis, Bricks in historical buildings of Toldeo City: characterisation and restoration, *Materials Characterisation*, **50** (2003), 59–68.

N.E.G. Lovestrom, F. Lucarelli, P.A. Mando, *et al.*, Galileo's writings: chronology by PIXE, *Nuclear Instruments and Methods in Physics Research B*, **95** (1995), 389–392.

M. Mader, D. Grambole, F. Herrmann, *et al.*, Non-destructive evaluation of glass corrosion states, *Nuclear Instruments and Methods in Physics Research B*, **136–138** (1998), 863–868.

M. Mader and C. Neelmeijer, Proton beam examination of glass – an analytical contribution for preventative conservation, *Nuclear Instruments and Methods in Physics Research B*, **226** (2004), 110–118.

M. Mader, D. Jembrih-Simburger, C. Neelmeijer and M. Schreiner, IBA of iridescent Art Nouveau glass – comparative studies, *Nuclear Instruments and Methods in Physics Research B*, **239** (2005), 107–113.

J.D. MacArthur, P. Del Carmine, F. Lucarelli and P.A. Mando, Identification of pigments in some colours on miniatures from the medieval age and early renaissance, *Nuclear Instruments and Methods in Physics Research B*, **45** (1990), 315–321.

P.A. Mando, M.E. Fedi, N. Grassi and A. Migliori, Differential PIXE for investigating the layer structure of paintings, *Nuclear Instruments and Methods in Physics Research B*, **239** (2005), 71–76.

M. Mantler and M. Schreiner, X-ray fluorescence spectrometry in art and archaeometry, *X-ray Spectrometry*, **29** (2000), 3–17.

P. Maravelaki-Kalaitzaki, V. Zafiropoulos and C. Fotakis, Excimer laser cleaning of encrustation on Pentelic marble: procedure and evaluation of the effects, *Applied Surface Science*, **148** (1999), 92–104.

P. Maravelaki-Kalaitzaki, Black crusts and patinas on Pentelic marble from the Parthenon and Erechtheum (Acripolis, Athens): characterisation and origin, *Analytica Chimica Acta*, **532** (2005), 187–198.

L. Maritan, L. Nodari, C. Mazzoli, *et al.*, Influence of firing conditions on ceramic products: experiments study of clay rich in organic matter, *Applied Clay Science*, **31** (2006), 1–15.

G.A. Mazzocchin, F. Agnoli, S. Mazzocchin and I. Colpo, Analysis of pigments from Roman wall paintings found in Vincenza, *Talanta*, **61** (2003a), 565–572.

G.A. Mazzocchin, F. Agnoli and I. Colpo, Investigation of Roman age pigments found on pottery fragments, *Analytica Chimica Acta*, **478** (2003b), 147–161.

G.A. Mazzocchin, D. Rudello, C. Bragato and F. Agnoli, A short note on Egyptian blue, *Journal of Cultural Heritage*, 5 (2004), 129–133.

R.J. McNeil, in *Archaeological Chemistry III* (ed J.B. Lambert), Advances in Chemistry Series 205, American Chemical Society, Washington DC (1984), pp 255–269.

M. Milazzo, Radiation applications in art and archaeometry. X-ray fluorescence applications to archaeometry. Possibility of obtaining non-destructive quantitative analysis, *Nuclear Instruments and Methods in Physics Research B*, **213** (2004), 683–692.

P. Mirti, X-ray microanalysis discloses the secrets of ancient Greek and Roman potters, *X-ray Spectrometry*, **29** (2000), 63–72.

L. Moens, A. von Bohlen and P. Vandenabeele, X-ray fluorescence, in *Modern Analytical Methods in Art and Archaeology* (eds E. Ciliberto and G. Spoto), John Wiley & Sons Inc., New York (2000), pp 55–79.

P. Moioli and C. Seccaroni, Analysis of art objects using a portable X-ray fluorescence spectrometer, *X-ray Spectrometry*, **29** (2000), 48–52.

H. Mommsen, T. Beier, H. Dittmann, *et al.*, X-ray fluorescence analysis with synchrotron radiation on the inks and paper of Incunabla, *Archaeometry*, **38** (1996), 347–357.

A. Moropoulou, A. Bakolas and K. Bisbikou, Characterisation of ancient, byzantine and later historic mortars by thermal and x-ray diffraction techniques, *Thermochimica Acta*, **269–270** (1995), 779–795.

N. Na, Q. Ouyang, H. Ma., *et al.*, Non-destructive and in situ identification of rice paper, seals and pigments by FTIR and XRF spectroscopy, *Talanta*, **64** (2004), 1000–1008.

C. Neelmeijer, W. Wagner and H.P. Schramm, Depth resolved ion beam analysis of objects of art, *Nuclear Instruments and Methods in Physics Research B*, **118** (1996), 338–345.

C. Neelmeijer, I. Brissaud, T. Calligaro, *et al.*, Paintings – a challenge for XRF and PIXE analysis, *X-ray Spectrometry*, **29** (2000), 101–110.

C. Neelmeijer and M. Mader, The merits of particle induced X-ray emission in revealing painting techniques, *Nuclear Instruments and Methods in Physics Research B*, **189** (2002), 293–302.

A.M.B. Olsson, T. Calligaro, S. Colinart, *et al.*, Micro-PIXE analysis of an ancient Egyptian papyrus: identification of pigments used for the 'Book of the Dead ', *Nuclear Instruments and Methods in Physics Research B*, **181** (2001), 707–714.

P.M. O'Neill, D.C. Creagh and M. Sterns, Studies of the composition of pigments used traditionally in Australian aboriginal bark paintings, *Radiation Physics and Chemistry*, **71** (2004), 841–842.

E. Pantos, W. Kockelmann, L.L. Chapman, *et al.*, Neutron and X-ray characterisation of the metallurgical properties of a 7th century BC Corinthian-type bronze helmet, *Nuclear Instruments and Methods in Physics Research B*, **239** (2005), 16–26.

E. Paparazzo, Surface and interface analysis of Roman lead pipe 'fistula': microchemistry of the soldering at the join, as seen by scanning Auger microscopy and X-ray photoelectron spectroscopy, *Applied Surface Science*, **74** (1994), 61–72.

E. Paparazzo and L. Moretto, Surface and interface microchemistry of archaeological objects studied with X-ray photoemission spectroscopy and scanning Auger microscopy, *Journal of Electron Spectroscopy and Related Phenomena*, **76** (1995), 653–658.

E. Paparazzo, L. Moretto, J.P. Northover, *et al.*, Scanning Auger microscopy and X-ray photoelectron spectroscopy studies of Roman bronzes, *Journal of Vacuum Science Technology A*, **13** (1995), 1229–1233.

E. Paparazzo and L. Moretto, X-ray photoemission study of soldered lead materials: relevance to the surface and interface chemical composition of Roman lead pipes fistulae, *Vacuum*, **49** (1998), 125–131.

E. Paparazzo and L. Moretto, X-ray photoelectron spectroscopy and scanning Auger microscopy studies of bronzes from the collections of the Vatican Museums, *Vacuum*, **55** (1999), 59–70.

E. Paparazzo, A.S. Lea, D.R. Bear and J.P. Northover, Scanning Auger microscopy studies of an ancient bronze, *Journal of Vacuum Science Technology A*, **19** (2001), 1126–1133.

L. Pappalardo, A portable PIXE system for the in situ characterisation of black and red pigments in Neolithic, copper age and bronze age pottery, *Nuclear Instruments and Methods in Physics Research B*, **150** (1999), 576–580.

L. Pappalardo, A.G. Karydas, N. Kotzamani, *et al.*, Complementary use of PIXE-alpha and XRF portable systems for the non-destructive and in situ characterisation of gemstones in museums, *Nuclear Instruments and Methods in Physics Research B*, **239** (2005), 114–121.

A.E. Pillay, C. Punyadeera, L. Jacobsen and J. Eriksen, Analysis of ancient pottery and ceramic objects using X-ray fluorescence spectrometry, *X-ray Spectrometry*, **29** (2000), 53–62.

A.M. Pollard and C. Heron, *Archaeological Chemistry*, Royal Society of Chemistry, Cambridge (1996).

C. Puyandeera, A.E. Pillart, L. Jacobson and G. Whitelaw, Application of XRF and correspondence analysis to provenance studies of coastal and inland archaeological pottery from the Mingeni River Area, South Africa, *X-ray Spectrometry*, **26** (1997), 249–256.

G. Querre, A. Bouquillon, T. Calligaro, *et al.*, PIXE analysis of jewels from a Achaemenid tomb (4th century BC), *Nuclear Instruments and Methods in Physics Research B* **109–110** (1996), 686–689.

S.J.B. Reed, *Electron Microprobe Analysis and Scanning Electron Microscopy in Geology*, 2nd edn, Cambridge University Press, Cambridge (2005).

F. Reiff, M. Bartels, M. Gastel and H.M. Ortner, Investigation of contemporary gilded forgeries of ancient coins, *Fresenius Journal of Analytical Chemistry*, **371** (2001), 1146–1153.

C. Remazeilles, V. Quillet, T. Calligaro, *et al.*, PIXE elemental mapping on original manuscripts with an external microbeam. Application to manuscripts damaged by iron–gall ink corrosion, *Nuclear Instruments and Methods in Physics Research B*, **181** (2001), 681–687.

M.P. Riccardi, P. Duminuco, C. Tomasi and P. Ferloni, Thermal, microscopic and X-ray diffraction studies on some ancient mortars, *Thermochimica Acta*, **321** (1998), 207–214.

C. Riotino, C. Sabbioni, N. Ghedini, *et al.*, Evaluation of atmospheric deposition on historic buildings by combined thermal analysis and combustion techniques, *Thermochimica Acta*, **321** (1998), 215–222.

J.C. Riviere, *Surface Analytical Techniques*, Oxford University Press, Oxford (1990).

J.D. Robertson, H. Neff and B. Higgins, Microanalysis of ceramics with PIXE and LA-ICP-MS, *Nuclear Instruments and Methods in Physics Research B*, **189** (2002), 378–381.

I. Rojas-Rodriquez, A. Herrara, C. Vazquez-Lopez, *et al.*, On the authenticity of eight Reales 1730 Mexican silver coins by X-ray diffraction and by energy dispersion spectroscopy techniques, *Nuclear Instruments and Methods in Physics Research B*, **215** (2004), 537–544.

J. Roque, T. Pradell, J. Molera and M. Vendrell-Saz, Evidence of nucleation and growth of metal copper and silver nanoparticles in lustre: AFM surface characterisation, *Journal of Non-Crystalline Solids*, **351** (2005), 568–575.

J. Roque, J. Molera, P. Sciau, *et al.*, Copper and silver nanocrystals in lustre lead glazes: development and optical properties, *Journal of the European Ceramic Society* (2006), **26** (2006), 3813–3824.

J.L. Ruvalcaba-Sil, M.A. Ontalba Salamanca, L. Manzanilla, *et al.*, Characterisation of pre-Hispanic pottery from Teotihuacan, Mexico, by a combined PIXE–RBS and XRD analysis, *Nuclear Instruments and Methods in Physics Research B*, **150** (1999), 591–596.

M. Schreiner, B. Fruhmann, D. Jembrih-Simburger and R. Linke, X-rays in art and archaeology: An overview, *Powder Diffraction*, **19** (2004), 3–11.

D.A. Scott, The application of scanning X-ray fluorescence microanalysis in the examination of cultural materials, *Archaeometry*, **43** (2001), 475–482.

D.A. Scott, The non-destructive investigation of copper alloy patinas, in *Non-Destructive Micro Analysis of Cultural Heritage Materials* (eds K. Jannsens and R. Van Grieken), Elsevier, Amsterdam (2004), pp 465–492.

D.A. Silva, H.R. Wenk and P.J.M. Monteiro, Comparative investigation of mortars from Roman Colosseum and cistern, *Thermochimica Acta*, **438** (2005), 35–40.

D.A. Skoog and J.J. Leary, *Principles of Instrumental Analysis*, Harcourt Brace College Publishers, Fort Worth (1992).

Z. Smit, P. Pelicon, G. Vidmar, *et al.*, Analysis of medieval glass by X-ray spectrometric methods, *Nuclear Instruments and Methods in Physics Research B*, **161–163** (2000), 718–723.

Z. Smit, K. Janssens, K. Proost and I. Langus, Confocal μ–XRF depth analysis of paint layers, *Nuclear Instruments and Methods in Physics Research B*, **219–220** (2004), 35–40.

Z. Smit, K. Janssens, E. Bulska, *et al.*, Trace element fingerprinting of façon-de-Venise glass, *Nuclear Instruments and Methods in Physics Research B*, **239** (2005a), 94–99.

Z. Smit, P. Pelicon, J. Simcic and J. Istenic, Metal analysis with PIXE: the case of Roman military equipment, *Nuclear Instruments and Methods in Physics Research B*, **239** (2005b), 27–34.

G. Spoto and E. Ciliberto, X-ray photoelectron spectroscopy and Auger electron spectroscopy in art and archaeology in *Modern Analytical Methods in Art and Archaeometry*, (eds E. Ciliberto and G. Spoto), John Wiley & Sons Inc., New York (2000), pp 363–404.

G. Spoto, E. Ciliberto, G.C. Allen, *et al.*, Chemical and structural properties of ancient metallic artefacts: multi-technique approach to the study of early bronzes, *British Corrosion Journal*, **35** (2000), 43–47.

M.C. Squarcialupi, G.P. Bernardini, V. Faso, *et al.*, Characterisation by XPS of the corrosion patina formed on bronzed surfaces, *Journal of Cultural Heritage*, **3** (2002), 199–204.

Z. Szokefalvi-Nagy, I. Demeter, A. Kocsonya and I. Kovacs, Non-destructive XRF analysis of paintings, *Nuclear Instruments and Methods in Physics Research B*, **226** (2004), 53–59.

M. Uda, K. Akiyoshi and M. Nakamura, Characterisation of ancient Chinese pottery decorated with a black pigment, *Nuclear Instruments and Methods in Physics Research B*, **150** (1999), 601–604.

M. Uda, M. Nakamura, S. Yoshimura, *et al.*, Amarna blue painted on ancient Egyptian pottery, *Nuclear Instruments and Methods in Physics Research B*, **189** (2002), 382–386.

M. Uda, A. Ishizaki, R. Satoh, *et al.*, 'Portable X-ray diffractometer equipped with XRF for archaeometry' *Nuclear Instruments and Methods in Physics Research B*, **239** (2005), 77–84.

M. Ursic, M. Budnar, J. Simcic and P. Pelicon, The influence of matrix composition and ink layer thickness on iron gall ink determination by the PIXE method, *Nuclear Instruments and Methods in Physics Research B*, **247** (2006), 342–348.

S. Valls del Barrio, M. Garcia-Valles, T. Pradell and M. Vendrell-Saz, The red–orange patina developed on a monumental dolostone, *Engineering Geology*, **63** (2002), 31–38.

G. Vittiglio, S. Bichlmeier. P. Klinger, *et al.*, A compact μ–XRF spectrometer for (in situ) analyses of cultural heritage and forensic materials, *Nuclear Instruments and Methods in Physics Research B*, **213** (2004), 693–698.

J. Vodopivec, M. Budnar and P. Pelicon, Application of the PIXE method to organic objects, *Nuclear Instruments and Methods in Physics Research B*, **239** (2005), 85–93.

T.J. Wess, M. Drakopoulos, A. Snigirev, *et al.*, The use of small-angle X-ray diffraction studies for the analysis of structural features in archaeological samples, *Archaeometry*, **43** (2001), 117–129.

W.P. Whitney, Electron microprobe analysis of glaze/glass–ceramic interface reactions, *Journal of Non-Crystalline Solids*, **38–39** (1980), 687–692.

P. Wobrauschek, G. Halmetschlager, S. Zamini, *et al.*, Energy-dispersive X-ray fluorescence analysis of Celtic glasses, *X-ray Spectrometry,fm* **29** (2000), 25–33.

C.T. Yap and V. Vijayakumar, Principal component analysis of trace elements from EDXRF studies, *Applied Spectroscopy*, **44** (1990), 1080–1083.

K.N. Yu and J.M. Miao, Non-destructive analysis of Jingdezhan blue and white porcelains of the Ming Dynasty using EDXRF, *X-ray Spectrometry*, **25** (1996), 281–285.

K.N. Yu, Attribution of antique Chinese blue-and-white porcelain using energy dispersive X-ray fluorescence (EDXRF), in *Radiation in Art and Archaeometry* (eds D.C. Creagh and D.A. Bradley), Elsevier, Amsterdam (2000), pp 317–246.

B. Zhang, H.S. Cheng, B. Ma, *et al.*, PIXE and ICP-AES analysis of early glass unearthed from Xinjiang (China), *Nuclear Instruments and Methods in Physics Research B*, **240** (2005), 559–564.

J. Zhu, J. Shan, P. Qiu, *et al.*, The multivariate statistical analysis and XRD analysis of pottery at Xigongqiao site, *Journal of Archaeological Science*, **31** (2004), 1685–1691.

A. Zucchiatti, F. Cardoni, P. Prati, *et al.*, PIXE analysis of pottery from the recovery of a renaissance wreck, *Nuclear Instruments and Methods in Physics Research B*, **136–138** (1998), 893–896.

A. Zucchiatti, A. Bouquillon, J. Castaing and J.R. Gaborit, Elemental analysis of a group of glazed terracotta angels from the Italian renaissance as a tool for reconstruction of a complex conservation history, *Archaeometry*, **45** (2003), 391–404.

7

Mass Spectrometry

7.1 INTRODUCTION

Mass spectrometry (MS) is a widely used technique for studying the masses of atoms, molecules or molecular fragments [De Hoffmann and Stroobant, 2001; Gross, 2004]. The technique is a powerful analytical tool as it provides detailed qualitative and quantitative analyses on even very small samples. In the basic experiment, a gaseous sample is bombarded with high energy electrons which cause one or more electrons to be ejected on impact. A magnetic field is used to separate the ions by their masses. A number of MS techniques have been developed to deal with a range of sample types. Molecular MS is used to characterise a broad range of compounds. The surfaces of materials may also be studied using MS techniques: secondary ion MS has proved very useful for the study of heritage materials. A form of atomic MS, ICP–MS, has emerged in recent years for the identification of elements in a range of heritage objects.

7.2 MOLECULAR MASS SPECTROMETRY

The fundamental components of a mass spectrometer are a sample inlet, an ion source, a measurement system and an ion detector under a high vacuum [De Hoffmann and Stroobant, 2001; Evershed, 2000; Gross, 2004; Harris, 2003; Harwood and Claridge, 1997; Skoog and Leary, 1992]. A basic MS instrument is a single focussing mass spectrometer; Figure 7.1 illustrates the layout of such an instrument. Volatile samples can be introduced by directly inserting them into an inlet reservoir, but

Analytical Techniques in Materials Conservation Barbara H. Stuart
© 2007 John Wiley & Sons, Ltd

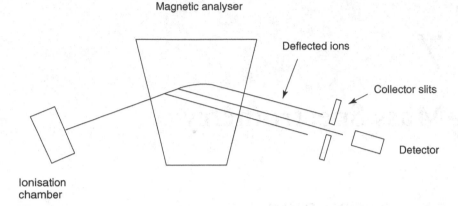

Figure 7.1   Schematic diagram of a single focussing mass spectrometer

less volatile substance can be introduced into an ionisation chamber. Electron impact (EI) ionisation is carried out using high energy electrons with energy of about 70 eV. The positively charged species produced are accelerated with a 4–8 keV potential and the beam is passed through a strong magnetic field at 90° to the beam. The ions are deflected according to their mass/charge (m/z) ratio and are focussed onto a detector by varying the field strength.

A double focussing mass spectrometer is also available, the layout of which is shown in Figure 7.2. The resolution of these instruments is enhanced by placing an electrostatic analyser in series with a magnetic analyser, which focuses the ion beam. Narrower slits are also used before

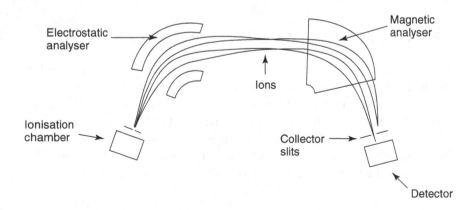

Figure 7.2   Schematic diagram of a double focussing mass spectrometer

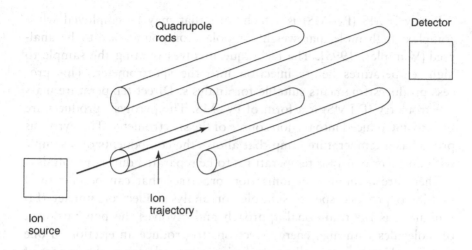

Figure 7.3   Schematic diagram of a quadrupole mass spectrometer

the detector to improve the resolution, allowing masses to be measured with an accuracy of 1–2 ppm.

Another type of spectrometer is the quadrupole mass spectrometer (Figure 7.3). Such instruments used four rods that produce a voltage along the path of the ions. When direct current and radio wave frequency voltages are applied to the rods, the ions oscillate. For each mass/charge ratio, a stable oscillation exists that allows the ions to travel the entire length of the rods without being lost, enabling only ions of a specific mass/charge ratio to be collected.

A time-of-flight (TOF) mass spectrometer can be used to measure high molecular weight substances. This type of instrument is based on the principle that lighter ions are accelerated faster than heavier ions, so will have a short time-of-flight over a specified distance. In a TOF spectrometer, the ions pass through a region with no field applied and the time to arrive at the detector is measured.

Mass spectrometers can be combined with other apparatus to further improve their analytical capabilities. Tandem MS or MS–MS links at least two stages of mass analysis. In MS–MS, a first spectrometer is used to isolate the component ions in a mixture. The ions are then introduced one at a time into a second spectrometer, where they are fragmented to produce a series of spectra. MS is commonly combined with gas or liquid chromatography to take advantage of the separation capabilities of these instruments. The use of these combined techniques is described in detail in Chapter 8.

Pyrolysis MS (Py–MS) is a technique that may be employed when complex, high molecular weight or polar compounds are to be analysed [Wampler, 1995]. The technique involves treating the sample to high temperatures before injection into the spectrometer. This process produces fragments suitable for analysis. Direct temperature mass spectrometry (DTMS) is a form of Py–MS. The pyrolysis products are directly introduced into the ion source of the spectrometer. The pyrolysis probe has a temperature ramp that allows the components of a sample with different pyrolysis temperatures to be separated during pyrolysis.

There are a number of ionisation processes that can be employed in MS to produce species suitable for analysis. Electron impact (EI) ionisation is the traditional approach and involves the bombardment of molecules with high energy electrons to produce an ejection of one or more electrons. As well as producing a radical cation, the impact of the electron may result in sufficient energy being produced to rupture a molecular bond and produce fragments. The positively charged ions are accelerated by an electric potential and passed through a magnetic field. The extent to which the ion is deflected depends upon the charge and the mass of the ion; measurement of the mass/charge ratio allows the mass of each charged species to be determined. The highest mass/charge value corresponds to the singly charged intact molecule known as the molecular ion of the parent ion.

Chemical ionisation (CI) is also a useful ionisation technique and involves the bombardment of the sample with positively charged atoms or molecules instead of electrons. This is a milder technique which enables molecules that are unable to cope with EI conditions to be investigated. Ammonia is commonly used as a source of primary ions. The addition of H^+ and NH_4^+ to a molecule produces quasi-molecular ion peaks at M+1 or M+18. As there is not a great excess of energy in this approach, fragmentation of ions is minimised. As such, a CI spectrum will usually have a strong base peak with fewer fragment peaks.

In field ionisation (FI) techniques, the ions are formed with a large electric field (up to 10^{10} Vm^{-1}). The sample is placed onto a fine wire on which carbon dendrites have been grown and along which the sample molecules are subjected to high potentials. Electrons from the sample molecules fill the orbitals of the metal and positive ions are produced. The ions are repulsed by the wire probe with little excess energy.

There are a number of desorption ionisation methods. In field desorption (FD) ionisation, a similar set-up to FI is used, but a current is passed through the wire and, as a consequence, thermal degradation may occur before the ionisation process is completed.

Laser desorption (LD) methods involve the use of a focussed laser beam to break weaker intermolecular bonds, such as hydrogen bonds, in a sample rather than breaking covalent bonds in a molecule. Matrix assisted laser desorption/ionisation (MALDI) is a particular LD method that is often used for macromolecules. In this technique, a sample solution is mixed with a solution of an UV-absorbing substance, such as 2,5-dihydroxybenzoic acid, which is the 'matrix'. Evaporation of the solution results in a mixture of fine crystals of the matrix and sample. When exposed to the laser, the matrix vaporises and passes into the gas phase and carries the sample. The presence of the matrix minimises association between the sample molecules and provides ionic species that transfer charge to the sample. After the ions expand into the source, a voltage pulse expels the ions into the spectrometer. MALDI is often used with a TOF spectrometer and provides excellent resolving power and mass accuracy.

Electrospray ionisation (ESI) is a milder technique that passes the sample solution through a fine needle at an electrical potential of about 4 kV. This allows the sample to be taken straight from a solution to ions in a gas. The solution is dispersed into a mist containing ionised droplets and the solvent is rapidly lost. An aerosol of ionised sample is then desorbed into the gas phase, which is then analysed by the spectrometer. The soft nature of ESI ionisation allows the structures of large molecules, such as proteins, to avoid decomposition.

Fast atom bombardment (FAB) sources are also useful for the study of high molecular weight samples. In FAB, the sample is often contained in a glycerol matrix, which favours ion formation, and is bombarded with energetic xenon or argon atoms. Ionisation occurs because of the transfer of the translational energy from the fast xenon atoms. Fragmentation is reduced by the minimal effect of vibrational energy. Another mild ionising technique for surfaces is secondary ion mass spectrometry (SIMS). This technique is described in Section 7.3 and applications of this technique are treated separately.

The appearance of a mass spectrum is illustrated in Figure 7.4. The spectrum is a plot of the relative abundance of the ions versus the m/z ratio. The base peak is the most abundant ion observed and is fixed at 100 % relative abundance. Positive ions are usually studied, but negative ions can also be examined. When EI ionisation is used, the most intense peak observed in the spectrum is due to the molecular ion ($M^{+\bullet}$), where the molecule under study has lost an electron to form a cation radical

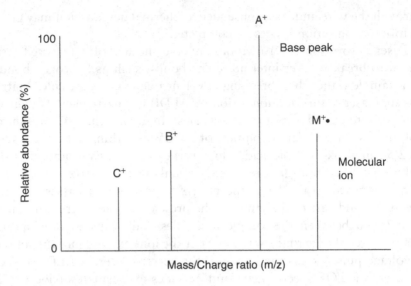

Figure 7.4 Schematic diagram of a mass spectrum

species. Any fragment ions produced will show peaks at lower mass values. The results are viewed as a series of peaks in a pyrogram in which the intensity of the compound detected is plotted as a function of time. The intensity is often quoted as the total ion current (TIC). The time axis can show the actual time, but a scan number is commonly used on this scale. For each peak in the pyrogram, a mass spectrum can be obtained.

If softer ionisation techniques such as CI are used, then other species such as MH^+ or MNH_4^+ are observed, depending on the technique used. The appearance of fragment ions in the spectrum provides help-ful information about the molecular structure as fragmentation usually produces specific stable fragments. Fragmentation often involves the breaking of one or two bonds and common fragment ions are observed in spectra. The interpretation of often complex spectra is aided by the collection of the details of common ions. Database libraries are available for use for MS enabling the peaks in a spectrum to be analysed.

The presence of isotopes should be considered when interpreting spectra and can, in fact, be exploited. Isotope ratio mass spectrometry (IRMS) is a specialist MS technique that is designed to measure isotopic ratios for elements including carbon, hydrogen, nitrogen, oxygen and sulfur [Benson *et al.*, 2006].

7.2.1 Paintings

MS techniques can be used to characterise the various components found in paintings. Materials used as binders, including polymers, oils and proteinaceous substances, can be identified using these techniques [Boon and Learner, 2002; Hynek *et al.*, 2004; Learner, 2004; Peris-Vincente, 2005; Tokarski *et al.*, 2006; van den Berg, 2002; van den Brink *et al.*, 2001; 2000]. Protein-based binders can be identified by determining the relative amount of amino acids released by an acid hydrolysis procedure. One study has demonstrated that the amino acids of proteinaceous binders, including gelatin, casein and egg, can be discriminated using direct infusion MS with positive ion ESI [Peris-Vincente *et al.*, 2005]. The results were analysed using LDA.

MS can also be employed to identify lipid binders. DTMS has been utilised for the characterisation of the drying oils used in oil paints [van den Berg, 2002]. Free fatty acids, cross-linked material, metal carboxylates and triacylglycerols may be differentiated by their mass spectra. MS is also effective for examining degraded lipids [van den Berg, 2002; van den Berg *et al.*, 2001; van den Brink *et al.*, 2001]. In Figure 7.5 the DTMS spectra of fresh and heat-treated linseed oil are illustrated [van den Berg *et al.*, 2004]. The inserts show the TIC. The main features of the spectrum of fresh linseed oil are that it shows fragments due to intact molecular ions (m/z 848–884), fragments due to the loss of one or two unsaturated fatty acids (m/z 570–620 and 310–344, respectively) and ions of carbon 16 and 18 fatty acids (m/z 239, 258–267). On first sight, the spectrum of the heat-treated oil appears very similar. However, inspection of the lower masses of each cluster reveals that heating results in the disappearance of highly unsaturated molecules relative to the less unsaturated species. This observation supports that cross-linking is induced on degradation.

DTMS has also been used to identify synthetic binders including acrylic, alkyds and PVA [Boon and Learner, 2002; Learner, 2004]. The characteristic mass fragments of the pyrolysis products that are observed in the mass spectra of some acrylic resins are listed in Table 7.1. A DTMS spectrum of a typical alkyd resin will show mass fragments associated with phthalic anhydride at m/z = 104, 148 and 76. Modified alkyds will show additional peaks which can be associated with the additional monomer (m/z = 43, 45 and 60) in a DTMS spectrum due to acetic acid, a major pyrolysis product resulting from side group elimination.

MS methods can be used to identify a range of pigments as these compounds produce very characteristic mass spectra [Boon and Learner, 2002; Grim and Allison, 2003; Kuckova *et al.*, 2005; Learner, 2004;

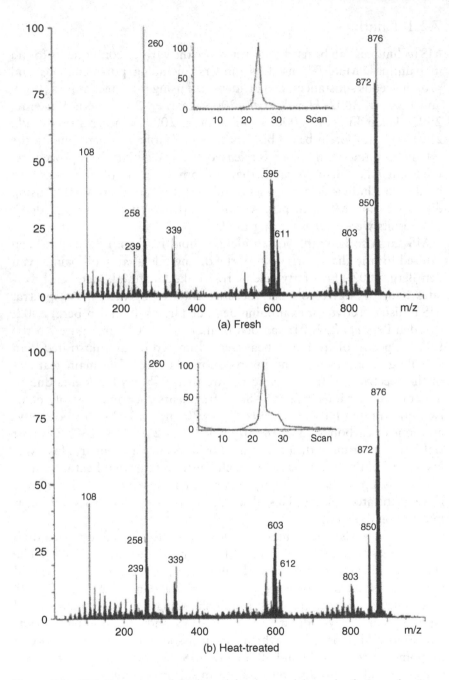

Figure 7.5 DTMS spectra of fresh and heat-treated linseed oils [van den Berg, 2002. Reproduced by permission of Wiley-VCH]

Table 7.1 DTMS results for some acrylic resins

Polymer	Mass/Charge ratio	Species
poly(butyl methacrylate)	87, 69, 56, 41	BMA
poly(ethyl methacrylate-methyl acrylate)	69, 114, 86, 41, 99	EMA
	55, 85	MA
	140, 113, 112	MA dimer
	126, 127, 155, 169, 95	EMA–MA dimer
	167, 194, 226, 227, 198	MA timer
	200, 149, 241, 255, 181, 180, 141, 208, 152, 153, 121	MA–MA–EMA trimer
poly(ethyl acrylate-methyl methacrylate)	55, 99	EA
	69, 100, 41, 85	MMA
	114, 143, 142, 115	EA sesquimer
	154, 155, 98, 126, 99, 127	EA dimer
	140, 168, 111, 95	EA–MMA
	255, 208, 254, 134, 226, 181, 180, 152, 153, 135, 106	EA trimer
	200, 222, 269, 167, 166, 194, 227	EA–EA–MMA trimers
poly(n-butyl acrylate-methyl methacrylate)	56	Butene
	55, 73	nBA
	100, 69, 41, 99, 85, 39	MMA
	115, 171	nBA sesquimer
	127, 126, 98	nBA dimer
	112, 141	nBA–MMA dimer
	236, 311, 181, 180, 310, 153, 152, 134, 384	nBA–nBA–MMA trimers
	256, 228, 250, 283, 282, 195, 194, 325, 356	nBA–nBA–MMA trimers

Maier *et al.*, 2004]. One study, for example, used MALDI and ESI–MS to analyse carminic acid in linseed oil [Maier *et al.*, 2004]. The use of these soft ionisation techniques in the negative ion mode proved to be effective for detecting the presence of carminic acid in the presence of binding media. MS can also be employed to pigment mixtures. For instance, a LDMS study of a commercial green pigment demonstrated that the pigment was a mixture of Prussian blue and lead chromate [Grim and Allison, 2003].

Paint varnishes can also be characterised using MS [Dietemann *et al.*, 2001; Scalarone *et al.*, 2003, 2005]. DTMS with EI and CI has been used to characterise five natural terpenoid resins: dammar, mastic, colophony, Manila copal and sandarac. These resins are mainly composed of cyclic di- and triterpenoids and information about the volatile and high molecular weight components are obtained from the spectra. Information

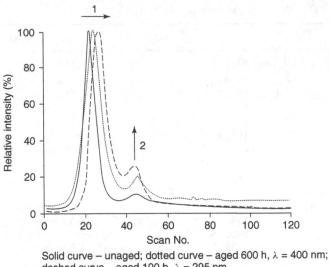

Figure 7.6 DTMS TICs for dammar. Solid curve – unaged; dotted curve – aged 600 h, λ = 400 nm; dashed curve – aged 100 h, λ = 295 nm. Journal of Mass Spectrometry, Scalarone *et al.*, 2003, John Wiley & Sons Limited. Reproduced with permission

about degradation reactions can also be obtained. The TICs for unaged and light-aged dammar resin are illustrated in Figure 7.6. The changes observed are a consequence of oxidation and cross-linking. The shift of the volatilisation peak (1) to higher scan numbers, and thus towards high temperature, provides evidence of the formation of oxidised tri- and diterpenoids. The increasing height of the pyrolysis peak (2) provides evidence of cross-linking reactions.

MS techniques can also be employed to examine the effects of laser cleaning of paints [Castillejo *et al.*, 2002; Teule *et al.*, 2003]. MS can be used to examine any degradation in binding media or pigments as a result of laser exposure.

7.2.2 Written Material

As for paintings, MS can be utilised to identify pigments used in illuminated manuscripts [Grim and Allison, 2004]. LDMS has been used to examine the colourants used in two manuscripts reported to be 17th and 19th century documents. This approach provides useful information about black inks, which are traditionally difficult to characterise chemically as they are complex mixtures. A mass spectrum can provide a more detailed analysis of such inks.

7.2.3 Natural Materials

Natural resins, gums and waxes can be characterised using MS [Garnier *et al.*, 2002; Regert and Rolando, 2002; Wright and Wheals, 1987]. While such materials are commonly examined using gas chromatography often coupled to a mass spectrometer (Chapter 8), it is also possible to gain information by using particular MS techniques which avoid the need for extraction and derivatisation procedures. This is particularly valuable when the material of interest is only available in minute quantities. Diterpenoid materials, including pine resin and birch bark tar, and beeswax are easily detected by comparison with the mass spectra obtained for contemporary reference samples.

7.3 SECONDARY ION MASS SPECTROMETRY

Secondary ion mass spectrometry (SIMS) is a technique involving the analysis of ions that are generated by the interaction of a ion beam in the keV range with a solid sample in a vacuum [Adriaens, 2000; Darque-Ceretti and Aucouturier, 2004; Dowsett and Adriaens, 2004]. The impact of the primary ion initiates a phenomenon known as a collision cascade in which the atoms in an approximately 10^3 nm^3 volume around the ion are in rapid motion. Some of the energy returns to the surface to break bonds and produce atomic and molecular species. A proportion of these are ionised during the emission process and are known as secondary ions. The secondary ions are collected by an electric field and focussed into a mass spectrometer.

There are two types of SIMS techniques. In static SIMS (SSIMS) each molecule is hit by only one primary ion so only the outer layer at the surface is analysed. Inert gas ions or metal ions such as gallium (Ga^+) and indium (In^+) are used. This technique provides compositional analysis for organic and inorganic species at the surface. This approach is regarded as virtually non-destructive as only the top one or two atomic layers are affected. Dynamic SIMS is used for depth profiling and bulk analysis. In this technique an intense beam of ions is used and rapidly erodes the surface. The beam can be rastered across a set area, removing one or more monolayers during each scan. When the process is repeated, a profile of the composition as a function of depth is created.

Sample preparation for SIMS is straightforward. The experiment can often be carried out directly on a solid surface. However, the results are better if the surface is flat and the sample must fit in the sampling compartment, the size of which will depend on the instrument type. The sample also needs to be conductive. This is not a problem for metal specimens, but for ceramics and glasses, electron flooding may be applied.

7.3.1 Metals

As metals are highly conductive materials and are amenable to high vacuums, they are very suitable for SIMS analysis. Historic metals, however, may cause some problems due to the presence of non-conducting patinas and corrosive products. Despite this a number of studies have been reported of successful SIMS analysis of heritage metals [Adriaens, 2000; Allen et al., 1995; Darque-Ceretti and Aucouturier, 2004; De Ryck et al., 2003; Dowsett and Adriaens, 2004; Dowsett et al., 2005; Griesser et al., 2005; Hallett et al., 2003; Kraft et al., 2004; Mayerhofer et al., 2005; Reiff et al., 2001; Sokhan et al., 2003; Spoto, 2000; Wouters et al., 1991; Yeung et al., 2000]. Studies of gold, silver, copper alloys, tin and lead artefacts have been reported.

SIMS can be used for provenance studies of metal artefacts based on isotopic analyses. As SIMS is sensitive to the atomic mass of elements, it is well suited to isotopic analysis. The lead isotope composition of bronze artefacts has been used for provenance studies [Darque-Ceretti and Aucouturier, 2004; Yeung et al., 2000]. Lead has four naturally occurring isotopes (^{204}Pb, ^{206}Pb, ^{207}Pb and ^{208}Pb) and the variation in the isotopic abundance of these vary depending on the source. The variation is large enough to identify the origin of a lead-containing bronze based on the measurement of lead isotope ratios.

SIMS provides a useful method for examining corrosion layers on the surface of metals. This technique can readily identify species in phases and inclusions in metals and the distributions of major and minor elements provides an understanding of a corrosion mechanism. A number of studies have used SIMS as a tool for characterising the dark patina that is commonly observed on bronze objects [Adriaens, 2000; Darque-Ceretti and Aucouturier, 2004; De Ryck et al., 2003; Spoto, 2000; Wouters et al., 1991]. Compounds such as cuprite (Cu_2O) and copper chlorides can be identified using SIMS and the distribution of compounds on a corroded surface can be determined.

7.3.2 Paintings

SIMS is a promising technique for identifying paint pigments [Darque-Ceretti and Aucouturier, 2004; Dowsett and Adriaens, 2004; Keune and

Boon, 2004, 2005; Spoto, 2000; Van Ham *et al.*, 2005]. One advantage of this approach is the potential to characterise both inorganic and organic components with one analysis. SSIMS coupled with a TOF spectrometer can provide improved sensitivity [Van Ham *et al.*, 2005]. Work on pure pigments in pressed pellets has proved successful. The mass spectra obtained from an examination of verdigris are illustrated in Figure 7.7 [Van Ham *et al.*, 2005]. Numerous high m/z ions can be detected and used for molecular identification. The clear identification of the acetate component of the pigment enables verdigris to be distinguished from another commonly used green pigment, malachite. However, the application of SSIMS to pigments in embedded paint fragments only provided elemental information because of charge build-up and contamination by other components.

7.3.3 Glass

SIMS has been used to examine corrosion in historic glass [Adriaens, 2000; Anderle *et al.*, 2006; Bertoncello *et al.*, 2002; Dal Bianco *et al.*, 2004, 2005; Daolio *et al.*, 1996; Darque-Ceretti and Aucouturier, 2004; Dowsett and Adriaens, 2004; Fearn *et al.*, 2005; Rogers *et al.*, 1993; Schreiner, 1991; Schreiner *et al.*, 1988; Schreiner *et al.*, 1999; Spoto, 2000]. As the depth resolution of SIMS is 50–100 Å, this technique is suited to the study of weathering of ancient glass. SIMS has been used to characterise the deterioration process in studies of medieval glass in Austria [Schreiner, 1991]. SIMS was used to quantitatively determine the elements present at varying depths into the surface of the glass. The surface of the glass sample was coated with a thin layer of gold in order to minimise the amount of charge. The SIMS depth profile for the glass is illustrated in Figure 7.8. Layer 1 is the gold layer, layer 2 is the outermost layer containing the corrosion products, layer 3 is the leached layer and layer 4 is the bulk material. The sputtering time is converted into a depth scale by measuring the depth of the crater created during the process. Elements such as potassium, calcium and barium are depleted in the leached layer compared to the bulk phase. Sodium appears constant in these layers and hydrogen is increased in the leached layer. All the elements are increased in the outer layer where the corrosion products are formed. It is well known that the network modifiers, such as alkaline elements, diffuse out of the glass as a result of weathering, and this is clearly shown by the SIMS results of this study of medieval glass.

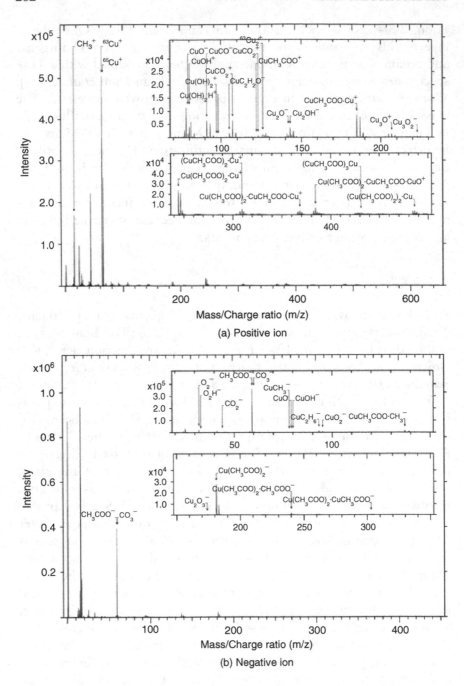

Figure 7.7 Positive and negative ion mass spectra of verdigris. With kind permission of Springer Science and Business Media, Analytical and Bioanalytical Chemistry, Vol 383, 2005, 991–997, R Van Ham *et al.*, figure 3

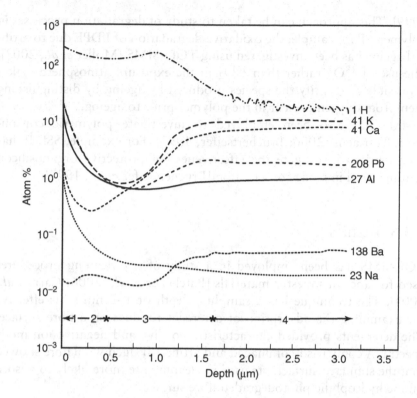

Figure 7.8 Depth distribution of elements in a medieval glass determined using SIMS. With kind permission of Springer Science and Business Media, Microchimica Acta, Vol 104, 1991, 255–264, M Schreiner, figure 4

7.3.4 Stone

SIMS can be used to date obsidian artefacts [Adriaens, 2000; Dowsett and Adriaens, 2004; Liritikis *et al.*, 2004; Patel *et al.*, 1998; Spoto, 2000]. The fact that artefacts made of obsidian tend to have smooth surfaces aids in the determination of surface composition using SIMS. Dating can be made using this technique because the thickness of a hydration layer on the surface layer of obsidian depends on how long the surface has been exposed to the atmosphere. The use of SIMS to measure the penetration depth of nitrogen in the hydrated surface also enables obsidian objects to be dated.

7.3.5 Synthetic Polymers

SIMS is a well established technique for the examination of polymer surfaces [Adriaens *et al.*, 1999; Briggs *et al.*, 1998; Van Vaeck *et al.*,

1999]. This approach can be taken to study of degradation processes in polymers. For example, the oxidative degradation of LDPE due to artificial ageing has been investigated using TOF–SIMS [Moller *et al.*, 2003]. The use of $^{18}O_2$ rather than $^{16}O_2$ in the exposure atmosphere makes it possible to identify the species produced by ageing by distinguishing them from oxygen present in the polymer prior to ageing.

SIMS has also been employed to investigate polymeric consolidants [Adriaens, 2000; Bruchertseifer, 1997]. For example, SSIMS has been used to investigate the effectiveness of protective organosilicon coatings used in stone conservation [Bruchertseifer *et al.*, 1997].

7.3.6 Textiles

TOF-SIMS has been employed in a study of the cleaning procedures used for ancient tapestry materials [Batcheller *et al.*, 2006; Carr *et al.*, 2004]. The technique has a sampling depth of 1–2 nm so is effective for examining the adsorption of detergent residues on a fibre surface. The detergents provided characteristic positive and negative ion mode spectra. A comparison of unaged and artificially aged wool fibre showed that the standard surfactants used in cleaning are more likely to adsorb on the hydrophilic photodegraded fibre surface.

7.4   ATOMIC MASS SPECTROMETRY

In atomic MS techniques the sample is atomised [Jeffries, 2004; Skoog and Leary, 1992; Young and Pollard, 2000] and the atoms converted to a stream of atoms, usually singly charged positive ions, which are separated on the basis of their m/z ratio. The most useful form of atomic MS is ICP–MS, which has become one of the most important techniques for elemental analysis. In most ICP–MS instruments, a liquid sample is dispersed into a stream of gas (usually argon or helium) and injected into the core of an ICP (described in Chapter 5). The layout of an ICP–MS instrument is illustrated in Figure 7.9. The sample is heated, vaporised and then ionised. The ions are passed through a core into a reduced pressure region. A number of ions pass through a second aperture into a high vacuum. Most instruments use a quadrupole mass analyser, through which the ions are detected. The transmitted ions are collected at a detector where the m/z values correspond to the element's natural isotope. The method is quantitative because the number of ions detected

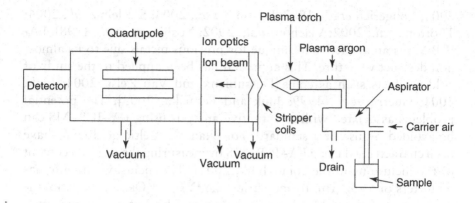

Figure 7.9   Schematic diagram of an ICP mass spectrometer

for each isotope will depend on the concentration of the element in the sample.

Milligram quantities of samples can be analysed by ICP–MS, and samples as small as 1 mg can be examined with sensitive instruments. A sensitivity of 1 part in 10^{15} is feasible using ICP–MS [Jeffries, 2004]. Contamination-free samples are extracted and completely dissolved in acid, usually nitric acid, hydrochloric acid or hydrofluoric acid. Metals are the simplest to dissolve in acids, while glasses and ceramics require more aggressive techniques involving higher temperatures and pressures.

Although most ICP–MS instruments are designed to analyse liquids, some instruments are designed to analyse solids via laser ablation (LA). LA–ICP–MS uses a laser to remove material from a small area of the sample. The particles produced are taken away from the sample in a flow of gas and injected into the plasma [Jeffries, 2004]. Although the ablation is destructive as such, the changes are on a microscope scale so visual destruction of an artefact is avoided. No sample preparation is required for LA–ICP–MS, but the sampling area should be reasonably flat. LA–ICP–MS provides good precision for samples of 1 ppm or greater using a 100 μm crater.

7.4.1 Metals

ICP–MS has been a popular technique for characterising metal artefacts. Manufacturing technology and provenance studies have been carried out on gold, silver, lead and copper alloy objects [Coustures *et al.*, 2003; Dussubieux and Van Zelst, 2004; Guerra *et al.*, 1999; Guerra and Calligaro, 2003; Hall *et al.*, 1998; Junk, 2001; Junk and Pernicka,

2003; Longerich *et al.*, 1987; Ponting *et al.*, 2003; S. Klein *et al.*, 2004; Thorton *et al.*, 2002; Vlachou *et al.*, 2002; Yoshinaga *et al.*, 1998]. LA–ICP–MS is an attractive technique for precious metals due to its almost non-destructive nature. This approach has been applied to the study of gold artefacts such as coins [Dussubieux and Van Zelst, 2004; Junk, 2001; Guerra *et al.*, 1999; Junk and Pernicka, 2003]. The potential problems associated with quantitative analysis using LA–ICP–MS can be avoided by use of a standard. For example, Celtic gold coins have been characterised using LA–ICP–MS by measuring the osmium content of the inclusions in the coins [Junk, 2001]. The inclusions usually has diameters of 5–30 μm. By measuring the ^{187}Os/^{188}Os isotope ratio, the ore source of the gold can be identified, as the isotopic composition depends on the source.

7.4.2 Glass

ICP–MS is an attractive technique for the examination of historic glass due to its practically non-destructive properties. The technique can effectively identify a broad range of elements found in such glasses [Brianese *et al.*, 2005; Casellato *et al.*, 2003; Gratuze *et al.*, 2001; Hartmann *et al.*, 1997; Smit *et al.*, 2005; Schulthesis *et al.*, 2004]. About 100–250 mg of sample is required if an ICP–MS method involving a solution is utilised. To determine the composition of the bulk glass the weathered outer layer should be removed before dissolution.

LA–ICP–MS is suitable for use when studying the surfaces of historic glass [Gratuze *et al.*, 2001; Schulthesis *et al.*, 2004; Smit *et al.*, 2005]. As LA–ICP–MS has a 10 μm depth resolution, this approach will measure the damaged surface. LA–ICP–MS has been used to characterise a sample of Venetian millefiori glass produced around 1880 [Gratuze *et al.*, 2001]. The specimen was made of white opaque glass decorated with a fine layer of purple, brown and red glasses. Each coloured part of the specimen was analysed using LA–ICP–MS (as well as SEM–EDS) and the results are listed in Table 7.2. In addition to the elements listed in Table 7.2, more than 30 other elements were determined using LA–ICP–MS and it was proposed that provenance studies of glass objects can be made.

7.4.3 Ceramics

ICP–MS is a popular technique for characterising and identifying ceramics [Habicht-Mauche *et al.*, 2002, 2000; Kennett *et al.*, 2002; Li *et al.*,

Table 7.2 LA-ICP-MS analysis of a Venetian millefiori glass

Compound	Brown glass	Red glass	Purple glass	White glass
Na_2O	21.2	20.3	22.2	17.3
MgO	0.98	1.11	0.84	1.90
Al_2O_3	0.26	0.65	0.29	0.25
SiO_2	65.2	67.8	70.5	55.1
P_2O_5	0.053	0.11	0.049	0.065
Cl	0.20	0.40	0.29	0.17
K_2O	0.23	1.58	0.11	0.39
CaO	3.4	4.8	3.1	6.28
MnO	0.16	0.19	2.3	0.035
Fe_2O_3	0.55	2.9	0.17	0.56
SnO_2	0.002	0.0064	0.0010	0.006
Sb_2O_3	0.99	0.036	0.0041	3.57
PbO	6.7	0.16	0.0045	14.4
CuO	0.03	2.3	0.003	0.11
As_2O_3	0.022	0.008	0.004	0.035
TiO_2	0.016	0.036	0.016	0.020

2005; Mallory-Greenough et al., 1998; M. Klein et al., 2004; Neff, 2003; Robertson et al., 2002; Zucchiatti et al., 2003]. An example of the application of ICP–MS to ceramic identification involved studies of Song dynasty porcelains from various kilns in northern China [Li et al., 2005]. The porcelain bodies were cleaned with Milli-Q water in an ultrasonic water bath. About 50–100 mg of sample was digested with hydrofluoric acid and nitric acid under high pressure. The quantitative analysis of 40 trace elements using ICP–MS allowed visually similar Ding-style white porcelains from three different kilns to be easily differentiated.

LA–ICP–MS lends itself to the analysis of precious ceramics [Neff, 2003; Robertson et al., 2002]. This approach allows the individual components in heterogeneous ceramics and surface materials, such as pigments and glazes, to be characterised. Despite the advantages of sensitivity and the capability of examining very small regions, quantitative LA–ICP–MS is more of a challenge. However, if one or more elements can be determined independently, then these can act as internal standards. For instance, in one study, the calcium concentrations were determined by EMPA and then used as internal standards for calibrating the LA–ICP–MS signals [Longerich et al., 1996].

7.4.4 Stone

ICP–MS has been utilised in a number of provenance studies of obsidian [Bellot-Gurlet et al., 2005; Gratuze, 1999; Gratuze et al., 2001;

Pereira et al., 2001; Tykot, 1997]. ICP–MS with solution nebulisation can be employed with the obsidian sample crushed and dissolved in acids such as hydrofluoric and nitric. LA–ICP–MS is also very effective. A broad range of elements can be analysed and multivariate analysis, especially PCA, can be applied to the data produced.

Precious stones, including opals and sapphires, have been characterised using LA–ICP–MS [Erel et al., 2003; Guillong and Gunther, 2001]. The use of laser ablation is particularly useful for such materials, where a substantially non-destructive technique is preferable to an approach requiring dissolution of the sample. One LA–ICP–MS study of opals has demonstrated that it is possible to distinguish artificial and natural opals through examination of the elements present [Erel et al., 2003]. Hf^+ and Zr^+ ions are specific to artificial opals, while Al^+, Ti^+, Fe^+, Rb^+, SiC_2^+ and SiC_2H^+ are detected in natural opals. In addition, natural Australian opals can be distinguished by the presence of Sr^+, Cs^+ and Ba^+.

7.4.5 Paintings

LA–ICP–MS has been applied recently to the analysis of artists' paints from different manufacturers to identify variation between the elements [Smith et al., 2005]. Trace element patterns have been established for oil, watercolour and acrylic paint pigments. For example, in a comparison of four different manufacturers of cadmium yellow light, the manufacturers could be easily distinguished by differences in the relative concentrations of ^{49}Ti, ^{55}Mn, ^{82}Se, ^{91}Zr, ^{93}Nb, ^{120}Sn, ^{181}Ta and ^{208}Pb. LA–ICP–MS thus shows potential as a technique that enables of comparison of the paints used by a particular artist with those used in questioned works with minimal damage.

7.4.6 Written Material

Several studies have reported the use of ICP–MS for the examination of ink corrosion of historic manuscripts [Bulska and Wagner, 2004; Wagner et al., 1999; Wagner and Bulska, 2004]. LA-ICP-MS has been used to investigate the distribution of certain elements, including iron and copper, in text written with iron-gall ink on medieval manuscripts. Such elements are believed responsible for accelerating the degradation of the manuscript paper. The distribution patterns of iron and copper on the surface of the manuscript can be determined.

REFERENCES

A. Adriaens, L. Van Vaeck and F. Adams, Static secondary ion mass spectrometry (SSIMS). Part 2: Applications in Materials Science, *Mass Spectrometry Reviews*, **18** (1999), 48–81.

A. Adriaens, The role of SIMS in understanding ancient materials, in *Radiation in Art and Archaeometry* (eds D.C. Creagh and D.A. Bradley), Elsevier, Amsterdam (2000), pp 180–201.

G. Allen, I.T. Brown, E. Ciliberto and G. Spoto, Scanning ion microscopy (SIM) and secondary ion mass spectrometry (SIMS) of early Iron Age bronzes, *European Journal of Mass Spectrometry*, **1** (1995), 493–497.

M. Anderle, M. Bersani, L. Vanzetti and S. Perderzoli, State of art in SIMS (secondary ion mass spectrometry) applications to archaeometry studies, *Macromolecular Symposia*, **238** (2006), 11–15.

J. Batcheller, A.M. Hacke, R. Mitchell and C.M. Carr, Investigation into the nature of historical tapestries using time of flight secondary ion mass spectrometry (TOF-SIMS), *Applied Surface Science*, **19** (2006), 7113–7116.

L. Bellot-Gurlet, G. Poupeau, J. Salomon, *et al.*, Obsidian provenance studies in archaeology: a comparison between PIXE, ICP-AES and ICP-MS, *Nuclear Instruments and Methods in Physics Research B*, **240** (2005), 583–588.

S. Benson, C. Lennard, P. Maynard and C. Roux, Forensic applications of isotope ratio mass spectrometry – a review, *Forensic Science International*, **157** (2006), 1–22.

R. Bertoncello, L. Milanese, U. Russo, *et al.*, Chemistry of cultural glasses: the early medieval glasses of Monselice's Hill (Padova, Italy), *Journal of Non-Crystalline Solids*, **306** (2002), 249–262.

J.J. Boon and T. Learner, Analytical mass spectrometry of artists' acrylic emulsion paints by direct temperature resolved mass spectrometry and laser desorption ionisation mass spectrometry, *Journal of Analytical and Applied Pyrolysis*, **64** (2002), 327–344.

N. Brianese, U. Casellato, F. Fenzi, *et al.*, Medieval and renaissance glass technology in Tuscany. Part 4: the XIVth sites of Santa Cristina (Gambassi-Firenze) and Poggio Imperiale (Siena), *Journal of Cultural Heritage*, **6** (2005), 213–225.

D. Briggs, D.R. Clarke, S. Suresh and I.M. Ward, *Surface Analysis of Polymers by XPS and Static SIMS*, Cambridge University Press, Cambridge (1998).

C. Bruchertseifer, K. Stoppek-Langner, J. Grobe, *et al.*, Examination of organosilicon impregnating mixtures by static SIMS and diffuse reflectance FTIR, *Fresenius Journal of Analytical Chemistry*, **358** (1997), 273–274.

E. Bulska and B. Wagner, A study of ancient manuscripts exposed to iron-gall ink corrosion, in *Non-Destructive Microanalysis of Cultural Heritage Materials* (eds K. Janssens and R. van Grieken), Elsevier, Amsterdam (2004), pp 755–788.

C.M. Carr, R. Mitchell and D. Howell, Surface chemical investigation into the cleaning procedures of ancient tapestry materials. Part 1, *Journal of Materials Science*, **39** (2004), 7317–7325.

U. Casellato, F. Fenzi, P. Guerriero, *et al.*, Medieval and renaissance glass technology in Valdelsa (Florence). Part 1: raw materials, sands and non-vitreous finds, *Journal of Cultural Heritage*, **4** (2003), 337–353.

M. Castillejo, M. Martin, M. Oujja, *et al.*, Analytical study of the chemical and physical changes induced by KrF laser cleaning of tempera paints, *Analytical Chemistry*, **74** (2002), 4662–4671.

M.P. Coustures, D. Beziat, F. Tollon, *et al.*, The use of trace element analysis of entrapped slag inclusions to establish ore-bar iron links: examples from two Gallo–Roman iron-making sites in France (Les Martys, Montagne Noire and Les Ferrys, Loitet), *Archaeometry*, 45 (2003), 599–613.

B. Dal Bianco, R. Bertoncello, L. Milanese and S. Barison, Glasses on the seabed: surface study of chemical corrosion in sunken Roman glasses, *Journal of Non-Crystalline Solids*, 343 (2004), 91–100.

B. Dal Bianco, R. Bertoncello, L. Milanese and S. Barison, Glass corrosion across the Alps: a surface study of chemical corrosion of glasses found in marine and ground environments, *Archaeometry*, 47 (2005), 351–360.

S. Daolio, C. Piccirillo, C. Pagura, *et al.*, Glass sample characterisation by secondary ion mass spectrometry, *Rapid Communications in Mass Spectrometry*, 10 (1996), 1286–1290.

E. Darque-Ceretti and M. Aucouturier, Secondary ion mass spectrometry. Application to archaeology and art objects, in *Non-Destructive Microanalysis of Cultural Heritage Materials* (eds K. Janssens and R. van Grieken), Elsevier, Amsterdam (2004), pp 397–461.

E. De Hoffmann and V. Stroobant, *Mass Spectrometry: Principles and Applications*, 2nd edn, John Wiley & Sons Ltd, Chichester (2001).

I. De Ryck, A. Adriaens, E. Pantos and F. Adams, A comparison of microbeam techniques for the analysis of corroded ancient bronze objects, *Analyst*, 128 (2003), 1104–1109.

P. Dietemann, M. Kalin, S. Zumbuhl, *et al.*, A mass spectrometry and electron paramagnetic resonance study of photochemical and thermal ageing of triterpenoid varnishes, *Analytical Chemistry*, 73 (2001), 2087–2096.

M. Dowsett and A. Adriaens, The role of SIMS in cultural heritage studies, *Nuclear Instruments and Methods in Physics Research B*, 226 (2004), 38–52.

M.G. Dowsett, A. Adriaens, M. Soares, *et al.*, The use of ultra-low-energy dynamic SIMS in the study of the tarnishing of silver, *Nuclear Instruments and Methods in Physics Research B*, 239 (2005), 51–64.

L. Dussubieux and L. Van Zelst, LA–ICP–MS analysis of platinum group elements and other elements of interest in ancient gold, *Applied Physics A*, 79 (2004), 353–356.

E. Erel, F. Aubriet, G. Finqueneisel and J.F. Muller, Capabilities of laser ablation mass spectrometry in the differentiation of natural and artificial opal gemstones, *Analytical Chemistry*, 75 (2003), 6422–6429.

R.P. Evershed, Biomolecular analysis by organic mass spectrometry, in *Modern Analytical Methods in Art and Archaeometry* (eds E. Ciliberto and G. Spoto), John Wiley & Sons Inc., New York (2000), pp 177–239.

S. Fearn, D.S. McPhail and V. Oakley, Moisture attack on museum glass measured by SIMS, *Physics and Chemistry of Glasses*, 46 (2005), 505–511.

N. Garnier, C. Cren-Olive, C. Rolando and M. Regert, Characterisation of archaeological beeswax by electron ionisation and electrospray ionisation mass spectrometry, *Analytical Chemistry*, 74 (2002), 4868–4877.

B. Gratuze, Obsidian characterisation by laser ablation ICP-MS and its application to prehistoric trade in the Mediterranean and the Near East: sources and distribution of obsidian within the Aegean and Anatolia, *Journal of Archaeological Science*, 26 (1999), 869–881.

B. Gratuze, M. Blet-Lemarquand and J.N. Barrandon, Mass spectrometry with laser sampling: a new tool to characterise archaeological materials, *Journal of Radioanalytical and Nuclear Chemistry*, 247 (2001), 645–656.

M. Griesser, R. Traum, K.E. Mayerhofer, *et al.*, Brown spot corrosion on historic gold coins and medals, *Surface Engineering*, 21 (2005), 385–392.

D.M. Grim and J. Allison, Identification of colourants as used in watercolour and oil paintings by UV laser desorption mass spectrometry, *International Journal of Mass Spectrometry*, 222 (2003), 85–99.

D.M. Grim and J. Allison, Laser desorption mass spectrometry as tool for the analysis of colourants: the identification of pigments used in illuminated manuscripts, *Archaeometry*, 46 (2004), 283–299.

J.H. Gross, *Mass Spectrometry: A Textbook*, Springer, Berlin (2004).

M.F. Guerra, C.O. Sarthe, A. Gondonneau and J.N. Barrandon, Precious metals and provenance enquiries using LA–ICP–MS, *Journal of Archaeological Science*, 26 (1999), 1101–1110.

M.F. Guerra and T. Calligaro, Gold cultural heritage objects: a review of studies of provenance and manufacturing technologies, *Measurement Science and Technology*, 14 (2003), 1527–1537.

M. Guilong and D. Gunther, Quasi non-destructive laser ablation inductively coupled plasma mass spectrometry fingerprinting of sapphires, *Spectrochimica Acta B*, 56 (2001), 1219–1231.

J.A. Habicht-Mauche, S.T. Glenn, H. Milford and A.R. Flegal, Isotopic tracing of prehistoric Rio Grande glaze paint production and trade, *Journal of Archaeological Science*, 27 (2000), 709–713.

J.A. Habicht-Mauche, S.T. Glenn, M.P. Schmidt, *et al.*, Stable lead isotope analysis of Rio Grande glaze paints and ores using ICP-MS: a comparison of acid dissolution and laser ablation techniques, *Journal of Archaeological Science*, 29 (2002), 1043–1053.

M.E. Hall, S.P. Brimmer, F.H. Li and L. Yablonsky, ICP–MS and ICP–OES studies of gold from a late Sarmatian burial, *Journal of Archaeological Science*, 25 (1998), 545–552.

K. Hallett, D. Thickett, D.S. McPhail and R.J. Chater, Application of SIMS to silver tarnish at the British Museum, *Applied Surface Science*, 203–204 (2003), 789–792.

D.C. Harris, *Quantitative Chemical Analysis*, 6th edn, W.H. Freeman, New York (2003).

G. Hartmann, I. Kappel, K. Grote and B. Arndt, Chemistry and technology of prehistoric glass from Lower Saxony and Hesse, *Journal of Archaeological Science*, 24 (1997), 547–559.

L.M. Harwood and T.D.W. Claridge, *Introduction to Organic Spectroscopy*, Oxford University Press, Oxford (1997).

R. Hynek, S. Kuckova, J. Hradilova and M. Kodicek, Matrix assisted laser desorption ionisation time-of-flight mass spectrometry as a tool for fast identification of protein binders in colour layers of paintings, *Rapid Communication in Mass Spectrometry*, 18 (2004), 1896–1900.

T.E. Jeffries, Laser ablation inductively coupled plasma mass spectrometry, in *Non-Destructive Microanalysis of Cultural Heritage Materials* (eds K. Janssens and R. van Grieken), Elsevier, Amsterdam (2004), pp 313–358.

S.A. Junk, Ancient artefacts and modern analytical techniques – usefulness of laser ablation ICPMS demonstrated with ancient gold coins, *Nuclear Instruments and Methods in Physics Research B*, 181 (2001), 723–727.

S.A. Junk and E. Pernicka, An assessment of osmium isotope ratios as a new tool to determine the provenance of gold with platinum group metal inclusions, *Archaeometry*, 45 (2003), 313–331.

D.J. Kennett, S. Sakai, H. Neff, R. Gossett and D.O. Larson, Compositional characterisation of prehistoric ceramics: a new approach, *Journal of Archaeological Science*, 29
(2002), 443–455.

K. Keune and J.J. Boon, Imaging secondary ion mass spectrometry of a paint cross
section taken from an early Netherlandish painting by Rogier can der Weyden,
Analytical Chemistry, 76 (2004), 1374–1385.

K. Keune and J.J. Boon, Analytical imaging studies clarifying the process of the darkening
of vermillion in paintings, *Analytical Chemistry*, 77 (2005), 4742–4750.

M. Klein, F. Jesse, H.U. Kasper and A. Golden, Chemical characterisation of ancient
pottery from Sudan by X-ray fluorescence spectrometry (XRF), electron microprobe analysis (EMPA) and inductively coupled plasma mass spectrometry (ICP–MS),
Archaeometry, 46 (2004), 339–356.

S. Klein, Y. Lahaye, G.P. Brey and H.M. Von Kaenel, The early Roman imperial AES
coinage II: tracing the copper sources by analysis of lead and copper isotopes – copper
coins of Augustus and Tiberius, *Archaeometry*, 46 (2004), 469–480.

G. Kraft, S. Flege, F. Reiff and H.M. Ortner, Investigation of contemporary forgeries of
ancient silver coins, *Mikrochimica Acta*, 145 (2004), 87–90.

S. Kuckova, I. Nemec, R. Hynck, *et al.*, Analysis of organic colouring and binding components in colour layer of art works, *Analytical and Bioanalytical Chemistry*, 382
(2005), 275–282.

T.J.S. Learner, *Analysis of Modern Paint*, The Getty Conservation Institute, Los Angeles
(2004).

B.P. Li, A. Greig, J.X. Zhao, *et al.*, ICPMS trace element analysis of Song dynasty
porcelains from Ding, Jiexiu and Guantai kilns, north China, *Journal of Archaeological
Science*, 32 (2005), 251–259.

I. Liritikis, M. Diakostamatiou, C.M. Stevenson, *et al.*, Dating of hydrated obsidian
surfaces by SIMS–SS, *Journal of Radioanalytical and Nuclear Chemistry*, 261 (2004),
51–60.

H.P. Longerich, B.J. Fryer and D.F. Strong, Trace analysis of natural alloys by inductively
coupled plasma mass spectrometry (ICP–MS): application to archaeological native
silver artefacts, *Spectrochimica Acta B* 1–2 (1987), 101–107.

H.P. Longerich, A.W. Jackson and D. Gunther, Laser ablation inductively coupled
plasma mass spectrometric transient signal data acquisition and analyte concentration
calculation, *Journal of Analytical Atomic Spectrometry*, 11 (1996), 899–904.

M.S. Maier, S.D. Parera and A.M. Seldes, Matrix-assisted laser desorption and electrospray ionisation mass spectrometry of carminic acid isolated from cochineal,
International Journal of Mass Spectrometry, 232 (2004), 225–229.

L.M. Mallory-Greenough, J.D. Greenough and J.V. Owen, New data for old pots:
trace element characterisation of ancient Egyptian pottery using ICP–MS, *Journal of
Archaeological Science*, 25 (1998), 85–97.

K.E. Mayerhofer, K. Piplits, R. Traum, *et al.*, Investigations of corrosion phenomena on
gold coins with SIMS, *Applied Surface Science*, 252 (2005), 133–135.

K. Moller, A. Jansson and P. Sjovall, Analysis of polymer oxidation using $^{18}O_2$ and
TOF-SIMS, *Polymer Degradation and Stability*, 80 (2003), 345–352.

H. Neff, Analysis of Mesoamerican plumbate pottery surfaces by laser ablation inductively coupled plasma mass spectrometry (LA–ICP–MS), *Journal of Archaeological
Science*, 30 (2003), 21–35.

S.B. Patel, R.E.M. Hedges and J.A. Kilner, Surface analysis of archaeological obsidian
by SIMS, *Journal of Archaeological Science*, 25 (1998), 1047–1054.

J. Peris-Vicente, E. Simo-Alfonso, J.V. Gimeno Adelantado and M.T. Domenech-Carbo, Direct infusion mass spectrometry as a fingerprint of protein-bonding media used in works of art, *Rapid Communications in Mass Spectrometry*, **19** (2005), 3463–3467.

C.E.B. Pereira, N. Miekeley, G. Poupeau and I.L. Kuchler, Determination of minor and trace elements in obsidian rock samples and archaeological artefacts by laser ablation inductively coupled plasma mass spectrometry using synthetic obsidian standards, *Spectrochimica Acta B*, **56** (2001), 1927–1940.

M. Ponting, J.A. Evans and V. Pashley, Fingerprinting of Roman mints using laser ablation MC–ICP–MS lead isotope analysis, *Archaeometry*, **45** (2003), 591–597.

M. Regert and C. Rolando, Identification of archaeological adhesives using direct inlet electron ionisation mass spectrometry, *Analytical Chemistry*, **74** (2002), 965–975.

F. Reiff, M. Bartels, M. Gastel and H.M. Ortner, Investigation of contemporary gilded forgeries of ancient coins, *Fresenius Journal of Analytical Chemistry*, **371** (2001), 1146–1153.

J.D. Robertson, H. Neff and B. Higgins, Microanalysis of ceramics with PIXE and LA–ICP–MS, *Nuclear Instruments and Methods in Physics Research B*, **189** (2002), 378–381.

P. Rogers, D. McPhail, J. Ryan and V. Oakley, A quantitative study of decay processes of Venetian glass in a museum environment, *Glass Technology*, **34** (1993), 67–68.

D. Scalarone, J. van der Horst, J.J. Boon and O. Chiantore, Direct temperature mass spectrometric detection of volatile terpenoids and natural terpenoid polymers in fresh and artificially aged resins, *Journal of Mass Spectrometry*, **38** (2003), 607–617.

D. Scalarone, M.C. Duursma, J.J. Boon and O. Chiantore, MALDI–TOF mass spectrometry on cellulosic surfaces of fresh and photo-aged di- and triterpenoid varnish resins, *Journal of Mass Spectrometry*, **40** (2005), 1527–1535.

G. Schultheis, T. Prohaska, G. Stingeder, *et al.*, Characterisation of ancient and art nouveau glass samples by Pb isotopic analysis using laser ablation coupled to a magnetic sector field inductively coupled plasma mass spectrometer (LA–ICP–SF–MS), *Journal of Analytical Atomic Spectrometry*, **19** (2004), 838–843.

M. Schreiner, M. Grasserbauer and P. March, Quantitative NRA and SIMS depth profiling of hydrogen in naturally weathered medieval glass, *Fresenius Journal of Analytical Chemistry*, **331** (1988), 428–432.

M. Schreiner, Glass of the past: the degradation and deterioration of medieval glass paintings, *Microchimica Acta*, **104** (1991), 255–264.

M. Schreiner, G. Woisetschlager, I. Schmitz and M. Wadsek, Characterisation of surface layers formed under natural environmental conditions on medieval stained glass and ancient copper alloys using SEM, SIMS and atomic force microscopy, *Journal of Analytical and Atomic Spectroscopy*, **14** (1999), 395–404.

D.A. Skoog and J.J. Leary, *Principles of Instrumental Analysis*, Harcourt Brace, Fort Worth (1992).

Z. Smit, K. Janssens, E. Bulska, *et al.*, Trace element fingerprinting of façon-de-Venise glass, *Nuclear Instruments and Methods in Physics Research B*, **239** (2005), 94–99.

K. Smith, K. Horton, R.J. Watling and N. Scoullar, Detecting art forgeries using LA–ICP–MS incorporating the in situapplication of laser-based collection technology, *Talanta*, **67** (2005), 402–413.

M. Sokhan, P. Gaspar, D.S. McPhail, *et al.*, Initial results on laser cleaning at the Victoria and Albert Museum, Natural History Museum and Tate Gallery, *Journal of Cultural Heritage*, **4** (2003), 230s–236s.

G. Spoto, Secondary ion mass spectrometry in art and archaeology, *Thermochimica Acta*, **365** (2000), 157–166.

R. Teule, H. Scholten, O.F van den Brink, *et al.*, Controlled UV laser cleaning of painted artworks: a systematic effect study on egg tempera paint samples, *Journal of Cultural Heritage*, **4** (2003), 209s–215s.

C.P. Thorton, C.C. Lamberg-Karlovsky, M. Liezers and S.M.M. Young, On pins and needles: tracing the evolution of copper-base alloying at Tepe Yahya, Iran, via ICP–MS analysis of common-place items, *Journal of Archaeological Science*, **29** (2002), 1451–1460.

C. Tokarski, E. Martin, C. Rolando and C Cren-Olive, Identification of proteins in renaissance paintings by proteomics, *Analytical Chemistry*, **78** (2006), 1494–1502.

R.H. Tykot, Characterisation of the Monte Arci (Sardinia) obsidian sources, *Journal of Archaeological Science*, **24** (1997), 467–479.

J.D.J. van den Berg, Analytical chemical studies on traditional linseed oil paints, PhD thesis, Universiteit van Amsterdam (2002).

O.F. van den Brink, G.B. Eijkel and J.J. Boon, Dosimetry of paintings: determination of the degree of chemical change in museum-exposed test paintings by mass spectrometry, *Thermochimica Acta*, **365** (2000), 1–23.

O.F. van den Brink, J.J. Boon, P.B. O'Connor, *et al.*, Matrix-assisted laser desorption/ionisation Fourier transform mass spectrometric analysis of oxygenated triglycerides and phosphatidylcholines in egg tempera paint dosimeters used for environmental monitoring of museum display conditions, *Journal of Mass Spectrometry*, **36** (2001), 479–492.

R. Van Ham, L. Van Vaeck, F. Adams and A. Adriaens, Feasibility of analysing molecular pigments in paint layers using TOF SSIMS, *Analytical and Bioanalytical Chemistry*, **383** (2005), 991–997.

L. Van Vaeck, A. Adriaens and R. Gijbels, Static secondary ion mass spectrometry (SSIMS). Part 1: Methodology and Structural Interpretation, *Mass Spectrometry Reviews*, **18** (1999), 1–47.

C. Vlachou, J.G. McDonnell and R.C. Janaway, The investigation of degradation effects in silvered copper alloy Roman coins (AD 250–350), *Conservation Science 2002*, pp 236–242.

B. Wagner, S. Garbos, E. Bulska and A. Hulanicki, Determination of iron and copper in old manuscripts by slurry sampling graphite furnace atomic absorption spectrometry and laser ablation inductively coupled plasma mass spectrometry, *Spectrochimica Acta B*, **54** (1999), 797–804.

B. Wagner and E. Bulska, One the use of laser ablation inductively coupled plasma mass spectrometry for the investigation of the written heritage, *Journal of Analytical Atomic Spectrometry*, **19** (2004), 1325–1329.

T.P. Wampler, *Applied Pyrolysis Handbook*, Marcel Dekker, New York (1995).

H.J. Wouters, L. Butaye and F.C. Adams, Application of SIMS in patina studies on bronze age copper alloys, *Fresenius Journal of Analytical Chemistry*, **342** (1991), 128–134.

M.M. Wright and B.B. Wheals, Pyrolysis mass spectrometry of natural gums, resins and waxes and its used for detecting such materials in ancient Egyptian mummy cases (cartonnages), *Journal of Analytical and Applied Pyrolysis*, **11** (1987), 195–211.

C.S.L. Yeung, R.W.M. Kwok, P.Y.K. Lam, *et al.*, SIMS analysis of lead isotope compositions in ancient Chinese metallic artefacts, *Surface and Interface Analysis*, **29** (2000), 487–491.

J. Yoshinaga, M. Yoneda, M. Morita and T. Suzuki, Lead in prehistoric, historic and contemporary study by ICP mass spectrometry, *Applied Geochemistry*, **13** (1998), 403–413.

S.M.M. Young and A.M. Pollard, Atomic Spectroscopy and Spectrometry, in *Modern Methods in Art and Archaeology* (eds E. Ciliberto and G. Spoto), John Wiley and Sons Inc., New York (2000), pp 21–53.

A. Zucchiatti, A. Bouquillon, J. Castaing and J.R. Gaborit, Elemental analyses of a group of glazed terracotta angels from the Italian renaissance as a tool for the reconstruction of a complex conservation history, *Archaeometry*, **45** (2003), 391–404.

8

Chromatography and Electrophoresis

8.1 INTRODUCTION

Chromatography is a group of techniques used to separate mixtures [Harris, 2003; Miller, 2004; Skoog and Leary, 1992]. Such techniques are important for separating complex mixtures, such as paint. They can be used to detect very small amounts of a component, making them excellent tools for characterising small samples of heritage material. There is a variety of chromatographic approaches, but all involve holding one phase in place while the other phase moves past. The mobile phase, the solvent moving through a column, can be a liquid or a gas. The stationary phase, the phase that does not move in the column, is commonly a liquid bonded to the inside of a capillary tube or onto the surface of the solid particles packed in a column. It is also possible to use solid particles as the stationary phase.

The gas or liquid that passes through the chromatography column is called the eluant and the fluid emerging from the column is called the eluate. The process itself is called elution. The columns can be packed with particles of the stationary phase or an open tubular column which is a narrow capillary with the stationary phase coated on the inside walls.

The solutes that are eluted from a chromatography column are observed with one of various detectors. The data is expressed as a chromatogram, which shows the detector responses as a function of elution time. The retention time (t_R) for each component is the time needed after the injection of the mixture into the column until the component

Analytical Techniques in Materials Conservation   Barbara H. Stuart
© 2007 John Wiley & Sons, Ltd

reaches the detector. The retention volume (V_R) is the volume of the mobile phase that is required to elute a solute from the column.

The effectiveness of a chromatographic technique depends on the separation of the compounds. One factor that reflects the separation is the difference in the elution times between peaks; the farther apart the peaks, the better the separation of compounds. Another factor that reflects separation is the broadness of the peaks. The wider the peaks, the poorer the separation of the compounds.

8.2   PAPER CHROMATOGRAPHY

Paper chromatography (PC) is the simplest chromatographic technique and, as the name implies, uses paper as the separation medium [Gasparic and Churacek, 1978; Zweig and Witaker, 1967]. A strip of filter paper is employed and a sample in solution is placed as a spot near one edge of the paper. The paper is placed in a suitable solvent and the solvent moves by capillary action through the paper and the sample. The sample components migrate at different characteristic rates that can be used to identify components. If the compounds are colourless, then reagents may be applied to reveal circular or elliptical spots. The method is simple, but there are several limitations. The sample cannot be volatile, the length of the migration path is limited and only qualitative analysis is possible. PC can also take some time and an experiment is often left for several hours to compete.

It is possible to identify compounds using PC by R_f values. The R_f value is:

$$R_f = \frac{\text{distance moved by compound}}{\text{distance moved by solvent front}} \qquad (8.1)$$

Although R_f values are referenced, experimental conditions do affect the values so a knowledge of the conditions used is required when comparing results.

PC has been superceded by other chromatographic techniques in terms of sensitivity, speed and the amount of information derived. However, as this technique is so straightforward and inexpensive, it can still be of assistance.

8.2.1 Paintings

Paint components have been examined using PC [Hey, 1958; Mills and Wilmer, 1952]. One early study identified various types of natural resins using paper chromatography [Mills and Wilmer, 1952]. The paper was wet with kerosene and a liquid phase of isopropanol/water/kerosene was used. The paper was sprayed with phenol in carbon tetrachloride and exposed to bromine vapour and the resulting colours varied according to the type and amount of resin present in the sample under examination. PC has also been applied to the analysis of proteins used in paint binders and adhesives [Brochwicz, 1970; Hey, 1958; Mills and White, 1994]. The protein needs to be hydrolysed into its component amino acids prior to analysis.

8.3 THIN LAYER CHROMATOGRAPHY

Thin layer chromatography (TLC) is another simple chromatographic method and superceded PC [Gasparic and Churacek, 1978; Striegel and Hill, 1996; Wall, 2005]. In this technique, the stationary phase (e.g. silica, alumina or cellulose) is coated as a thin layer onto a glass or plastic plate. The sample in solution is applied as a spot (0.1–1.0 μl) on the bottom of the plate, which is then placed in a tank containing the mobile phase, a mixture of solvents. The separation occurs as the solvent is carried up the plate by capillary action. The separated components are usually detected by spraying the plate with a suitable visualising agent such as a fluorescent dye or iodine vapours. The appearance of a TLC plate is illustrated in Figure 8.1. TLC has the advantage over PC in that

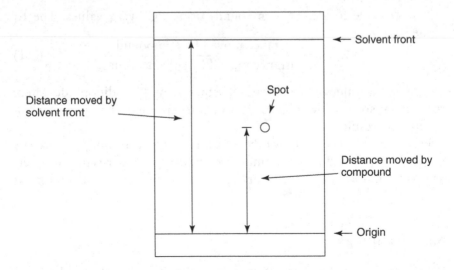

Figure 8.1 Separation on a TLC plate

the separation process is much faster and the separation zone is sharper. As with PC, R_f values for different compounds can be used to identify the components in unknown samples. TLC requires mg quantities of sample and precise quantitative analysis is not possible.

8.3.1 Paintings

Binding media containing carbohydrates can also be identified with the use of TLC [Kharbade and Joshi, 1995; Striegel and Hill, 1996; Vallance, 1997]. Hydrolysis with acid, such as hydrochloric acid or trifluoroacetic acid, is required to break down the carbohydrates into the component monosaccharides. Several TLC methods for carbohydrate analysis have been reported. One study used silica gel plates with an acetonitrile/water solvent system and an aminohippuric acid detection reagent [Striegel and Hill, 1996]. Another study has reported a technique using silica gel and a solvent of 1-propanol/water/ammonium hydroxide, with a reagent of *p*-anisidine phthalate [Kharbade and Joshi, 1995]. Commercially available monosaccharides including arabinose, fucose, galactose, galacturonic acid, glucose, glucuronic acid, mannose, rhamnose, ribose and xylose can be used as reference sugars.

Protein-based paint binders, such as gelatin, casein, egg white and egg yolk, can be identified using TLC via the determination of amino acids present in a sample [Broekman-Bokstijn *et al.*, 1970; Masschelein-Kleiner, 1974; Mills and White, 1994; Striegel and Hill, 1996; Tomek and Pechova, 1992; White, 1984]. The protein in a sample must be broken down into its component amino acids by hydrolysis. Liquid or vapour phase acid hydrolysis can be carried out under vacuum [Striegel and Hill, 1996]. Cellulose or silica gel may be used for the stationary phase for TLC of amino acids. Commercially available pure amino acids can be used for reference. One study used a butanol/acetic acid/water solvent system and a ninhydrin detection reagent to identify the proteins [Striegel and Hill, 1996]. In another study, the dansyl derivatives of amino acids were separated using polyamide plates and a water/formic acid solvent [Tomek and Pechova, 1992]. These derivatives can be detected by the fluorescence under UV light.

For identifying waxes by TLC, the samples should be dissolved in chloroform [Striegel and Hill, 1996]. Silica gel plates can be used in a petroleum ether/diethyl ether/acetic acid solvent system. To identify the components, an anisaldehyde detection reagent is applied, the plate heated and examined under UV light.

The identification of paint resins by TLC is more difficult because of the number of components present [Striegel and Hill, 1996]. Samples can be dissolved in ethyl acetate and applied to a silica gel plate in a benzene/methanol solvent system. It is recommended that the plates are developed two or three times in the same solvent system. An antimony trichloride detection reagent is used with UV light to visualise the pattern of fluorescence spots produced. One study used TLC to identify the lining adhesive on 'Water Lilies' by Claude Monet [Striegel and Hill, 1996]. Prior to analysis, the adhesive was thought to be a wax–resin mixture. TLC was carried out on a chloroform solution (1.3 mg in 50 µl) and 12 reference waxes were analysed on the same chromatographic plate. Beeswax was identified as the adhesive.

8.3.2 Textiles

TLC has been applied for identifying dyes in textiles of historical interest [Karadag and Dolen, 1997; Kharbade and Agrawal, 1985; Masschelein-Kleiner, 1967; Mills and White, 1994; Rai and Shok, 1981; Schweppe, 1979, 1989; Timar-Balazsy and Eastop, 1998; Wallert and Boynter, 1996]. TLC is useful if a natural dye contains several red coloured substances, and particularly so for identifying hydroxyflavone and hydroxyanthraquinone dyes. Polyamide powder is recognised as a good material for the stationary phase, with mobile phases usually involving mixtures of methanol with acids such as formic acid when separating natural dyes [Mills and White, 1994; Schweppe, 1979]. An example of the use of TLC was the investigation of a piece of satin made in Italy at the beginning of the 16th century [Schweppe, 1979]. TLC on polyamide was used to show that both carminic and kermesic acids were present in the dye used on the satin. As the only dye known to contain both these acids is Polish kermes, it was determined to be the dye used for this piece of satin.

8.4 GAS CHROMATOGRAPHY

Gas chromatography (GC) involves the introduction of gaseous or vaporised samples into a long column when the sample components are separated [Harris, 2003; Miller, 2004; Skoog and Leary, 1992]. The components are flushed sequentially from the column to a detector and each component may be identified by measuring the elution time. The output is a chromatogram which shows when each solute was eluted

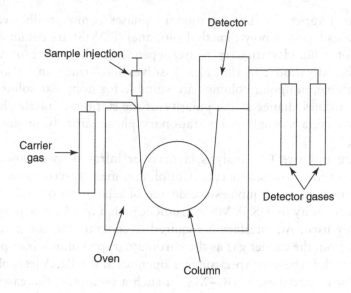

Figure 8.2 Schematic diagram of a gas chromatograph

and the peaks areas provide information about how much of each component is present. The identity of the component can be determined by comparing its elution time with a database of known compounds.

A schematic diagram of a gas chromatograph is illustrated in Figure 8.2. The volatile liquid or gaseous sample is injected via a septum, a rubber disk, into a heated port. The injection process may be automated. The vapour is carried through the column by a carrier gas such as helium, nitrogen or hydrogen. The column temperature is accurately controlled and temperature programming can be employed. In the past, most GC was carried out using packed columns. However, most analyses now employ narrow open tubular columns, commonly fused silica coated with polyimide. Different types of open tubular columns, which result in different performance properties, are available. Wall-coated columns have the liquid stationary phase on the inside wall of the column. Support-coated columns have the stationary phase coated onto a solid support attached to the inside wall of the column. Porous layer columns have a solid stationary phase on the inside of the wall of the column. The inside diameter of the column is usually 0.10–0.53 mm and typical lengths are 15–100 cm.

Gas liquid chromatography is the most widespread form of GC and is based upon the partition of the analyte between a gaseous mobile phase and a liquid phase immobilised on the surface of an inert solid.

There is a variety of liquid stationary phases commercially available and phases based in poly(dimethyl siloxane) (PDMS) are common. The choice of a liquid stationary phase depends on the nature of analyte under investigation and the 'like dissolves like' rule can be applied. For example, nonpolar columns are suitable for nonpolar solutes. The retention times change as the polarity of the stationary phase changes. For instance, a strongly polar stationary phase strongly retains polar solutes.

For quantitative GC analysis, a mass or infrared spectrometer can be used to identify components. Coupling a mass spectrometer with a gas chromatograph combines the degree of separation of GC with the analytical ability of MS. A MS instrument with a quadrupole mass filter is usually used. An interface is required to separate the compounds of interest from the carrier gas as the chromatograph column is at positive pressure while the mass spectrometer operates at 0.1 Pa. A jet molecular separator can be used in GC–MS. In such a separator, the gases from the GC pass from an outlet onto a collector nozzle a short distance from the outlet. The lighter components of the gas diffuse more rapidly away from the jet. Thus, the gas is not collected while the heavier components pass to the spectrometer for analysis.

In gas chromatography – infrared spectroscopy (GC–IR), a common method for coupling a gas chromatograph to a FTIR spectrometer is to use a light pipe: a heated flow cell that allows the continuous scanning of the effluent emerging for the GC column [White, 1990]. The nature of this technique requires that interferograms are collected in short time intervals. Data can be displayed in real time and are commonly monitored as the changing spectrum of the GC effluent and the changing infrared absorption as a function of time. The latter is called a Gram–Schmidt chromatogram.

Quantitative GC analysis involves the measurement of the area of a chromatographic peak. Such analysis is mostly carried out by adding a known quantity of an internal standard to the unknown sample. The response factor is measured using standard mixtures and the amount of unknown may be calculated using the following equation:

$$\frac{A_x}{[X]} = F\frac{A_s}{[S]} \tag{8.2}$$

where A_x and A_s are the areas of the unknown analyte and standard signals, respectively, [X] and [S] are the concentrations of the analyte and the standard, respectively, and F is the response factor.

There are several commonly used GC detectors. Thermal conductivity detectors have been popular because they are simple and respond to all analytes, but they are not sensitive enough for very small samples in narrow bore columns. Such detectors measure the change in voltage resulting when the analyte emerges from the column. In a flame ionisation detector, the eluate is burned in a hydrogen and air mixture to produce CHO^+ in the flame. Electrons flow from the anode to the cathode where they neutralise CHO^+ in the flame and the current is the detector signal. The detection limit for a flame ionisation detector is about 100 times smaller than that for thermal conductivity detectors. There are other detectors, such as an electron capture detector, which respond to particular classes of compounds.

Before injecting a GC sample, some form of sample preparation will be required. An extraction procedure may be required to extract an analyte from a more complex system which cannot be examined using GC. The concentration of the analyte is an important factor – the concentration must be high enough to measure. It is also common to chemically transform an analyte via derivatisation. This allows non-volatile compounds to be converted to volatile derivatives suitable for GC analysis. Although solutions are commonly injected into the chromatograph, solid-phase microextraction (SPME) provides a simple method of extracting compounds without using a solvent. This method utilises a fused silica fibre coated with a film of non-volatile liquid stationary phase attached to a syringe which can be injected into the GC.

Different injection set-ups can be used to introduce a sample into a GC. Split injection delivers only a small amount of the sample to the column and is useful if the analyte is known to be at a concentration of greater than 0.1 %. Splitless injection is useful for analytes at trace concentrations.

Pyrolysis GC (Py–GC) can be useful for examining complex, high molecular weight and/or polar compounds which are difficult to examine using traditional GC [Chiavari and Prati, 2003]. The technique involves a thermal pre-treatment at high temperatures (greater than 600 °C) of a sample in an inert atmosphere before injection. The compounds are fragmented to produce volatile molecule. A programme is produced which displays peaks due to the decomposition products which can be used as a fingerprint for the sample. Pyrolysis–capillary GC provides an even more sensitive approach [Wampler, 1999]. The use of capillary column chromatography instead of packed columns results in increased resolution. Pyrolysis GC–MS (Py–GC–MS) is widely used as this enables a mass spectrum for each peak to be obtained and more information provided about the sample.

8.4.1 Paintings

A variety of binders used in paintings can be analysed using GC techniques [Colombini and Modugno, 2004; del Cruz-Canizaries *et al.*, 2004; Domenech-Carbo *et al.*, 2001; Gimeno-Adelantado *et al.*, 2001, 2002; Mateo-Castro *et al.*, 1997, 2001; Pitthard *et al.*, 2006; Schilling *et al.*, 1996; Schilling and Khanjian, 1996; Vallance, 1997]. Although a destructive technique, a sample of less than 0.1 mg can be removed from the painting for GC analysis. Proteinaceous binders have been widely identified using GC methods, particularly GC–MS [Aruga *et al.*, 1999; Bersani *et al.*, 2003; Casoli *et al.*, 1996; Colombini *et al.*, 1999a, 1999b, 2000a; Lleti *et al.*, 2003; Marinach *et al.*, 2004; Mateo-Castro *et al.*, 2001; Pitthard *et al.*, 2004; Rampazzi *et al.*, 2002; Schneider and Kenndler, 2001; Scott *et al.*, 2001; Tsakalof *et al.*, 2004]. Hydrolysis is the first step in the study of proteinaceous binding media – hydrolysis using enzymatic or chemical catalysis breaks the peptide bonds to produce free amino acids. A derivatisation procedure is also carried out on proteinaceous binding media to make them suitable for analysis. Consideration must be given to the potential interference of pigments during the analysis of protein binders. Pigments, particularly lead, copper, calcium and manganese pigments, complex the amino acids obtained as a product of the hydrolysis of proteins [Mateo-Castro *et al.*, 1997]. However, pigment interference can be suppressed by incorporating a step prior to the derivatisation procedure which complexes the metals [Colombini and Modugno, 2004]. Several derivatisation methods have been successfully used for the analysis of protein based components in paint samples: the transformation of amino acids in their N-trifluoroacetyl methyl esters or into the analogous propyl or butyl esters; silylation of both the acidic and aminic groups with the formation of trimethylsilyl derivatives or tert-butyldimethylsilyl derivatives; the derivatisation with ethyl chloroformate in ethanol/water/pyridine [Colombini and Modugno, 2004]. Generally proteinaceous binders are identified according to the amount of certain amino acids [Colombini and Modugno, 2004; de la Cruz-Canizaries *et al.*, 2004; Domenech-Carbo *et al.*, 2001; Galletti *et al.*, 1996; Gimeno-Adelantado *et al.*, 2002; Mateo-Castro *et al.*, 1997, 2001; Schilling *et al.*, 1996; Schilling and Khanjian, 1996; Vallance, 1997]. A straightforward approach for the identification of proteinaceous binders is to use characteristic amino acid values [Casoli *et al.*,

1996; Schilling *et al.*, 1996; Schilling and Khanjian, 1996]. For instance, animal glue contains a significantly higher amount of glycine compared to egg and milk and casein contains higher amount of glutamic acid compared to egg. Thus, glycine and glutamic acid are always included in the ratios measured. The identification of collagen can be confirmed by the presence of hydroxyproline. Multivariate analysis can also be applied on the relative amino acid percentages [Aruga *et al.*, 1999; Colombini *et al.*, 1998, 1999a, 1999b, 2000a; Lleti *et al.*, 2003].

An alternative approach to the hydrolysis and derivatisation procedures prior to GC-MS analysis is thermal pyrolysis. The pyrolysis can be carried out using a simple thermal degradation process or by thermally assisted hydrolysis methylation (THM), mainly with tetramethylammonium hydroxide or silylation. Py–GC–MS has been used in a number of studies involving proteinaceous binding media [Bocchini and Traldi, 1998; Carbini *et al.*, 1994, 1996; Chiavari *et al.*, 1993, 1995a; Chiavari and Prati, 2003; Zang *et al.*, 2001]. Egg, casein and animal glues have been characterised using this approach. Animal glues are identified using fragments of hydroxyproline such as pyyrole and diketodipyrrole [Chiavari *et al.*, 1991, 1993, 1998; Chiavari and Prati, 2003]. Such glues have also been shown to have a characteristic distribution of compounds which contain proline and other amino acids, such as proline-glycine and proline-proline diketopiperazines that are detectable using Py–GC–MS. Casein and milk are difficult to detect by pyrolysis.

Lipid binders, such as oils and waxes, have also been analysed using GC and GC–MS [Bonaduce and Colombini, 2004; Casoli *et al.*, 1996, 1998; Chiavari *et al.*, 2005; Colombini *et al.*, 1999a, 2000a, 2002a; Favaro *et al.*, 2005; Gimeno-Adelantado *et al.*, 2001; Marinach *et al.*, 2004; Mateo-Castro *et al.*, 2001, 1997; Peris-Vincente *et al.*, 2006; Pitthard *et al.*, 2004, 2005; Plater *et al.*, 2003; Prati *et al.*, 2004; Rampazzi *et al.*, 2002; Schilling, 2005; van den Berg *et al.*, 2001, 2002]. Such binders require a derivatisation procedure to make them amenable to GC analysis. The most common derivatives used are trimethylsilyl, N-tert-butyldimethylsilyl or alkyl esters. For instance, the fatty acid content of paint oils can be determined by methylation to convert free fatty acids into methyl esters, which may be separated in GC. One study found it effective to use the ratios between GC peak areas of each fatty acid ethyl ester observed for linseed, sunflower and poppyseed oils [Mateo-Castro *et al.*, 2001]. This method allows changes associated with ageing to be monitored. A chromatogram of a N-methyl-N-tert-butyldimethylsilyl (MTBS) derivative of a sample of a paint layer used in a GC–MS study of

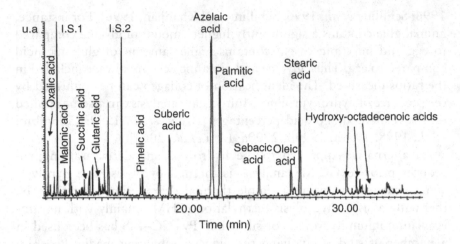

Figure 8.3 TIC chromatogram for a MTBS derivative of a paint sample. Reprinted from Microchemical Journal, Vol 73, Colombini *et al.*, 175–185, 2002 with permission from Elsevier

the deterioration of lipid binders is illustrated in Figure 8.3 [Colombini *et al.*, 2002a]. Films of whole egg and linseed oil were formed to examine the effects of accelerated ageing. One experiment involved the effect of UV radiation at 365 nm on the paint film. The relative percentages of the fatty acids for unaged and aged samples were determined and are listed in Table 8.1. The ratios of palmitic/stearic acids (P/S), azelaic/palmitic acids (A/P) and oleic/stearic acids (O/S) were calculated as a means of monitoring changes and these values are also listed in Table 8.1. The consistency

Table 8.1 Fatty acid composition of unaged and aged paint films

Fatty acid	Unaged	Aged
Lauric	0.2	0.4
Myristic	0.4	0.7
Suberic	4.1	8.0
Azelaic	22.4	37.6
Palmitic	19.4	25.9
Sebacic	2.8	4.5
Oleic	38.2	6.0
Stearic	13.1	16.9
P/S	1.5	1.5
A/P	1.2	1.5
O/S	2.9	0.4

of the P/S ratio on ageing (also noted in further ageing experiments) provided a means of identifying lipid binders. The decrease in the O/S ratio indicated a degradation of the oleic acid.

Py–GC–MS has also been effectively used to characterise lipid components [Bocchini and Traldi, 1998; Cappitelli et al., 2002; Carbini et al., 1994; Challinor, 1996; Chiavari et al., 1993, 1995a, 2005; Chiavari and Prati, 2003; Favaro et al., 2005; Learner, 2004; Scalarone et al., 2001; van den Berg, 2002]. Drying oils show complex pyrograms, but several peaks can be assigned. The main products observed are palmitic and stearic acids with a number of shorter chain acids also observed. In Py–GC–MS, the P/S ratio has been found to be not very reproducible, so it is not a reliable indicator of the oil type [Learner, 2004]. When normal pyrolysis conditions are employed, a pyrogram similar to that of drying oils is observed for the lipid component of egg binder. However, there are several low intensity markers in egg that can potentially be used to differentiate these lipids. For instance, the presence of azelaic acids allows the lipid fraction of egg to be distinguished from a drying oil such as linseed oil [Chiavari and Prati, 2003].

Gum media have also been characterised using GC–MS methods [Schilling, 2005; Schneider and Kenndler, 2001; Vallance et al., 1998]. Samples can be hydrolysed under acidic conditions and then silylated with HMDS. The chromatograms of sugar based compounds can be complicated by the presence of multiple peaks for each monosaccharide resulting from the presence of different structural forms, but oxime derivatives can be used for classification [Vallance et al., 1998]. One study carried out GC–MS on standard gum media [Valance et al., 1998]. The results were compared with suspected gum media samples taken from tempera paintings by William Blake at the Tate Gallery, London. The GC–MS results indicated the presence of a mixed gum media including gums tragacanth, karaya and arabic, with added cane sugar, raising questions about the gums available to artists of the period.

Py–GC–MS can also be used to identify gum media [Chiavari and Prati, 2003; Stevanato et al., 1997]. Curie-point Py–GC–MS has been used to examine a range of polysaccharide binders [Stevanato et al., 1997]. The pyrograms of gum arabic and gum tragacanth are shown in Figures 8.4 and 8.5, respectively [Stevanato et al., 1997]. It can be seen that some of the pyrolysis products are common to both these gums due to the presence of the same sugar and uronic acid components in the original sample, indicated by peaks at the same retention times in the pyrograms. However, it is also seen that there are differences in certain peaks due to the presence of characteristic components and to

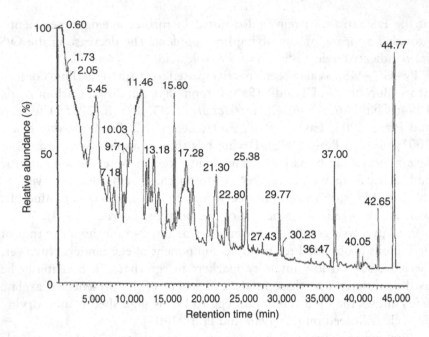

Figure 8.4 Py–GC–MS pyrogram of gum arabic. Rapid Communications in Mass Spectrometry, Stevenato *et al.*, 1997, John Wiley & Sons Limited with permission

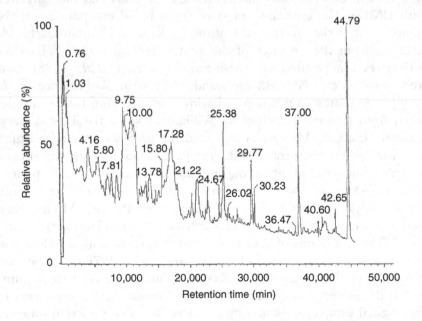

Figure 8.5 Py–GC–MS pyrogram of gum tragacanth. Rapid Communications in Mass Spectrometry, Stevenato *et al.*, 1997, John Wiley & Sons Limited with permission

the different quantities of the original components that make up the gum media. Thus, Py–GC–MS can provide characteristic patterns for different gums while avoiding derivatisation procedures.

GC–MS has been successfully applied to the identification of resins used as paint varnishes and binding media, particularly terpenoid resins [Colombini *et al.*, 2000b; de la Cruz-Canizares *et al.*, 2005; Osete-Cortina *et al.*, 2004; Prati *et al.*, 2004; Regert, 2004; Scalarone *et al.*, 2003; van den Berg *et al.*, 2000; van der Doelen *et al.*, 1998a, 1998b]. GC analysis of terpenic compounds is commonly based on the formation of methyl esters from the carboxylic groups using a variety of derivatisation agents such as diazomethane, methyl chloroformate and direct methylation. The formation of trimethyl ethers has also been employed. The gas chromatograms of methylated dammar and mastic resins are illustrated in Figure 8.6 [van der Doelen *et al.*, 1998a]. Both

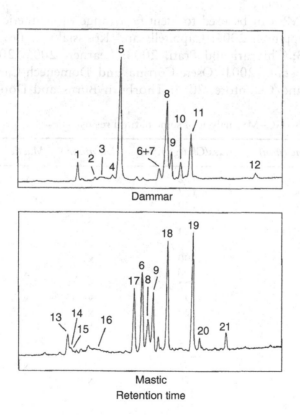

Figure 8.6 Gas chromatograms of methylated dammar and mastic resins [van der Doelen *et al.*, 1998a]

of these resins consist mainly of triterpenoids, with a smaller amount of polymeric material. The polymeric fraction of the resins is too large to be analysed by GC, so methanolic extracts containing the triterpenoid fraction were analysed using GC–MS.

Py–GC–MS has also been applied to the identification of natural resins used in paintings [Bocchini and Traldi, 1998; Chiavari *et al.*, 1993, 1995b; Chiavari and Prati, 2003; de la Cruz-Canizares *et al.*, 2005; Osete-Cortina and Domenech-Carbo, 2005a, 2005b; Scalarone *et al.*, 2002; Stevenato *et al.*, 1997; van den Berg *et al.*, 1998a, 1998b]. Specific pyrolysis products have been identified as markers for the identification of resins. Table 8.2 summarises the components that can aid in the identification of some common resins: dammar, mastic and sandarac [Stevenato *et al.*, 1997]. The data show that it is easy to distinguish mastic resin from the dammar and sandarac resins by examination of the pyrolysis products with retention times of 4.02, 5.48, 7.20, 15.9, 24.9 and 25.7 min.

Py–GC–MS can be used to identify a range of synthetic polymeric binders [Cappitelli, 2004; Cappitelli and Koussiaki, 2006; Chiantore *et al.*, 2003; Chiavari and Prati, 2003; Learner, 2005, 2004, 2001; Nakamura *et al.*, 2001; Osete-Cortina and Domenech-Carbo, 2006; Scalarone and Chiantore, 2004; Thorburn Burns and Doolan, 2000;

Table 8.2 Py–GC–MS analysis of some natural resins

Retention time (min)	Mass/Charge ratio	Dammar	Mastic	Sandarac
4.01	172		x	
5.48	172		x	
7.2	150		x	
8.2	–	x		x
9.3	174			x
11.6	176	x		x
12.7	190	x		x
15.9	200		x	
17.1	172	x		x
21.8	228	x	x	x
22.9	222	x		x
24.3	242	x		x
24.9	272		x	
25.7	272		x	
27.6	256	x	x	x
29	270	x	x	x
31	284	x	x	x
31.7	280		x	
38.7	302			x
43.5	264	x		x

Table 8.3 Py–GC–MS data for acrylic emulsions

Acrylic emulsion	Scan number	Mass/Charge ratio	Identity
Ethyl methacrylate	63	86	MA
copolymer	96	100	MMA
	160	114	EMA
	774	172	MA dimer
	882	200	MA–EMA
	1334	258	MA timer
	1377	286	(MA)$_2$–EMA
	1403	286	(MA)$_2$–EMA
	1453	286	(MA)$_2$–EMA
Ethyl acrylate	44	46	Ethanol
copolymer	62	88	Ethyl ethanoate
	90	100	EA
	97	100	MMA
	155	114	EMA
	841	188	EA–MMA sesquimer
	881	200	EA–MMA
	891	188	EA sesquimer
	954	200	EA dimer
	1428	300	EA–EA–MMA
	1455	300	EA–EA–MMA
	1511	300	EA trimer
	1526	300	EA–EA–MMA
	1898	400	EA tetramer
	1905	400	EA tetramer
	2218	500	EA pentamer
n-butyl acrylate –	42	56	n-butene
methyl	77	74	n-butanol
methacrylate	97	100	MMA
copolymer	319	128	nBA
	447	142	NBMA
	1075	216	nBA–MMA sesquimer
	1106	228	nBA–MMA
	1329	244	nBA sesquimer
	1375	256	nBA dimer
	1737	356	nBA–nBA–MMA
	1761	356	nBA–nBA–MMA
	1843	356	nBA–nBA–MMA
	1953	384	nBA trimer

(MA methyl acrylate; MMA methyl methacrylate; EMA ethyl methacrylate; EA ethyl acrylate; BA butyl acrylate)

Wilcken and Schulten, 1996]. Characteristic pyrograms can be obtained for acrylic, PVA and alkyd, the main types of synthetic binders. In Table 8.3 the main pyrolysis products determined by Py–GC–MS for several acrylic emulsions are listed [Learner, 2004]. PVA resins produce quite different pyrograms – they show peaks due to the formation of

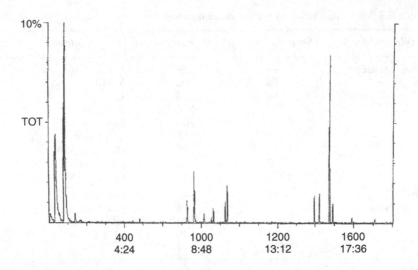

Figure 8.7   Py–GC–MS pyrogram of red paint from Hockney's 'Mr and Mrs Clark and Percy'. Reproduced by permission from The Analysis of Modern Paints, Thomas J S Learner, 2004

acetic acid and benzene produced via a side group elimination reaction. Alkyd resins can be readily identified by a peak due phthalic anhydride. In Figure 8.7 an illustration of a pyrogram of a modern paint is provided [Learner, 2004]. The pyrogram was obtained for a sample of red paint from the palette used by David Hockney for the painting 'Mr and Mrs Clark and Percy'. Comparison with Table 8.2 demonstrates that the paint was based on an ethyl acrylate – methyl methacrylate acrylic emulsion. Additional peaks between scan numbers 700 and 100 can be attributed to pigments.

It is also possible to identify pigments using Py–GC–MS [Learner, 2004; Prati *et al.*, 2004; Sonoda *et al.*, 1993; Sonoda, 1999]. Red, orange and yellow azo pigments show characteristic pyrograms. Certain degradation products of azo pigments are suitably volatile to pass through a GC column. Other types of organic pigments, such as phthalocyanines and quinacridones, do not undergo suitable reactions on pyrolysis to products with a volatility capable of passing through a GC column.

8.4.2 Natural Materials

A number of natural materials used in paintings have already been described in the previous section. Such materials are used in many other artefacts of cultural significance and GC techniques have been used to characterise natural resins, waxes, organic residues, wood and bituminous materials. For instance, Py–GC–MS has been successfully employed to determine the geographical origin of amber [Bocchini and Traldi, 1998; Carlson et al., 1997; Shedrinsky et al., 1993, 2004]. The presence of succinic acid or its anhydride can be readily detected using MS and enables amber to be differentiated from the synthetic substitutes for amber.

GC–MS has been used in many studies of organic residues and coatings on artefacts [Agozzino et al., 2001; Asperger et al., 1999; Bonaduce and Colombini, 2004; Chiavari et al., 1991; Chiavari and Prati, 2003; Colombini et al., 2005; Copley et al., 2005a, 2005b; Dudd et al., 1999; Evershed et al., 1990, 1997, 1999, 2002; Evershed, 1993; Mejanelle et al., 1997; Regert et al., 2003, 2005; Shredinsky et al., 1989, 1991; Tchapla et al., 2004]. Fats can be characterised by examining the ratio of palmitic to stearic acid and the type of mono-, di- and triacylglycerols present [Dudd et al., 1999l; Evershed et al., 1997, 1999, 2002]. Beeswax has been identified using GC–MS methods via the detection of odd-numbered linear hydrocarbon (C_{21}–C_{33}), even-numbered free fatty acids (C_{22}–C_{30}) and long-chain palmitate esters in the range C_{40}–C_{52} [Asperger et al., 1999; Bonaduce and Colombini, 2004; Chiavari and Prati, 2003; Regert et al., 2003]. One problem associated with wax is the need for a caustic saponification step prior to analysis. To overcome a difficult sample pre-treatment step, Py–GC–MS has been successfully applied with an in-situ derivatisation with HMDS [Bonaduce and Colombini, 2004]. In Figure 8.8 an example of a pyrogram of beeswax taken from an Egyptian sarcophagus is illustrated [Chiavari and Prati, 2003].

8.4.3 Written Material

GC-MS can be used to identify the components of inks used in the production of historic documents [Arpino et al., 1977; Bleton et al., 1996; Casas-Catalan and Domenech-Carbo, 2005; Ferrer and Sistach, 2005]. For example, this technique has been applied to the identification of natural dyes, including indigo and saffron, used for illuminating manuscripts [Casas-Catalan and Domenech-Carbo, 2005]. It was noted in this study that Py–GC–MS is a suitable technique for characterising samples in which indigo is believed to be present. The procedure used, which included an on-line derivatisation of the dye using HMDS, enabled pyrolysis products to be used to identify indigotin and indirubin molecules which are markers for indigo dye.

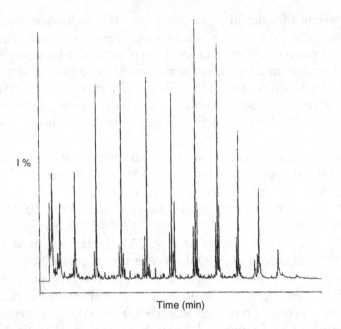

Figure 8.8 Py–GC–MS pyrogram of beeswax. Reproduced by permission from Chromatographia, G Chiavari & S Prati, Vol 58, 543–544, 2003

8.4.4 Stone

The crusts that form on heritage building stones can be characterised using GC–MS methods [De Angelis *et al.*, 2002, 1999; Rampazzi *et al.*, 2004; Saiz-Jimenez *et al.*, 1991; Saiz-Jimenez, 1993]. A knowledge of the composition of these products, including black crusts and gypsum deposits, can provide an understanding of the decay processes. Py–GC–MS and SPME coupled with GC–MS have been used to identify the components of deposits. Although the compositions of such crusts are complex, valuable information can be obtained regarding the stone environment. Pollution of petrogenic origin can be determined by the presence of polycyclic aromatic hydrocarbons, such as indene, naphthalene and phenanthrene, as well as the oxygenated analogues. Wood combustion is indicated by the presence of odd carbon number hydrocarbons, resin terpenoids and polyphenols. Additionally, microbial metabolism is shown by n-alkanoic acids as methyl esters with 14–18 carbon atoms. Multivariate analysis can be applied to decipher the complex data produced by GC–MS analysis [De Angelis *et al.*, 2002].

8.4.5 Synthetic Polymers

Apart from the analysis of the polymeric components of paints, GC-MS methods have been used to study other polymers such as those used in sculptures [Quye, 1995; Quye and Williamson, 1999]. GC–MS proves useful when examining polymer mixtures, additives or degradation products. Where it is difficult to dissolve a polymer in a suitable solvent, Py–GC–MS is helpful. However, it has been observed that cellulose nitrate and cellulose acetate do not produce unambiguous results using Py-GC-MS [Quye and Williamson, 1999].

8.4.6 Textiles

GC-MS has been used to analyse anthraquinone dyes by employing a silylating reagent [Henriksen and Kjosen, 1983]. Py–GC–MS has also been applied to the analysis of dyes [Casas-Catalan and Domenech-Carbo, 2005; Fabbri *et al.*, 2000]. The methylation of pyrolysates with tetramethylammonium hydroxide (TMAH) has been used for the analysis of dyes [Fabbri *et al.*, 2000]. Another study used Py–GC–MS to identify natural dyes, including madder, curcuma and saffron, which are used on textiles [Casas-Catalan and Domenech-Carbo, 2005]. On-line derivatisation of the dyes using HMDS was found to be an effective procedure for the unambiguous identification of the dyes.

8.4.7 Museum Environments

The development of SPME has enabled GC-MS to be used in the measurement of volatile organic compounds that can cause deterioration in museum objects [Godoi *et al.*, 2005; Lattuati-Derieux *et al.*, 2006; Ryhl-Svendsen and Glastrup, 2002]. This approach has been used to measure concentrations of acetic acid and formic acid in museum environments.

8.5 HIGH PERFORMANCE LIQUID CHROMATOGRAPHY

Liquid chromatography (LC) has a mobile phase that is a liquid and proves useful when the compounds under investigation are not sufficiently volatile for GC. High performance liquid chromatography (HPLC) employs high pressure to force the solvent through columns

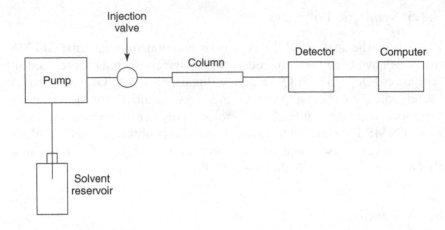

Figure 8.9 Schematic diagram of a HPLC apparatus

containing fine particles that produce a high resolution separation [Lough and Wainer, 1996; Meyer, 2004]. In Figure 8.9 a schematic diagram of the layout of an apparatus used for HPLC is illustrated. The HPLC system consists of a solvent delivery system, a sample injection valve, a high pressure column, a detector, a computer and often an oven for temperature control of the column.

In LC, packed columns are used, as diffusion in liquids is much slower and open columns would be impractical. The efficiency of a packed column improves with decreasing size of the stationary phase particles and typical particle sizes in HPLC are 3–10 μm. However, with a small particle size comes an increased resistance to solvent flow and HPLC requires pressures of about 7–40 MPa to obtain flow rates of the order of 0.5–5 ml min^{-1}. The columns are usually manufactured from steel or polymer and are from 5–30 cm in length with an inner diameter of 1–5 mm. Heating the column will decrease the solvent viscosity and so reduce the required pressure or increase the flow. The most common column support is silica. Very pure solvent must be used in order to avoid the introduction of damaging impurities.

In absorption chromatography, the solvent molecules compete with the solute molecules for sites on the stationary phase. Elution occurs when the solvent displaces the solute from the stationary phase. Absorption chromatography on silica can use a polar stationary phase and a less polar solvent and is known as normal phase chromatography. Reversed phase (RP) chromatography has a stationary phase that is nonpolar or weakly polar and a more polar solvent. RP chromatography improves the signal peak because the stationary phase has few sites that can strongly absorb a solute and it is also less sensitive to polar impurities. Isocratic elution is carried out with a single or constant solvent, but if this does not provide a suitably rapid elution, then gradient elution can be employed. Gradient elution involves adding increasing amounts of a second solvent to the first solvent to produce a continuous gradient.

There are various detectors that may be used for HPLC. UV detectors are common and qualitative information about each analyte can be provided by a photodiode array detector (DAD). Although a refractive index detector is able to provide a universal response, it is regarded as not very sensitive. Fluorescence and electrochemical detectors are very sensitive but are selective. MS provides both qualitative and quantitative detection of the substance eluted from the column.

The advantage of LC–MS is that it may be used to study components that are too thermally unstable or involatile to be analysed by GC–MS. However, it has proved more difficult to combine MS with LC than with GC because of the large volume of solvent generated by this method. However, suitable instrumentation is currently being developed and thermospray interface can be utilised to remove the solvent.

Liquid chromatography may also be used in conjunction with infrared spectroscopy [White, 1990]. The eluant from a liquid chromatograph may be passed through a liquid flow-through cell. Supercritical fluid chromatography (SFC), where supercritical carbon dioxide is commonly used as a mobile phase, is used with FTIR to improve detection limits.

8.5.1 Textiles

HPLC is widely used to study dyes from textiles and has been successfully applied to colourants used in historic textiles [Ackacha *et al.*, 2003; Balakina *et al.*, 2006; Halpine, 1996; Halpine, 1998; Karapanagiotis *et al.*, 2005; Karapanagiotis and Chryssoulakis, 2005; Novotna *et al.*, 1999; Orska-Gawrys *et al.*, 2003; Puchalska *et al.*, 2004; Surowiec *et al.*, 2003, 2004a, 2005; Szostek *et al.*, 2003; Trojanowicz *et al.*, 2004; Wouters, 1985; Wouters *et al.*, 1990; Wouters and Roario-Chirinos, 1992; Wouters and Verhecken, 1989, 1991; Zhang and Laursen, 2005]. The use of HPLC with a DAD has proven to be particularly useful for identifying the components of many natural dyes. DAD detection of the component peaks allows the components to be identified based on the UV-visible spectra. A promising method for the identification of dye components is LC–MS as the information provided by the MS spectra can elucidate the structure of the component where DAD is not sufficient.

HPLC systems that combine both types of detectors (LC–DAD–MS) provide even more useful information.

Before dyes can be analysed using HPLC, they must be extracted from the fibres and as most textiles are mordanted with metal ions, the extraction process must be able to disrupt the dye–metal complex. An established dye extraction process for heritage textiles is hydrolysis with acidified methanol. This procedure provides an efficient extraction of anthraquinone and flavanoids dyes from textiles and hydrolysis of their glycosidic forms to aglycones. However, the efficiency of this approach for indigoids has been less acceptable [Surowiec *et al.*, 2005; Szostek *et al.*, 2003]. It has been shown that blue indigoid dyes are better extracted with hot pyridine, pyridine with water or with dimethylformamide.

The identification of dyes from historical textiles is usually based on comparison with known references. Where UV-visible detection is used with HPLC, a combination of the elution and the UV-visible peaks can be used to identify dye components. If a mass spectrometer is being employed, the parent mass and fragmentation patterns can be used to elucidate structure. In Table 8.4 the UV-visible maxima and major mass spectrum peaks of a number of dyes that can be used in the identification process are listed [Balakina *et al.*, 2006]. There can also be variation in the UV-visible bands due to the solvent, so references should be recorded under the same experimental conditions. Retention times have not been included because these can vary considerably based on the experimental conditions used. It is possible that the sample under investigation may not match the references employed due to degradation processes, an inefficient extraction process or limited references.

8.5.2 Paintings

A knowledge of the composition of binders used in paintings can be obtained via HPLC analysis. Proteinaceous binding media can be

Table 8.4 Characteristics of dyes detected using HPLC

Dye	UV-visible absorbance maxima (nm)	Mass/Charge ratio of major ions
Alizarin	279, 330, 430	241
Carminic acid	274, 312, 495	493
Indigo	617	263
Kermesic acid	272, 310, 490	331
Purpurin	255, 294, 482	257

analysed quantitatively using amino acid analysis [Colombini *et al.*, 2004; Colombini and Modugno, 2004; Galletti *et al.*, 1996; Grzywacz, 1994; Halpine, 1998; Peris-Vincente *et al.*, 2005a; Vallance, 1997]. Electrochemical, UV-visible or fluorescence detection can be used to analyse amino acids, but fluorescence is usually chosen because it is more sensitive. Samples in quantities of the order of μg can be examined. As amino acids do not have fluorescent properties, a derivatisation procedure is required prior to analysis. There are a number of derivative reagents that have been used for amino acids, including *o*-phthalaldehyde, 9-fluorenylmethylchloroformate, carbazole-N-(2-methyl)acetyl chloride, 4-dimethylamminoazobenzene-4'-sulfonyl chloride, 6-aminoquinolyl-N-hydroxysucinimidylcarbamate, 4-fluoro-7-nitro-2,1,3-benzoxadiazole and phenylisothiocyanate. Consideration must be given to the possibility of interference by pigments in the analysis: the presence of metallic cations in certain pigments can form complexes with some amino acids, changing the amount of free amino acid measured. However, ethylenediaminetetraacetic acid (EDTA) can be used to minimise this effect.

Oils used in paintings have also been studied using HPLC [Peris-Vicente *et al.*, 2005a, 2005b, 2004; Shibayama *et al.*, 1999; van den Berg *et al.*, 2004]. HPLC can be used to measure the fatty acid content of the oils as the amount of each fatty acid is characteristic of the type of oil. Changes to the composition associated with ageing may also be determined. Hydrolysis is carried out on the triglycerides present in the oil to produce the fatty acid derivatives. UV-visible detection has been used to determine the fatty acid content in paint oil, but a derivatisation process is required to produce a UV-visible absorbing agent. Reagents such as 2-nitrophenylhydrazides, *p*-phenazophencicyl, 1-chlormethylsatin, phenacyl, naphtacyl, *p*-methoxyanilines, 2-(phthalimino)ethyl and nitrobenyl have been used in derivatisation procedures. UV-visible detection is usually less sensitive and less selective than fluorescent detection, so the latter is the preferred approach. As fatty acids do not possess fluorescent groups, a derivatisation procedure with a fluorescent reagent is required. A number of reagents can be used: 9-(2-hydroxyethyl)-carbazole, 9-(hydroxymethyl)anthracene, 2-(4-hydrazinocarbonylphenyl)-4,5-diphenylimidazole, 9-anthyldiazomethane, N-(1-napthyl)ethylenediamine, 5-(4-pyridyl)-2-thiophenmethanol, 2-(2,3-naphthalimino)ethyltrifluoromethane sulfonate. RP columns using methanol and water mixtures are usually employed in the separation of fatty acid derivatives. The elution volumes of the derivatives are affected by the number of carbon atoms and the number of unsaturated bonds in the fatty acid chain. Oils are usually best characterised by the ratio of palmitic to stearic acid. Although oils can be distinguished by this ratio even after ageing, materials such as waxes may be present and will affect the value of the ratio.

HPLC–MS has been utilised to examine aged triterpenoid varnishes used in paintings [van der Doelen *et al.*, 1998a, 1998b]. HPLC was identified as a more effective technique than GC–MS for the elucidation of ageing products – HPLC can resolve more degradation products. A range of oxidised triterpenoid components were characterised using this approach.

8.5.3 Stone

HPLC has been used for the analysis of heritage building stone [Cardiano *et al.*, 2004; Marinoni *et al.*, 2002; Schroeder *et al.*, 1990]. For example, the HPLC analysis of organic acids on the surface of various building stones, including limestone, quartzite and sandstone, from historical sites in Germany, has been reported [Schroeder *et al.*, 1990]. The quantitative determination of carboxylic acids was carried out in order to measure the presence of potentially destructive microorganisms.

8.6 SIZE EXCLUSION CHROMATOGRAPHY

Size exclusion chromatography (SEC) involves separating of molecules according to their size [Harris, 2003; Mori and Barth, 1999]. This technique is also known by the names gel permeation chromatography (GPC), gel filtration chromatography or molecular exclusion chromatography. SEC is mainly used to separate molecules of different molecular weights and may be employed to estimate the average molecular weight of a sample.

In Figure 8.10 a schematic diagram shows the movement of molecules through a SEC column. Larger molecules cannot penetrate the pores of the stationary phase and are eluted by a solvent volume equal to the volume of the mobile phase. Smaller molecules require a larger volume for elution. There is a number of gels used in SEC including Sephadex (dextran cross-linked by glycerin) and Bio-Gel (polyacrylamide cross-linked by N,N′-methylenebisacrylamide). There are different grades of these materials with differing pore sizes. In highly cross-linked gels with the smallest pore sizes, molecules with a molecular weight greater

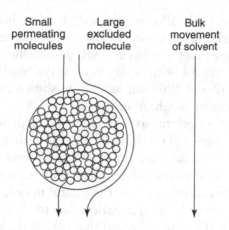

Figure 8.10 Schematic diagram of the movement of molecules through a SEC column

than 700 are excluded. The largest pore sizes exclude molecules with molecular weights greater than 10^8. UV detection with a photochemical array detector or a refractometer detector may be used to detect the output.

While SEC can be used to purify samples, the application to molecular weight distribution is probably of most interest to materials conservation. For each stationary phase, there is a range over which there is a logarithmic relationship between the molecular weight and the elution volume. The molecular weight of an unknown material may be estimated by comparing its elution volume with those of a set of standards. However, care must be taken when interpreting the results of such experiments as molecules with the same molecular weight, but with different shapes, can show different elution properties. When studying proteins, a high ionic strength should be used to eliminate any electrostatic adsorption of the solute by charged sites on the gel.

8.6.1 Written Material

SEC has been used as a means of characterising degradation in paper [Dupont, 2002; Rudolph et al., 2004; Stol et al., 2002]. This approach can determine the molecular weight distribution of cellulose. SEC has been employed to investigate the artificial ageing of paper samples [Stol et al., 2002] and the effects of laser treatment on paper [Rudolph et al., 2004]. In the latter research, the average molecular weight of cellulose was determined on samples that had been modified by reaction with phenyl isocyanate. This allowed the cellulose to become soluble in tetrahydrofuran, the solvent used as the mobile phase. The study was used to demonstrate that the molecular weight of cellulose is significantly decreased when treated with a laser of wavelength 355 nm, while treatment with 532 and 1064 nm light were shown to have a minimal effect on the molecular weight of cellulose.

The degradation of gelatin used as a sizing agent in paper has also been investigated using SEC [Dupont, 2002]. 17th and 18th century and artificially aged modern gelatin sized papers were examined in this study. Sodium dodecyl sulfate (SDS), an anionic surfactant, was used to ensure a rod-shape conformation of the gelatin in order to minimise the error in molecular weight determination due to differences in volume with the standards. The study showed that gelatin undergoes hydrolysis upon ageing, affecting the molecular weight.

8.6.2 Paintings

SEC can be utilised to characterise resins used in paintings [de la Rie, 1987; Halpine, 1998; Maines and de la Rie, 2005; Scalarone *et al.*, 2003; van der Doelen and Boon, 2000]. The effects of ageing of resins, such as copals, sandarac and dammar, may be determined via molecular weight determination. For instance, an increase in molecular weight of resins calculated using SEC provides evidence of cross-linking in the sample, which is associated with undesirable physical properties such as brittleness and darkening.

8.6.3 Textiles

The deterioration of textile fibres including silk, linen and cotton, can be monitored using SEC [Tse and Dupont, 2001]. Consideration should be given to the shape of the protein molecules studied in such materials. For example, globular proteins are usually used as standards which differs from the random coil conformation that silk is assumed to possess in solution. Lithium thiocyanate is regarded as the least degrading solvent for silk, so is preferred for SEC measurements for the fibre.

8.6.4 Synthetic Polymers

SEC is a commonly used technique to determine the average molecular weight and molecular weight distribution of polymers [Dawkins, 1989].

This technique has been used to characterise acrylic copolymers with potential as consolidants [Princi et al., 2005]. SEC also has great potential for the study of degraded polymer objects [Ballany et al., 2001; Quye and Williamson, 1999]. For instance, the degradation of cellulose acetate objects may be monitored using the molecular weight data gained from SEC. The molecular weight of polymers can be used as an indication of brittleness.

8.7 ION CHROMATOGRAPHY

In ion chromatography (IC), retention is based on an attraction between solute ions and the charged sites bound to the stationary phase [Fritz and Gjerde, 2000; Harris, 2003]. This technique is used to separate charged compounds. The stationary phase is usually an ion exchange resin that carries charged functional groups which interact with oppositely charged groups of the compound to be retained. A positively charged anion exchanger interacts with anions, while a negatively charged cation exchange interacts with cations. The bound compounds can be eluted from the column by gradient elution or by isocratic elution with a change in pH or salt concentrations.

Ion exchange resins are amorphous particles of organic materials such as PS. PS resins are made of copolymers consisting of styrene and divinylbenzene, the relative amounts of which result in different degrees of cross-linking in the resin. The benzene rings can be modified to contain SO_3^- groups to produce a cation exchange resin, or modified with NR_3^+ groups to produce an anion exchange resin.

Any electrolytes that might interfere with the analysis can be removed using suppressed ion chromatography prior to detection by electrical conductivity. In suppressed ion anion chromatography, the solution passes through a suppressor in which the cations are replaced by H^+ to convert the eluant to water. Suppressed ion cation chromatography has a suppressor which replaces the anion with OH^-.

8.7.1 Stone

IC has been widely used to measure the concentrations of ions present on the surfaces of heritage stones and mortars [Backbier et al., 1993; Fassina et al., 2002; Ghendini et al., 2000; Gobbi et al., 1995, 1998; Halpine, 1998; Perry and Duffy, 1997; Riotino et al., 1998; Sabbioni et al., 2001; Schroeder et al., 1990; Vleugels et al., 1992; Warke and Smith, 2000]. The most commonly encountered salts on building stone are sulfates, chlorides and nitrates and the anions of these compounds may be detected using IC. One study used IC to detect for these ions in limestone samples taken from cathedrals in the Netherlands [Backbier *et al.*, 1993]. Depth profiling at 1 cm intervals up to 14 cm into the surface of the stones was carried out. The top centimetre of the stones showed a significantly higher value for all three anions. IC was also used in this study to examine the after effects of the application of a poultice to the stone surface. The treatment was found to be effective, with a notable decrease in the concentration of all the anions at the stone surface detected using IC.

8.7.2 Synthetic Polymers

IC can be employed to determine the presence of ions in polymers that are indicators of degradation. This technique has been used to examine the degradation in cellulose acetate artefacts [Ballany *et al.*, 2001]. IC was used to investigate the presence of acetate, formate, chloride, nitrate, sulfate and oxalate in the artefacts. An increase in acetate levels is an indication that deacetylation is occurring, oxalate indicates chain scission and formate is a result of oxidative degradation. Chloride and sulfate ions are from residual chemicals used during the manufacturing process. Increased nitrate concentrations are used as an indication of a mixture with cellulose nitrate. The IC results for a range of samples found that there is no correlation between the residual sulfate or chloride and the extent of degradation. An increase in oxalate in older samples was indicative that chain scission is a process that occurs later in the degradation process.

8.7.3 Paintings

IC can be used to identify plants gums used in paintings [Colombini *et al.*, 2002b]. Anion exchange chromatography can be used to determine the sugar components in the form of their oxoanions. The sugars are produced by hydrolysis with trifluoroacetic acid. The sugars are detected by measuring the electrical current generated by their oxidation at an gold electrode. The paint sample is prepared for anion exchange by separating other components, such as pigments, by passing through a cation exchange resin. This method was used to separate the following sugars: fucose, rhamnose, arabinose, galactose, glucose, mannose and xylose.

8.7.4 Metals

Information regarding metal corrosion can be gained using IC. For example, in a study of corrosion products from lead roofs on historic buildings in the UK, IC was used to determine the concentrations of acetic and formic acids [Edwards *et al.*, 1997]. Timber used in roofs can be a source of these acids, which are aggressive towards lead. Acetic and formic acid from soluble lead salts which can be extracted into an aqueous solution. The concentration of acetate and formate ions can then be determined by ion exchange chromatography.

8.8   CAPILLARY ELECTROPHORESIS

Electrophoresis involves the migration of ions in solution under the influence of an electric field [Baker, 1995; Harris, 2003; Skoog and Leary, 1992; Weinberger, 2000]. This technique has been used since the 1930s, particularly for separating of proteins and nucleic acids. Molecules are separated by injecting a sample in a buffer into a narrow tube, paper or gel. When a direct current potential is applied across the buffer using electrodes, the ions of the sample migrate toward one of the electrodes. The rate of migration depends on the charge and the size of the molecule.

Capillary electrophoresis (CE) is an instrumental version of electrophoresis and has emerged in recent decades as a more efficient technique than slab electrophoresis. The use of a capillary allows much higher electric fields to be applied, resulting in better resolution and a shorter analytical time. In Figure 8.11 the layout of an apparatus used to conduct CE experiments is illustrated. A voltage of about 30 kV separates the components of a solution inside a fixed silica capillary tube. The tube is about 50 cm in length and has an inner diameter of 25–75 μm. CE requires only very small sample volumes (0.1–10 nl) compared to microlitre volumes required for slab electrophoresis. The separated compounds are eluted from one end of the capillary. CE uses UV-visible or conductometric detectors, which are more effective than the traditional staining methods of slab electrophoresis. CE can also be combined with MS, enabling large molecules such as proteins or nucleic acids to be identified.

There are a number of types of CE experimental set-ups, including capillary zone electrophoresis (CZE) and capillary gel electrophoresis (CGE). In CZE, the buffer composition is constant throughout the separation region. The different ionic components migrate according to

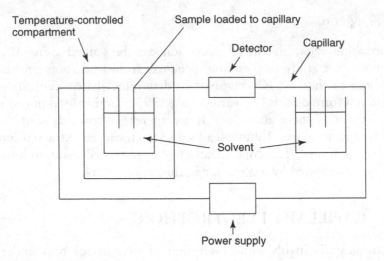

Figure 8.11   Schematic diagram of a CE apparatus

their mobility and separate into zones that may be resolved. CGE is carried out using a porous polymer gel. The pores of the gel contain a buffer mixture in which the separation is carried out.

8.8.1 Paintings

CE can be used to identify different types of binding media in paintings. Animal glues, plant gums and drying oils have all been successfully characterised using this technique [Grossl et al., 2005; Harrison et al., 2005; Kaml et al., 2004; Surowiec et al., 2004b]. Animal glues, including egg, collagen (parchment glue), casein, bone, fish and rabbit glues, can be identified using CZE by determining the characteristic amino acids [Harrison et al., 2005; Kaml et al., 2004]. Acid hydrolysis was first carried out on the proteins to prepare the amino acids for analysis. No derivatisation step was required. The profiles of the amino acids were recorded using a conductivity detector. The glues can be identified by examining the relative peak area of the amino acids. Electropherograms obtained for parchment glue, casein and egg white after acid hydrolysis are illustrated in Figure 8.12 [Kaml et al., 2004]. The main markers are hydroxyproline, proline, glycine, glutamic acid, serine and valine. It has been determined that the presence of gums and oils in a sample does not affect successful identification of animal glues used as binders [Harrison et al., 2005].

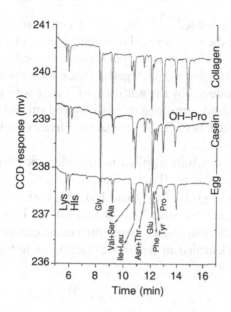

Figure 8.12 Electropherograms for parchment glue, casein and egg white binders [Kaml *et al.*, 2004. Reproduced by permission of Wiley-VCH]

Plant gums, such as gum arabic and gum tragacanth, can be identified using CZE [Grossl *et al.*, 2005]. The component sugars are first produced by hydrolysis with trifluoroacetic acid. When analysed using CE, plant gums reveal a typical composition in the resulting electropherogram. For example, a peak of glucuronic acid together with that of rhamnose, is indicative of gum arabic.

CE can also be used to investigate the drying oils used in paintings [Surowiec *et al.*, 2004b]. The fatty acids of the oils and the dicarboxylic acids formed by oxidation can be identified in their underivatised forms. The sampling methods is simple – a common hydrolysis followed by extraction from acidic solution with diethylether. Despite this benefit, this method has a lower separation efficiency than capillary GC.

8.8.2 Textiles

CZE, in combination with MS, has proved to be an effective technique for investigating historic cellulose textiles [Kouznetsov *et al.*, 1994, 1996]. Linen samples dated between 1200 BC and 1500 AD were studied using this approach in order to determine the age and the origin of such textiles. CZE is used to compare the cellulose sequences through a fractionation of the low molecular weight compounds, including β-D-glucose and cellulose. Enzymated hydrolysis is carried out on the textiles. While carbohydrates are not charged species under normal conditions, a separation is produced by the addition of borate ions to the separation medium. This method forms negatively charged sugar-borate complexes that can be separated by CZE. MS is used to identify each electrophoretic fraction.

CE has been successfully applied to the identification of textile dyes containing anthraquinones [Puchalska *et al.*, 2003]. CE has been coupled with both a UV-visible DAD and ESI-MS, with the latter proving to be more selective and sensitive ($0.1-0.5\mu g\ ml^{-1}$). This method allowed anthraquinone components, alizarin, purpurin, carminic acid and laccaic acid, to be identified in natural dyestuffs including cochineal and lac dye.

REFERENCES

M.A. Ackacha, K. Polec-Pawlak and M. Jarosz, Identification of anthraquinone colouring matters in natural red dyestuffs by high performance liquid chromatography with ultraviolet and electrospray mass spectrometric detection, *Journal of Separation Science*, **26** (2003), 1028–1034.

P. Agozzino, G. Avellone, I.D. Donato and F. Filizzola, Mass spectrometry for cultural heritage knowledge has chromatographic/mass spectrometric analysis of organic remains in Neolithic potsherds, *Journal of Mass Spectrometry*, **36** (2001), 443–444.

P. Arpino, J.P. Moreau, C. Oruezabal and F. Fliedler, Gas chromatographic – mass spectrometric analysis of tannin hydrolysates from the ink of ancient manuscripts (XIth to XVIth century), *Journal of Chromatography*, **134** (1977), 433–439.

R. Aruga, P. Mirti, A. Casoli and G. Palla, Classification of ancient proteinaceous painting media by the joint used of pattern recognition and factor analysis on GC–MS data, *Fresenius Journal of Analytical Chemistry*, **365** (1999), 559–566.

A. Asperger, W. Engewald and G. Fabian, Advances in the analysis of natural waxes provided by thermally assisted hydrolysis and methylation (THM) in combination with GCMS, *Journal of Analytical and Applied Pyrolysis*, **52** (1999), 51–63.

L. Backbier, J. Rousseau and J.C.J. Bart, Analytical study of salt migration and efflorescence in a medieval cathedral, *Analytica Chimica Acta*, **283** (1993), 855–867.

D.R. Baker, *Capillary Electrophoresis*, Wiley–Interscience, New York (1995).

G.G. Balakina, V.G. Vasiliev, E.V. Karpova and V.I. Mamatyuk, HPLC and molecular spectroscopic investigations of the red dye obtained from an ancient Pazyryk textile, *Dyes and Pigments*, **71** (2006), 54–60.

J. Ballany, D. Littlejohn, R.A. Pethrick and A. Quye, Probing the factors that control degradation in museum collections of cellulose acetate artefacts, in *Historic Textiles, Papers and Polymers in Museums* (eds J. M. Cardamone and M.T. Baker), American Chemical Society, Washington (2001), pp 145–165.

D. Bersani, G. Antonioli, P.P. Lottici and A. Casoli, Raman microspectroscopic investigation of wall paintings in S. Giovanni Evangelista Abbey in Parma: a comparison between two artists of the 16th century, *Spectrochimica Acta A*, 59 (2003), 2409–2417.

J. Bleton, P. Mejanelle, J. Sansoulet, S. Goursaud and A. Tchapla, Characterisation of neutral sugars and uronic acids after methanolysis and trimethylsilyation for recognition of plant gums, *Journal of Chromatography A*, 720 (1996), 27–49.

P. Bocchini and P. Traldi, Organic mass spectrometry in our cultural heritage, *Journal of Mass Spectrometry*, 33 (1998), 1053–1062.

I. Bonaduce and M.P. Colombini, Characterisation of beeswax in works of art by gas chromatography – mass spectrometry and pyrolysis gas chromatography – mass spectrometry, *Journal of Chromatography A*, 1028 (2004), 297–306.

Z. Brockwicz, Investigation on the identification of casein and egg media in murals by chromatographic separations on filter paper discs, *Materialy Zachodnio–Pomorskie*, 16 (1970), 601–637.

M. Broekman-Bokstijn, J.R.J. Van Asperen De Boer, E.H. Van Thul-Ehrnreich and C.M. Verdvyn-Groen, The scientific examination of the polychromed sculpture in the Herlin Alterpiece, *Studies in Conservation*, 15 (1970), 370–400.

F. Cappitelli, T. Learner and O. Chiantore, An initial assessment of thermally assisted hydrolysis and methylation gas chromatography – mass spectrometry for the identification of oils from dried paint films, *Journal of Analytical and Applied Pyrolysis*, 63 (2002), 339–348.

F. Cappitelli, THM–GC–MS and FTIR for the study of binding media in Yellow Islands by Jackson Pollock and Break Point by Fiona Banner, *Journal of Analytical and Applied Pyrolysis*, 71 (2004), 405–415.

F. Cappitelli and F. Koussiaki, THM–GC–MS and FTIR for the investigation of paints in Picasso's Still Life, Weeping Woman and Nude Woman in a Red Chair from the Tate Collection, London, *Journal of Analytical and Applied Pyrolysis*, 75 (2006), 200–204.

M. Carbini, S. Volpin and P. Traldi, Curie-point pyrolysis gas chromatography – mass spectrometry in the identification of paint media, *Organic Mass Spectrometry*, 29 (1994), 561–565.

M. Carbini, R. Stevano, M. Rovea *et al.*, Curie-point pyrolysis gas chromatography – mass spectrometry in the art field. 2 – The characterisation of proteinaceous binders, *Rapid Communications in Mass Spectrometry*, 10 (1996), 1240–1242.

P. Cardiano, S. Ioppolo, C. De Stefano, *et al.*, Study and characterisation of the ancient bricks of monastery of 'San Filippo di Fragalà' in Frazzanò (Sicily), *Analytica Chimica Acta*, 519 (2004), 103–111.

L. Carlson, A. Feldthus, T. Klarskov and A. Shedrinsky, Geographical classification of amber based on pyrolysis and infrared spectroscopy data, *Journal of Analytical and Applied Pyrolysis*, 43 (1997), 71–81.

M.J. Casas-Catalan and M.T. Domenech-Carbo, Identification of natural dyes used in works of art by pyrolysis-gas chromatography/mass spectrometry combined with in situ trimethylsilylation, *Analytical and Bioanalytical Chemistry*, 382 (2005), 259–268.

A. Casoli, P.C. Musini and G. Palla, Gas chromatographic – mass spectrometric approach to the problem of characterising binding media in paintings, *Journal of Chromatography A*, 731 (1996), 237–246.

A. Casoli, G. Palla and J. Tavlaridis, Gas chromatography – mass spectrometry of works of art: characterisation of binding media in post-Byzantine icons, *Studies in Conservation*, 43 (1998), 150–158.

J.M. Challinor,   A rapid simple pyrolysis derivatisation gas chromatography – mass spectrometry method for profiling of fatty acids in trace quantities of lipids, *Journal of Analytical and Applied Pyrolysis*, **17** (1996), 185–197.

O. Chiantore, D. Scalarone and T. Learner,   Characterisation of artists' acrylic emulsion paints, *International Journal of Polymer Analysis and Characterisation*, **8** (2003), 67–82.

G. Chiavari, S. Ferretti, G.C. Galletti and R. Mazzeo,   Analytical pyrolysis as a tool for the characterisation of organic substances in artistic and archaeological objects, *Journal of Analytical and Applied Pyrolysis*, **20** (1991), 253–261.

G. Chiavari, G.C. Galletti, G. Lanterna and R. Mazzeo,   The potential of pyrolysis gas chromatography – mass spectrometry in the recognition of ancient painting media, *Journal of Analytical and Applied Pyrolysis*, **24** (1993), 227–242.

G. Chiavari, D. Fabbri, G.C. Galletti and R. Mazzeo,   Use of analytical pyrolysis to characterise Egyptian painting layers, *Chromatographia*, **40** (1995a), 594–600.

G. Chiavari, D. Fabbri, R. Mazzeo, *et al.*,   Pyrolysis gas chromatography – mass spectrometry of natural resins used for artistic objects, *Chromatographia* **41** (1995b) 273–281.

G. Chiavari, G.C. Galletti, P. Bocchini, *et al.*,   Thermally assisted hydrolysis methylation gas chromatography mass spectrometry in painting layers from the Cupola of Santa Maria del Fiore Cathedral in Florence, *Science and Technology for Cultural Heritage*, **7** (1998), 19–25.

G. Chiavari and S. Prati,   Analytical pyrolysis as diagnostic tool in the investigation of works of art, *Chromatographia*, **58** (2003), 543–554.

G. Chiavari, D. Fabbri and S. Prati,   Effect of pigments on the analysis of fatty acids in siccative oils by pyrolysis methylation and silylation, *Journal of Analytical and Applied Pyrolysis*, **74** (2005), 39–44.

M.P. Colombini, B. Muscatello, R. Fuoco and A. Giacomelli,   Characterisation of proteinaceous binders in wall painting samples by microwave assisted acid hydrolysis and GC–MS determination of amino acids, *Studies in Conservation*, **43** (1998), 33–41.

M.P. Colombini, F. Modugno, M. Giacomelli and S. Francesconi,   Characterisation of proteinaceous binders and drying oils in wall painting samples by gas chromatography – mass spectrometry, *Journal of Chromatography A*, **846** (1999a), 113–124.

M.P. Colombini, F. Modugno and A. Giacomelli,   Two procedures for suppressing interference from inorganic pigments in the analysis by gas chromatography – mass spectrometry of proteinaceous binders in paintings, *Journal of Chromatography A*, **846** (1999b), 101–111.

M.P. Colombini, F. Modugno, E. Menicagli, *et al.*,   GC–MS characterisation of proteinaceous and lipid binders in UV aged polychrome artefacts, *Microchemical Journal*, **67** (2000a), 291–300.

M.P. Colombini, F. Modugno, S. Giannarelli, *et al.*,   GC-MS characterisation of paint varnishes, *Microchemical Journal*, **67** (2000b), 385–396.

M.P. Colombini, F. Modugno, R. Fuoco and A. Tognazzi,   A GC–MS study on the deterioration of lipidic paint binder, *Microchemical Journal*, **73** (2002a), 175–185.

M.P. Colombini, A. Ceccarini and A. Carmignani,   Ion chromatography characterisation of polysaccharides in ancient wall paintings, *Journal of Chromatography A*, **968** (2002b), 79–88.

M.P. Colombini and F. Modugno,   Characterisation of proteinaceous binders in artistic paintings by chromatographic techniques, *Journal of Separation Science*, **27** (2004), 147–160.

M.P. Colombini, A. Carmignani, F. Modugno, *et al.*, Integrated analytical techniques for the study of ancient Greek polychromy, *Talanta*, **63** (2004), 839–848.

M.P. Colombini, G. Giachi, F. Modugno and E. Ribechini, Characterisation of organic residues in pottery vessels of the Roman age from Antinoe (Egypt), *Microchemical Journal*, **79** (2005), 83–90.

M.S. Copley, R. Berstan, V. Straker, *et al.*, Dairying in antiquity. II Evidence from absorbed lipid residues dating to the British Bronze Age, *Journal of Archaeological Science*, **32** (2005a), 505–521.

M.S. Copley, R. Berstan, A.J. Mukherjee, *et al.*, Dairying in antiquity. III Evidence from absorbed lipid residues dating to the British Neolithic, *Journal of Archaeological Science*, **32** (2005b), 523–546.

J.V. Dawkins, Size exclusion chromatography, in *Comprehensive Polymer Science*, Vol. 1 (eds C. Booth and C. Price), Pergamon Press, Oxford (1989), pp 231–258.

F. De Angelis, A. Di Tullio, G. Mellerio, *et al.*, Investigation by solid-phase micro-extraction and gas chromatography/mass spectrometry of organic films on stone monuments, *Rapid Communications in Mass Spectrometry*, **13** (1999), 895–900.

F. De Angelis, A. Di Tullio, R. Ceci, *et al.*, Application of multivariate analysis for recognition of organic patinas on stone monuments, *Journal of Separation Science*, **25** (2002), 29–36.

J. de la Cruz-Canizares, M.T. Domenech-Carbo, J.V. Gimeno-Adelantado, *et al.*, Suppression of pigment interference in the gas chromatographic analysis of proteinaceous binding media in paintings with EDTA, *Journal of Chromatography A*, **1025** (2004), 277–285.

J. de la Cruz-Canizares, M.T. Domenech-Carbo, J.V. Gimeno-Adelantado, *et al.*, Study of Burseraceae resins used in binding media and varnishes from artworks by gas chromatography-mass spectrometry and pyrolysis gas chromatography-mass spectrometry, *Journal of Chromatography A*, **1093** (2005), 177–194.

E.R. de la Rie, The influence of varnishes on the appearance of paintings, *Studies in Conservation*, **32** (1987), 1–13.

M.T. Domenech-Carbo, M.J. Casas-Catalan, A. Domenech-Carbo, *et al.*, Analytical study of canvas painting collection from the Basilica de la Virgen de los Desamparados using SEM/EDX, FTIR, GC and electrochemical techniques, *Fresenius Journal of Analytical Chemistry*, **369** (2001), 571–575.

S. Dudd, R.P. Evershed and A.M. Gibson, Evidence for varying patterns exploitation of animal products in different prehistoric pottery traditions based on lipids preserved in surface and absorbed residues, *Journal of Archaeological Science*, **26** (1999), 1473–1482.

A.L. Dupont, Study of the degradation of gelatin in paper upon ageing using aqueous size exclusion chromatography, *Journal of Chromatography A*, **950** (2002), 113–124.

R. Edwards, W. Bordass and D. Farrell, Determination of acetic and formic acid in lead corrosion products by ion exchange chromatography, *Analyst*, **122** (1997), 1517–1520.

R.P. Evershed, C. Heron and L.J. Goad, Analysis of organic residues of archaeological origin by high temperature gas chromatography and gas chromatography – mass spectrometry, *Analyst*, **115** (1990), 1339–1342.

R.P. Evershed, Biomolecular archaeology and lipids, *World Archaeology*, **25** (1993), 74–93.

R.P. Evershed, H.R. Mottram, S.N. Dodd, *et al.*, New criteria for the identification of animal fats preserved in archaeological pottery, *Naturwissenschaften*, **84** (1997), 402–406.

R.P. Evershed, S.N. Dudd, S. Charters, *et al.*, Lipids as carriers of anthropogenic signals from prehistory, *Philosophical Transactions of the Royal Society*, **354** (1999), 19–31.

R.P. Evershed, S.N. Dudd, M.S. Copley, *et al.*, Chemistry of archaeological animal fats, *Accounts of Chemical Research*, **35** (2002), 660–668.

D. Fabbri, G. Chiavari and H. Ling, Analysis of anthraquinoid and indigoid dyes used in ancient artistic works by thermally assisted hydrolysis and methylation in the presence of tetramethylammonium hydroxide, *Journal of Analytical and Applied Pyrolysis* **56**, (2000), 167–178.

V. Fassina, M. Favaro, A. Naccari and M. Pigo, Evaluation of compatibility and durability of a hydraulic lime-based plaster applied on brick wall masonry of historical buildings affected by rising damp phenomena, *Journal of Cultural Heritage*, **3** (2002), 45–51.

M. Favaro, P.A. Vigato, A. Andeotti and M.P. Colombini, La Medusa by Caravaggio: characterisation of the painting technique and evaluation of the state of conservation, *Journal of Cultural Heritage*, **6** (2005), 295–305.

N. Ferrer and M.C. Sistach, Characterisation by FTIR spectroscopy on ink components in ancient manuscripts, *Restaurator*, **26** (2005), 105–117.

J.S. Fritz and D.T. Gjende, *Ion Chromatography*, Wiley–VCH, Weinheim (2000).

G.C. Galletti, P. Bocchini, G. Chiavari and D. Fabbri, High performance liquid chromatography versus gas chromatography in the analysis of proteinaceous painting ligands, *Fresenius Journal of Analytical Chemistry*, **354** (1996), 381–383.

J. Gasparic and J. Churacek, *Laboratory Handbook of Paper and Thin Layer Chromatography*, Ellis Horwood, New York (1978).

N. Ghendini, G. Gobbi, C. Sabbioni and G. Zappia, Determination of elemental and organic carbon on damaged stone monuments, *Atmospheric Environment*, **34** (2000), 4383–4391.

J.V. Gimeno-Adelantado, R. Mateo-Castro, M.T. Domenech-Carbo, *et al.*, Identification of lipid binders in paintings by gas chromatography. Influence of pigments, *Journal of Chromatography A*, **922** (2001), 385–390.

J.V. Gimeno-Adelantado, R. Mateo-Castro, M.T. Domenech-Carbo, *et al.*, Analytical study of proteinaceous binding media in works of art by gas chromatography using alkyl chloroformates as derivatising agents, *Talanta*, **56** (2002), 71–77.

G. Gobbi, G. Zappia and G. Sabbioni, Anion determination in damage layers of stone monuments, *Atmospheric Environment*, **29** (1995), 703–707.

G. Gobbi, G. Zappia and C. Sabbioni, Sulfite quantification on damaged stones and mortars, *Atmospheric Environment*, **32** (1998), 783–789.

A.F.L. Godoi, L. Van Vaeck and R. Van Grieken, Use of solid-phase microextraction for the detection of acetic acid by ion-trap gas chromatography – mass spectrometry and application to indoor levels in museums, *Journal of Chromatography A*, **1067** (2005), 331–336.

M. Grossl, S. Harrison, I. Kaml and E. Kenndler, Characterisation of natural polysaccharides (plant gums) used as binding media for artistic and historic works by capillary zone electrophoresis, *Journal of Chromatography A*, **1077** (2005), 80–89.

C.M. Grzywacz, Identification of proteinaceous binding media in paintings by amino acid analysis using 9-fluorenylmethyl chloroformate derivatisation and reversed phase high performance liquid chromatography, *Journal of Chromatography A*, **676** (1994), 177–183.

S.M. Halpine, An improved dye and lake pigment analysis method using high performance liquid chromatography and diode array detector, *Studies in Conservation*, **41** (1996), 76–94.

S.M. Halpine, HPLC applications in art conservation, in *Handbook of HPLC* (ed E. Katz), Marcel Dekker, New York (1998), pp 903–927.

D.C. Harris, *Quantitative Chemical Analysis*, 6th edn, W.H. Freeman, New York (2003).

S.M. Harrison, I. Kaml, V. Prokoratova, *et al.*, Animal glues in mixtures of natural binding media used in artistic and historic objects: identification by capillary zone electrophoresis, *Analytical and Bioanalytical Chemistry*, **382** (2005), 1520–1526.

L.M. Henriksen and H. Kjosen, Derivatisation of natural anthraquinones by reductive silylation for gas chromatographic and gas chromatographic–mass spectrometric analysis, *Journal of Chromatography*, **258** (1983), 252–257.

M. Hey, The analysis of paint media by paper chromatography, *Studies in Conservation*, **3** (1958), 183–193.

I. Kaml, K. Vcelakova and E. Kenndler, Characterisation and identification of proteinaceous binding media (animal glues) from their amino acid profile by capillary zone electrophoresis, *Journal of Separation Science*, **27** (2004), 161–166.

R. Karadag and E. Dolen, Examination of historical textiles with dyestuff analyses by TLC and derivative spectrophotometry, *Turkish Journal of Chemistry*, **21** (1997), 126–133.

I. Karapanagiotis and Y. Chryssoulakis, Investigation of red natural dyes used in historical objects by HPLC–DAD–MS, *Annali di Chimica*, **95** (2005), 75–84.

I. Karapanagiotis, A. Tsakalof, S. Danilia and Y. Chyssoulakis, Identification of red natural dyes in post-Byzantine icons by HPLC, *Journal of Liquid Chromatography and Related Technology*, **28** (2005), 739–749.

B.V. Kharbade and O.P. Agrawal, Identification of natural red dyes in old Indian textiles. Evaluation of thin layer chromatographic systems, *Journal of Chromatography*, **347** (1985), 447–454.

B.V. Kharbade and G.P. Joshi, Thin layer chromatographic and hydrolysis methods for the identification of plant gums in art objects, *Studies in Conservation*, **40** (1995), 93–102.

D.A. Kouznetsov, A.A. Ivanov and P.R. Veletsky, Detection of alkylated cellulose derivatives in several archaeological linen textile samples by capillary electrophoresis/mass spectrometry, *Analytical Chemistry*, **66** (1994), 4359–4365.

D.A. Kouznetsov, A.A. Ivanov and P.R. Veletsky, Analysis of cellulose chemical modification: a potentially promising technique for characterising cellulose archaeological textiles, *Journal of Archaeological Sciences*, **23** (1996), 23–34.

A. Lattuati-Derieux, S. Bonnassies-Termes and B. Laverdrine, Characterisation of compounds emitted during natural and artificial ageing of a book. Use of headspace solid phase microextraction gas chromatography-mass spectrometry, *Journal of Cultural Heritage*, **7**, (2006) 123–133.

T. Learner, The analysis of synthetic paints by pyrolysis gas chromatography – mass spectrometry (Py–GC–MS), *Studies in Conservation*, **46** (2001), 225–241.

T.J.S. Learner, *Analysis of Modern Paints*, Getty Conservation Institute, Los Angeles (2004).

T. Learner, Modern paints, in *Scientific Examination of Art: Modern Techniques in Conservation and Analysis*, National Academies Press, Washington DC (2005), pp 137–151.

R. Lleti, L.A. Sarabia, M.C. Ortiz, *et al.*, Application of the Kohonen artificial neural network in the identification of the proteinaceous binders in samples of panel painting using gas chromatography – mass spectrometry, *Analyst*, **128** (2003), 281–286.

W.J. Lough and I.M. Wainer (eds), *High Performance Liquid Chromatography: Fundamental Principles and Practice*, Chapman and Hall, London (1996).

C.A. Maines and E. Rene de la Rie, Size exclusion chromatography and differential scanning calorimetry of low molecular weight resins used as varnishes for paintings, *Progress in Organic Coatings*, **52** (2005), 39–45.

C. Marinach, M.C. Papillon and C. Pepe, Identification of binding media in works of art by gas chromatography – mass spectrometry, *Journal of Cultural Heritage*, **5** (2004), 231–240.

N. Marinoni, A. Pavese, R. Bugini and G. Di Silvestro, Black limestone used in Lombard architecture, *Journal of Cultural Heritage*, **3** (2002), 241–249.

L. Masschelein-Kleiner, Microanalysis of hydroxyquinones in red lakes, *Mikrochimica Acta*, **6** (1967), 1080–1085.

L. Masschelein-Kleiner, An improved method for the thin layer chromatography of media in tempera paintings, *Studies in Conservation*, **19** (1974), 207–211.

R. Mateo-Castro, M.T. Domenech-Carbo, V. Peris-Martinez, *et al.*, Study of binding media in works of art by gas chromatographic analysis of amino acids and fatty acids derivatised with ethyl chloroformate, *Journal of Chromatography A*, **778** (1997), 373–381.

R. Mateo-Castro, J.V. Gimeno-Adelantado, F. Bosch-Reig, *et al.*, Identification by GC–FID and GC–MS of amino acids, fatty and bile acid in binding media used in works of art, *Fresenius Journal of Analytical Chemistry*, **369** (2001), 642–646.

P. Mejanelle, J. Bleton, S. Goursaud and A. Tchapla, Identification of phenolic acids and inositols in balms and tissues from an Egyptian mummy, *Journal of Chromatography A*, **767** (1997), 177–186.

V.R. Meyer, *Practical High Performance Liquid Chromatography*, Wiley–VCH, Weinheim (2004).

J.M. Miller, *Chromatography: Concepts and Contrasts*, 2nd edn, Wiley–Interscience, New York (2004).

J.S. Mills and E.A. Wilmer, Paper chromatography of natural resins, *Nature*, **169** (1952), 1064.

J.S. Mills and R. White, *The Organic Chemistry of Museum Objects*, 2nd edn, Butterworth-Heinemann, Oxford (1994).

S. Mori and H.G. Barth, *Size Exclusion Chromatography*, Springer, Berlin (1999).

S. Nakamura, M. Takino and S. Daishima, Analysis of waterborne paints by gas chromatography – mass spectrometry with a temperature-programmable pyrolyser, *Journal of Chromatography A*, **912** (2001), 329–334.

P. Novotna, V. Pacakova, Z. Bosakova and K. Stulik, High performance liquid chromatographic determination of some anthraquinone and naphthoquinone dyes occurring in historical textiles, *Journal of Chromatography A*, **863** (1999), 235–241.

J. Orska-Gawrys, I. Surowicz, J. Kehl, *et al.*, Identification of natural dyes in archaeological Coptic textiles by liquid chromatography with diode array detection, *Journal of Chromatography A*, **989** (2003), 239–248.

L. Osete-Cortina, M.T. Domenech-Carbo, *et al.*, Identification of diterpenes in canvas painting varnishes by gas chromatography – mass spectrometry with combined derivatisation *Journal of Chromatography A*, **1024** (2004), 187–194.

L. Osete-Cortina and M.T. Domenech-Carbo, Analytical characterisation of diterpenoid resins present in pictorial varnishes using pyrolysis gas chromatography–mass spectrometry with on line trimethylsilylation, *Journal of Chromatography A*, **1065** (2005a), 265–278.

L. Osete-Cortina and M.T. Domenech-Carbo, Study of the effects of chemical cleaning on pinaceae resin-based varnishes from panel and canvas paintings using pyrolysis gas chromatography – mass spectrometry, *Journal of Analytical and Applied Pyrolysis*, **76** (2005b), 144–153.

L. Osete-Cortina and M.T. Domenech-Carbo, Characterisation of acrylic resins used for restoration of artworks by pyrolysis–silylation–gas chromatography/mass spectrometry with hexamethyldisilazane, *Journal of Chromatography A*, **1127** (2006) 228–236.

J. Peris-Vicente, J.V. Gimeno Adelantado, *et al.*, Identification of drying oils used in pictorial works of art by liquid chromatography of the 2-nitrophenylhydrazides derivatives of fatty acids, *Talanta*, **64** (2004), 326–333.

J. Peris-Vicente, J.V. Gimeno Adelantado, M.T. Domenech-Carbo, *et al.*, Characterisation of proteinaceous glues in old paintings by separation of the o-phthaldehyde derivatives of their amino acids by liquid chromatography with fluorescence detection, *Talanta*, **68** (2005a), 1648–1654.

J. Peris-Vicente, J.V. Gimeno Adelantado, M.T. Domenech-Carbo, *et al.*, Identification of lipid binders in old oil paintings by separation of 4-bromomethyl-7-methoxycoumarin derivatives of fatty acids by liquid chromatography with fluorescence detection, *Journal of Chromatography A*, **1076** (2005b), 44–50.

J. Peris-Vicente, J.V. Gimeno Adelantado, M.T. Domenech-Carbo, *et al.*, Characterisation of waxes used in pictorial artworks according to their relative amount of fatty acids and hydrocarbons by gas chromatography, *Journal of Chromatography A*, **1101** (2006), 254–260.

S.H. Perry and A.P. Duffy, The short-term effects of mortar joints on salt movement in stone, *Atmospheric Environment*, **31** (1997), 1297–1305.

V. Pitthard, P. Finch and T. Bayerova, Direct chemolysis – gas chromatography – mass spectrometry for analysis of paint materials, *Journal of Separation Science*, **27** (2004), 2000–208.

V. Pitthard, S. Stanek, M. Griesser and T. Muxeneder, Gas chromatography – mass spectrometry of binding media from early 20th century paint samples from Arnold Schonberg's palette, *Chromatographia*, **62** (2005), 175–182

V. Pitthard, M. Griesser, S. Stanek and T. Bayerova, Study of complex organic binding media systems on artworks applying GC–MS analysis: selected examples from the Kunsthistorisches Museum, Vienna, *Macromolecular Symposia*, **238** (2006), 37–45.

M.J. Plater, B. De Silva, T. Gelbrich, *et al.*, The characterisation of lead fatty acid soaps in 'protrusions' in aged traditional oil paint, *Polyhedron*, **22** (2003), 3171–3179.

S. Prati, S. Smith and G. Chiavari, Characterisation of siccative oils, resins and pigments in art works by thermochemolysis coupled to thermal desorption and pyrolysis GC and GC-MS, *Chromatographia*, **59** (2004), 227–231.

E. Princi, S. Vicini, E. Pedemonte, *et al.*, New polymeric materials for paper and textile conservation. I. Synthesis and characterisation of acrylic copolymers, *Journal of Applied Polymer Science*, **98** (2005), 1157–1164.

M. Puchalska, M. Orlinska, M.A. Ackacha, *et al.*, Identification of anthraquinone colouring matters in natural red dyes by electrospray mass spectrometry coupled to capillary electrophoresis, *Journal of Mass Spectrometry*, 38 (2003), 1252–1258.

M. Puchalska, K. Polec-Pawiak, I. Zadrozna, *et al.*, Identification of indigoid dyes in natural organic pigments used in historical art objects by high performance liquid chromatography coupled to electrospray ionisation mass spectrometry, *Journal of Mass Spectrometry*, 39 (2004), 1441–1449.

A. Quye, Historical plastics come of age, *Chemistry in Britain*, 31 (1995), 617–620.

A. Quye and C. Williamson, *Plastics: Collecting and Conserving*, NMS Publishing, Edinburgh (1999).

P.P. Rai and M. Shok, Thin layer chromatography of hydroxyanthraquinones in plant extracts, *Chromatographia*, 14 (1981), 599–600.

L. Rampazzi, F. Cariati, G. Tanda and M.P. Colombini, Characterisation of wall paintings in the Sos Furrighesos necropolis (Anela, Italy), *Journal of Cultural Heritage*, 3 (2002), 237–240.

L. Rampazzi, A. Andreotti, I. Bonaduce, *et al.*, Analytical investigation of calcium oxalate films in marble monuments, *Talanta*, 63 (2004), 967–977.

M. Regert, N. Garnier, O. Decavallas, *et al.*, Structural characterisation of lipid constituents from natural substances preserved in archaeological environments, *Measurement Science and Technology*, 14 (2003), 1620–1630.

M. Regert, Investigating the history of prehistoric glues by gas chromatography – mass spectrometry, *Journal of Separation Science*, 27 (2004), 244–254.

M. Regert, J. Longlois and S. Colinart, Characterisation of wax works of art by gas chromatographic procedures, *Journal of Chromatography A*, 1091 (2005), 124–136.

C. Riotino, C. Sabbioni, N. Ghedini, *et al.*, Evaluation of atmospheric deposition on historic buildings by combined thermal analysis and combustion techniques, *Thermochimica Acta*, 321 (1998), 215–222.

P. Rudolph, F.J. Ligterink, J.L. Pedersoli, *et al.*, Characterisation of laser-treated paper, *Applied Physics*, 79 (2004), 181–186.

M. Ryhl-Svendsen and J. Glastrup, Acetic acid and formic acid concentrations in the museum environment measured by SPME–GC–MS, *Atmospheric Environment*, 26 (2002), 3909–3916.

C. Sabbioni, G. Zappia, C. Riotino, *et al.*, Atmospheric deterioration of ancient and modern hydraulic mortars, *Atmospheric Environment*, 35 (2001), 539–548.

C. Saiz-Jimenez, B. Hermosin, J.J. Ortega-Calvo and G. Gomez-Alarcon, Applications of analytical pyrolysis to the study of stony cultural properties, *Journal of Analytical and Applied Pyrolysis*, 20 (1991), 239–251.

C. Saiz-Jimenez, Deposition of airborne organic pollutants on historic buildings, *Atmospheric Environment*, 27B (1993), 77–85.

D. Scalarone, M. Lazzari and O. Chiantore, Thermally assisted hydrolysis and methylation pyrolysis gas chromatography – mass spectrometry of light-aged linseed oil, *Journal of Analytical and Applied Pyrolysis*, 58–59 (2001), 503–512.

D. Scalarone, M. Lazzari and O. Chiantore, Ageing behaviour and pyrolytic characterisation of diterpenic resins used as art materials: colophony and Venice turpentine, *Journal of Analytical and Applied Pyrolysis*, 64 (2002), 345–361.

D. Scalarone, M. Lazzari and O. Chiantore, Ageing behaviour and analytical pyrolysis characterisation of diterpenic resins used as art materials: Manila copal and sandarac, *Journal of Analytical and Applied Pyrolysis*, 68–69 (2003), 115–126.

D. Scalarone and O. Chiantore, Separation techniques for the analysis of artists' acrylic emulsion paints, *Journal of Separation Science*, **27** (2004), 263–274.

M.R. Schilling, H.P. Khanjian and L.A.C. Souza, Gas chromatographic analysis of amino acids as ethyl chloroformate derivatives, *Journal of the American Institute for Conservation*, **35** (1996), 45–59.

M.R. Schilling and H.P. Khanjian, Gas chromatographic analysis of amino acids as ethyl chloroformate derivatives, *Journal of the American Institute for Conservation*, **35** (1996), 123–144.

M.R. Schilling, Paint media analysis, in *Scientific Examination of Art: Modern Techniques in Conservation and Analysis*, National Academies Press, Washington DC (2005), pp 186–205.

U. Schneider and E. Kenndler, Identification of plant and animal glues in museum objects by GC–MS, after catalytic hydrolysis of the proteins by the use of a cation exchanger, with simultaneous separation from the carbohydrates, *Fresenius Journal of Analytical Chemistry*, **371** (2001), 81–87.

B. Schroeder, J. Mangels, K. Selke, *et al.*, Comparative analysis of organic acids located on the surface of natural building stones by high performance liquid, gas and ion chromatography, *Journal of Chromatography*, **514** (1990), 241–251.

H. Schweppe, Identification of dyes on old textiles, *Journal of the American Institute for Conservation*, **19** (1979), 14–23.

H. Schweppe, Identification of red madder and insect dyes by thin layer chromatography, in *Historic Textile and Paper Materials II* (eds S.H. Zeronian and H.L. Needles), American Chemical Society, Washington (1989), pp 188–219.

D.A. Scott, N. Khandekar, M.R. Schilling, *et al.*, Technical examination of a 15th century German illuminated manuscript on paper: a case study in the identification of materials, *Studies in Conservation*, **46** (2001), 93–108.

A.M. Shedrinsky, T.P. Wampler, N. Indicator and N.S. Baer, Application of analytical pyrolysis to problems in art and archaeology: a review, *Journal of Analytical and Applied Pyrolysis*, **15** (1989), 393–412.

A.M. Shedrinsky, R.E. Stone and N.S. Baer, Pyrolysis gas chromatographic studies on Egyptian archaeological specimens: organic patinas on the 'Three Princesses' gold vessels, *Journal of Analytical and Applied Pyrolysis*, **20** (1991), 229–238.

A.M. Shedrinsky, D.A. Grimaldi, J.J. Boon and N.S. Baer, Application of pyrolysis gas chromatography and pyrolysis gas chromatography – mass spectrometry to the unmasking of amber forgeries, *Journal of Analytical and Applied Pyrolysis*, **25** (1993), 77–95.

A.M. Shedrinsky, T.P. Wampler and K.V. Chugunov, The examination of amber beads from the collection of the state hermitage museum found in Arzhan-2 burial memorial site, *Journal of Analytical and Applied Pyrolysis*, **71** (2004), 69–81.

N. Shibayama, S.Q. Lomax, K. Sutherland and E.R. de la Rie, Atmospheric pressure chemical ionisation liquid chromatography mass spectrometry and its application to conservation: analysis of triacylglycerols, *Studies in Conservation*, **44** (1999), 253–268.

D.A. Skoog and J.J. Leary, *Principles of Instrumental Analysis*, Harcourt Brace College Publishers, Fort Worth (1992).

N. Sonoda, J.P. Rioux and A.R. Duval, Identification des matériaux synthétiques dans les peintures modernes. II Pigments organiques et matière picturale, *Studies in Conservation*, **38** (1993), 99–127.

N. Sonoda, Characterisation of organic azo-pigments by pyrolysis-gas chromatography, *Studies in Conservation*, **44** (1999), 195–208.

R. Stevanato, M. Rovea, M. Carbini, *et al.*, Curie-point pyrolysis gas chromatography – mass spectrometry in the art field. Part 3: The characterisation of some non-proteinaceous binders, *Rapid Communications in Mass Spectrometry*, 11 (1997), 286–294.

R. Stol, J.L. Pedersoil, H. Poppe and W.T. Kok, Application of size exclusion electrochromatography to the microanalytical determination of the molecular mass distribution of celluloses from objects of cultural and historical value, *Analytical Chemistry*, 74 (2002), 2314–2320.

M. Striegel and J. Hill, *Thin Layer Chromatography for Binding Media Analysis*, Getty Conservation Institute, Los Angeles (1996).

I. Surowiec, J. Orska-Gawrys, M. Biesaga, *et al.*, Identification of natural dyestuff in archaeological Coptic textiles by HPLC with fluorescence detection, *Analytical Letters*, 36 (2003), 1211–1229.

I. Surowiec, W. Nowik and M. Trojanowicz, Identification of insoluble red dyewoods by high performance liquid chromatography – photodiode array detection (HPLC–PDA) fingerprinting, *Journal of Separation Science*, 27 (2004a), 209–216.

I. Surowiec, I. Kaml and E. Kenndler, Analysis of drying oils used as binding media for objects of art by capillary electrophoresis with indirect UV and conductivity detection, *Journal of Chromatography A*, 1024 (2004b), 245–254.

I. Surowiec, A. Quye and M. Trojanowicz, Liquid chromatography determination of natural dyes in extracts from historical Scottish textiles excavated from peat bogs, *Journal of Chromatography A*, 1112 (2005), 209–217.

B. Szostek, J. Orska-Gawrys, I. Surowiec and M. Trojanowicz, Investigation of natural dyes occurring in historical Coptic textiles by high performance liquid chromatography with UV-vis and mass spectrometric detection, *Journal of Chromatography A*, 1012 (2003), 179–192.

A. Tchapla, P. Mejanelle, J. Bleton and S. Goursaud, Characterisation of embalming materials of a mummy of the Ptolemaic era. Comparison with balms from mummies of different eras, *Journal of Separation Science*, 27 (2004), 217–234.

D. Thorburn Burns and K.P. Doolan, A comparison of pyrolysis gas chromatography – mass spectrometry and Fourier transform infrared spectroscopy for the analysis of a series of modified alkyd paint resins, *Analytica Chimica Acta*, 422 (2000), 217–230.

A. Timar-Balazsy and D. Eastop, *Chemical Principles of Textile Conservation*, Butterworht-Heinemann, Oxford (1998).

J. Tomek and D. Pechova, A note of the thin layer chromatography of media in paintings, *Studies in Conservation*, 37 (1992), 39–41.

M. Trojanowicz, J. Orska-Gawrys, I. Surowiec, *et al.*, Chromatographic investigation of dyes extracted from Coptic textiles from the National Museum in Warsaw, *Studies in Conservation*, 49 (2004), 115–130.

A.K. Tsakalof, K.A. Bairachtari, I.S. Asiani, *et al.*, Impact of biological factors on binding media identification in art objects: identification of animal glue in the presence of Aspergillus niger, *Journal of Separation Science*, 27 (2004), 167–173.

S. Tse and A.L. Dupont, Measuring silk deterioration by high performance size exclusion chromatography, viscometry and electrophoresis, in *Historic Textiles, Papers and Polymers in Museums* (eds J. M. Cardamone and M.T. Baker, American Chemical Society, Washington (2001), pp 98–114.

S.L. Vallance, Application of chromatography in art conservation: techniques used for the analysis and identification of proteinaceous and gum binding media, *Analyst*, **122** (1997), 75R–81R.

S.L. Vallance, B.W. Singer, S.M. Hitchen and J.H. Townsend, The development and initial application of a gas chromatographic method for the characterisation of gum media, *Journal of the American Institute for Conservation*, **37** (1998), 294–311.

K.J. van den Berg, J. van der Horst, J.J. Boon and O.O. Sudmeijer, *Cis*-1,4-poly-β-myrcene; the structure of the polymeric fraction of mastic resin (Pistacia lentiscus L.) elucidated, *Tetrahedron Letters*, **39** (1998a), 2645–2648.

J.D.J. van den Berg, J.J. Boon and K.J. van den Berg, Identification of an original non-terpenoid varnish from the early 20th century oil painting 'The White Horse' (1929) by H. Menzel, *Analytical Chemistry*, **70** (1998b), 1823–1830.

K.J. van den Berg, J.J. Boon, I. Pastorova and L.F.M. Spetter, Mass spectrometric methodology for the analysis of highly oxidised diterpenoid acids in Old Master paintings, *Journal of Mass Spectrometry*, **35** (2000), 512–533.

J.D.J. van den Berg, K.J. van den Berg and J.J. Boon, Determination of the degree of hydrolysis of oil paint samples using a two-step derivatisation method and on-column GC–MS, *Progress in Organic Coatings*, **41**, (2001) 143–155.

J.D.J. van den Berg, Analytical chemical studies on traditional linseed oil paints, PhD thesis, Universiteit van Amsterdam (2002).

J.D.J. van den Berg, K.J. van den Berg and J.J. Boon, Identification of non-cross-linked compounds in methanolic extracts of cured and aged linseed oil-based paint films using gas chromatography – mass spectrometry, *Journal of Chromatography A*, **950** (2002), 195–211.

J.D.J. van den Berg, N.D. Vermist, L. Carlyle, *et al.*, Effects of traditional processing methods of linseed oil on the composition of its triacylglycerols, *Journal of Separation Science*, **27** (2004), 181–199.

G.A. van der Doelen, K.J. van den Berg and J.J. Boon, Comparative chromatographic and mass spectrometric studies of triterpenoid varnishes: fresh material and aged samples from paintings, *Studies in Conservation*, **43** (1998a), 249–264.

G.A. van der Doelen, K.J. van den Berg, J.J. Boon, *et al.*, Analysis of fresh triterpenoid resins and aged triterpenoid varnishes by high performance liquid chromatography – atmospheric pressure chemical ionisation (tandem) mass spectrometry, *Journal of Chromatography A*, **809** (1998b), 21–37.

G.A. van der Doelen and J.J. Boon, Artificial ageing of varnish triterpenoids in solution, *Journal of Photochemistry and Photobiology A: Chemistry*, **134** (2000), 45–57.

G. Vleugels, E. Roekens, A. Van Put, *et al.*, Analytical study of the weathering of the Jeronimos Monastery in Lisbon, *The Science of the Total Environment*, **120** (1992), 225–243.

P.E. Wall, *Thin Layer Chromatography: A Modern Practical Approach*, Royal Society of Chemistry, Cambridge (2005).

A. Wallert and R. Boytner, Dyes from the Tumilaca and Chiribaya cultures, South Coast of Peru, *Journal of Archaeological Science*, **23** (1996), 853–861.

T.P. Wampler, Introduction to pyrolysis–capillary gas chromatography, *Journal of Chromatography A*, **842** (1999), 207–220.

P.A. Warke and B.J. Smith, Salt distribution in clay-rich weathered sandstone, *Earth Surface Processes and Landforms*, **25** (2000), 1333–1342.

R. Weinberger, *Practical Capillary Electrophoresis*, 2nd edn, Academic Press, San Diego (2000).

R. White, The characterisation of proteinaceous binders in art objects, *National Gallery Technical Bulletin*, **8** (1984), 5–14.

R. White, *Chromatography/Fourier Transform Infrared Spectroscopy and its Applications*, Marcel Dekker, New York (1990).

H. Wilken and H.R. Schulten, Differentiation of resin-modified paints by pyrolysis gas chromatography – mass spectrometry and principal component analysis, *Fresenius Journal of Analytical Chemistry*, **355** (1996), 157–163.

J. Wouters, High performance liquid chromatography of anthraquinones: analysis of plant and insect extracts and dyes textiles, *Studies in Conservation*, **30** (1985), 119–128.

J. Wouters and A. Verhecken, The coccid insect dyes: HPLC and computerised diode array analysis of dyed yarns, *Studies in Conservation*, **34**, (1989) 189–200.

J. Wouters, L. Maes and R. Germer, The identification of haematitie as red colourant on an Egyptian textile from the second millennium BC, *Studies in Conservation*, **35** (1990), 89–92.

J. Wouters and A. Verhecken, HPLC of blue and purple indigoid natural dyes, *Journal of the Society of Dyers and Colourists*, **7** (1991), 266–269.

J. Wouters and N. Rosario-Chirinos, Dye analysis of pre-Columbian Peruvian textiles with high performance liquid chromatography and diode array detection, *Journal of the American Institute for Conservation*, **31** (1992), 237–255.

X. Zang, J.C. Brown, J.D.H. van Heemst, A. Palumbo and P.G. Hatcher, Characterisation of amino acids and proteinaceous materials using on-line tetramethylammonium hydroxide (TMAH) thermochemolysis and gas chromatography-mass spectroscopy technique, *Journal of Analytical and Applied Pyrolysis*, **61** (2001), 181–193.

X. Zhang and R.A. Laursen, Development of mild extraction methods for the analysis of natural dyes in textiles of historical interest using LC–diode array detector–MS, *Analytical Chemistry*, **77** (2005), 2022–2025.

G. Zweig and J.R. Whitaker, *Paper Chromatography and Electrophoresis*, Academic Press, New York (1967).

9

Thermal and Mechanical Analysis

9.1 INTRODUCTION

Thermal analysis involves the measurement of physical and chemical changes that a material undergoes as it is heated. These changes can include a release or absorption of energy, a weight loss or gain, or a change in dimension or strength. Such changes occur at temperatures that are characteristic to a particular material and its thermal history. Thermal methods that may be potentially useful to the conservation scientist are differential scanning calorimetry, differential thermal analysis, thermogravimetric analysis, dynamic thermal analysis and thermal mechanical analysis.

The mechanical properties of materials are dependent upon their structural properties. For instance, a degraded sample will respond quite differently to the application of a force to a sample which shows no sign of deterioration. There are a number of standard deformation tests which provide information about materials such as tensile, flexural, fatigue, impact and hardness. Of course, these approaches are generally destructive so are of more use when plenty of sample is available or model ageing experiments are being carried out.

9.2 THERMOGRAVIMETRIC ANALYSIS

Thermogravimetric analysis (TGA) is a thermal method that involves the measurement of the weight loss of a material as a function of temperature

Analytical Techniques in Materials Conservation   Barbara H. Stuart
© 2007 John Wiley & Sons, Ltd

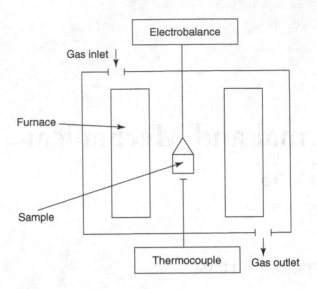

Figure 9.1 Schematic diagram of a TGA instrument

or time [Haines, 1995; Sandler *et al.*, 1998]. It can be used to quantify the mass change in a material associated with transitions or degradation processes. TGA data provide characteristic curves for a given material because each material will show a unique pattern of reactions at specific temperatures.

In TGA, a sample is placed in a furnace while suspended from one arm of a sensitive balance. The change in sample mass is recorded while the sample is maintained at the required temperature or subjected to a programmed heating. The instrument layout of TGA is shown schematically in Figure 9.1. The thermobalance can detect to 0.1 μg and calibration can be made using standard masses. Calcium oxalate monohydrate ($CaC_2O_4.H_2O$) is commonly used for calibration. A heater allows temperatures of up to 2800 °C to be obtained and temperature rates can be varied from 0.1 to 300 °C min^{-1}, with both heating and cooling of the samples possible. Reactions studied in the thermobalance can be carried out in different atmospheres, such as nitrogen, argon, helium or oxygen.

The TGA curve may be plotted as the sample mass loss as a function of temperature, or alternatively, in a differential form where the change in sample mass with time is plotted as a function of temperature. The forms of TGA curves are illustrated in Figure 9.2, which shows how TGA may be used to determine the mass loss. The mass loss associated with an initial step, such as solvent evaporation, is w_0-w_1. The TGA

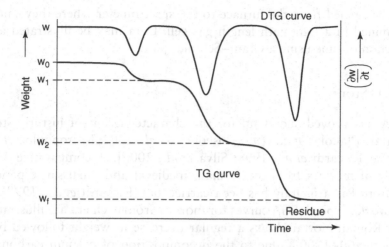

Figure 9.2 Characteristics of TGA curves

curve illustrated shows two degradation processes, but the number will depend on the individual sample. The second step represents the first degradation process and the mass loss is w_1-w_2. The third step, representing a second degradation process, shows a mass loss w_2-w_f, where w_f is the residue that does not decompose in the temperature range covered by the experiment. The derivative curve (DTG) shows a peak associated with each separate step that represents the maximum rate of mass loss.

TGA has been combined with other techniques to supplement the information gained from an experiment. TGA and DSC/DTA can be combined to study a sample. The instruments are designed with a modified thermobalance that weighs both the reference and sample and measures each temperature. To gain more information about thermal transitions it is sometimes necessary to identify the products generated during the process. This is carried out via evolved gas analysis (EGA), where the gases involved in the thermal process are detected for further analyses. TGA can be combined with MS or GC-MS to identify the gases formed on heating. Infrared spectrometers may also be combined with thermal analysis instrumentation. It is possible to combine thermal analysis apparatus with an infrared spectrometer to obtain a complete picture of the chemical and physical changes occurring in thermal processes [Haines, 1995; Hellgeth, 2002]. The most common approach is to combine a FTIR spectrometer with a thermal method such as TGA to obtain an EGA. The evolved gases produced during a TGA experiment can be carried from the furnace to the spectrometer where they can be examined in a long path length gas cell. Data may be illustrated as a function of time using a Gram–Schmidt plot.

9.2.1 Stone

TGA has proved successful for the characterisation of historic stone mortars [Bakolas *et al.*, 1998; Biscontin *et al.*, 2002; Moropoulou *et al.*, 1995a; Riccardi *et al.*, 1998; Silva *et al.*, 2005]. A comparative TGA study of mortars from the Roman, medieval and renaissance periods all from Pavia in Italy has been carried out [Riccardi *et al.*, 1998]. In Figure 9.3 typical TGA curves for mortars from each era are illustrated. The Roman mortar shows a regular decrease in weight followed by a step at 900–1000 K due to the decomposition of calcium carbonate. The medieval mortar shows a step near 400 K, which is probably due to the loss of structural water. A large step occurs near 420 K is observed for the renaissance mortar. The total mass loss for the renaissance mortar may be related to the loss of 1.5 molecules of water from calcium sulfate dihydrate ($CaSO_4.2H_2O$) (the hydrated salt that represent approximately 60 % of the initial mass). This is followed by the slow loss of the remaining water from calcium sulfate hemihydrate ($CaSO_4.0.5\ H_2O$) between 500 and 900 K.

TGA is also very useful for examining of the degradation products found on stone surfaces [Friolo *et al.*, 2003, 2005; Ghendini *et al.*, 2003; Riotino *et al.*, 1998]. TGA–DTA has been used as a tool in the analysis

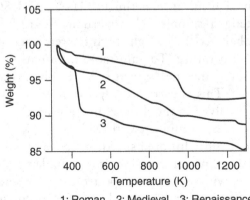

1: Roman   2: Medieval   3: Renaissance

Figure 9.3   TGA data for mortars from different eras. Reprinted from Thermochimica Acta, Vol 321, Riccardi *et al.*, 207–214, 1998 with permission from Elsevier

of black crust deposited on buildings, including the Leaning Tower of Pisa in Italy, due to atmospheric pollutants [Ghendini *et al.*, 2003]. The thermal analysis data was able to distinguish between carbonate, organic and elemental carbon. This approach is also able to estimate the quantities present, which aids in the understanding of the weathering processes occurring to the stone.

9.2.2 Ceramics

TGA can be used to investigate the technologies used to produce ceramics [Drebushchak *et al.*, 2006; Moropoulou *et al.*, 1995b; Wiedemann and Baya, 1992]. This technique has been used to study a number of Byzantine and medieval ceramics [Moropoulou *et al.*, 1995b]. The ceramics show characteristic DTG peaks that can be attributed to the decomposition of clay minerals.

9.2.3 Synthetic Polymers

TGA of polymers can provide a means of studying thermal degradation [Haines, 1995]. Polymers exhibit a range of degradation processes. For instance, polymers such as poly(methyl methacrylate) (PMMA) and polystyrene (PS) depolymerise. Polyethylene (PE) will produce unsaturated hydrocarbons from chain segments of varying lengths. The temperatures of degradation of polyalkenes are affected by substitution. For instance, polytetrafluoroethylene (PTFE) decomposes at a much higher temperature than PE because of fluorine substitution. By comparison, polypropylene (PP) decomposes at a lower temperature than PE because of the substitution of a methyl group. Poly(vinyl chloride) (PVC) and polyacrylonitrile (PAN) can eliminate small molecules initially and form unsaturated links and cross-linking before eventually degrading via complex reactions to char which oxidises in air. Polyamides may absorb moisture, the loss of which can be observed below 100°C, sometimes in stages. Polymers such as cellulose, polyester resins and phenol-formaldehyde possess complex decomposition schemes. The decomposition processes often eliminate small molecules which may be flammable or toxic. In an oxidising atmosphere these reactions are further complicated.

For phthalate ester plasticisers, a sharp peak near 300°C in the derivative TGA curve due to evaporation of the plasticiser will be present. In the case of a calcium carbonate filler, the TGA curve shows a decomposition to calcium oxide near 800 °C. TGA can also be used to quantitatively analyse polymer additives.

TGA has been employed as part of characterisation studies of polymers used in conservation [Cocca *et al.*, 2004; D'Orazio *et al.*, 2001]. The technique was used to evaluate the thermal stability of copolymers of ethyl acrylate – methyl methacrylate used as stone consolidants [Cocca *et al.*, 2004]. Inspection of the TGA curves provided evidence of a single step degradation process and that the polymers were thermally stable until 340 °C.

9.2.4 Natural Materials

TGA has been used to characterise ivory of different origin: mammoth and elephant [Burragato *et al.*, 1998]. This approach requires only 2–3 mg of sample, which means that minimal damage to an object is possible. Different thermal behaviour is shown in the DTG curves produced by mammoth and elephant ivories. Mammoth ivory shows a characteristic DTG peak in the temperature range 600–900 °C due to the different ageing conditions – mammoth ivory was frozen without changes to the amount of organic material.

Different types of resins have been studied using TGA [Ragazzi *et al.*, 2003; Rodgers and Curne, 1999]. A study of a range of fossil resins, including amber, used information from the DTG curves to classify the resins [Ragazzi *et al.*, 2003]. All the samples demonstrated a main exothermal event and the temperature at which the DTG peak associated with this process occurred varied among the samples. An increasing value of the temperature was able to be correlated with the age of the specimen.

9.2.5 Paintings

TGA provides a useful means of quantitatively determining the composition of paint samples [Odlyha, 2000; Odlyha *et al.*, 2000b, 1999]. For example, TGA can be used to estimate the relative medium to pigment concentration in a paint sample. One TGA study of an azurite tempera paint was able to determine the amount of egg medium in the paint [Odlyha *et al.*, 2000b]. From the TGA data obtained by heating samples in oxygen up to 700 °C, the weight losses of the pigment and medium, the pigment alone and the medium alone were determined. The medium concentration was calculated using the ratio:

$$\frac{\text{pigment residual mass} - \text{paint tempera residual mass}}{\text{pigment residual mass} - \text{binder residual mass used}}$$
$$\text{to study supports used in paintings}$$

For this calculation, it was determined that the paint contained 13 wt-% medium.

TGA can also be used to study supports used in paintings. For example, this method has been applied in a study of new, artificially aged and archival linen used in 19th century paintings [Carr *et al.*, 2003]. Notable differences in the TGA data of the archival samples were noted when compared to the new samples, including a reduction in the initiation temperature for the major degradation process.

9.2.6 Written Material

TGA provides a useful means of identifying writing materials such as parchment, paper, leather and papyrus [Budrugeac *et al.*, 2004, 2003; Franceschi *et al.*, 2004; Marcolli and Weidemann, 2001; Wiedemann *et al.*, 1996]. This technique has been used effectively to distinguish Japanese papers of different ages [Wiedemann *et al.*, 1996]. A comparison of mass loss data obtained for 1820 and 1952 paper shows significant differences in the thermal decomposition of lignins. Such differences can be use to quantitatively estimate the age of paper samples.

9.3   DIFFERENTIAL SCANNING CALORIMETRY/ DIFFERENTIAL THERMAL ANALYSIS

Differential scanning calorimetry (DSC) and differential thermal analysis (DTA) are two thermal methods that may be introduced together as they are often used to study the same phenomena [Charsley and Warrington, 1992; Haines, 1995]. DSC is a technique that records the energy necessary to establish a zero temperature difference between a sample and a reference material as a function of time or temperature. The two specimens are subjected to identical temperature conditions in an environment heated or cooled at a controlled rate. By comparison, DTA involves measuring the difference in temperature between the sample and the reference material as a function of time or temperature.

The layout of the apparatus used for DSC/DTA experiments is illustrated in Figure 9.4. A small sample (milligrams) is contained in a

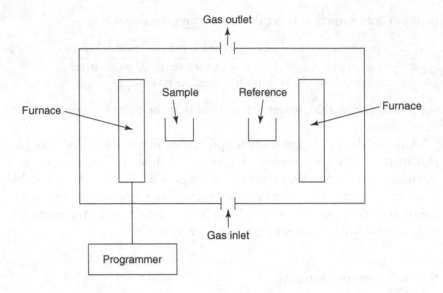

Figure 9.4 Schematic diagram of a DSC/DTA apparatus

crucible and placed in a furnace. The crucible is usually an aluminium pan and lid which is pressed and weighed. Powders, film or fibres can be examined. The reference material is often alumina (Al_2O_3). The furnace is electrically heated with an electronic control. Heating rates up to 100 K min^{-1} can be used, but a normal rate is 10 K min^{-1}. Temperatures below room temperature can be measured by using a coolant such as liquid nitrogen. The sample chamber can be purged with a gas, such as nitrogen, to control the reactions under study. Thermocouples are used as sensors for many DSC/DTA apparatus. A computer is used to control the experimental parameters and to convert the data into DSC or DTA curves.

DSC curves are plotted with heat flow as a function of temperature (or time) at a constant rate of heating. A shift in the baseline results from the change in heat capacity of the sample. The basic equation used for DSC is:

$$\Delta T = \frac{qC_p}{K} \tag{9.1}$$

where ΔT is the difference in temperature between the reference material and the sample, q is the heating rate, C_p is the heat capacity and K is a calibration factor for the instrument. DSC can also be used for measuring enthalpy in transitions. The peak area between the curve and the baseline is proportional to the enthalpy change (ΔH) in the sample. ΔH can be determined from the area of the curve peak (A) using:

$$\Delta H \, m = K \, A \qquad\qquad (9.2)$$

where m is the mass of the sample and K is a calibration coefficient dependent on the instrument.

A DTA curve is usually a plot of the temperature difference (ΔT) versus temperature, or sometimes time. An endothermic process is shown by a downward peak. DSC/DTA curves provide valuable information regarding physical changes such as melting temperatures and/or chemical reactions occurring in the sample of interest.

9.3.1 Synthetic Polymers

DSC is widely used to characterise the thermal properties of polymers [Haines, 1995]. A typical DSC curve for a polymer and illustrates the main transitions observed is shown in Figure 9.5 [Cheng, 2002; Stuart, 2002]. Transitions due to specific thermal processes can be identified. The glass transition temperature (T_g) is the temperature at which an amorphous polymer ceases to be brittle and glassy and becomes less rigid and rubbery. As a polymer is heated up to the glass transition temperature, the molecular rotation about single bonds becomes significantly easier. The glass transition temperature of a polymer can be detected using DSC as an endothermic shift from the baseline is observed at the glass transition temperature in the traces of crystallisable polymers. Such a change results from an increase in heat capacity due to the increased molecular motions in the polymer. The glass transition temperature is affected by a number of factors, including the nature of substituent groups attached to the polymer backbone, the type of secondary bonding between chains, cross-linking, molecular weight and the presence of plasticisers [Schilling, 1989; Stuart, 2002]. The glass transition temperatures for a number of polymers encountered in conservation science are listed in Table 9.1. It should be noted that the glass transition temperature will also depend greatly on the heating and cooling rates used in a DSC run. As the glass transition occurs over a temperature range and not at a single temperature, there may be some difficulty in defining glass transition temperature. Although the beginning of the transition is often defined as the glass transition temperature, the standard procedure involves the use of regression lines [Schilling, 1989], an illustration of which is shown in Figure 9.6. The point of inflection is commonly used as the glass transition temperature value.

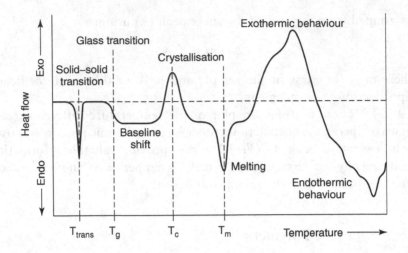

Figure 9.5   Typical DSC curve for a polymer

Table 9.1   Glass transition temperatures (T_g) for some common polymers

Polymer	T_g (°C)
LDPE	−110
PTFE	−97
HDPE	−90
PP	−15
PMMA	12
PVA	29
Nylon 6,6	57
PVC	87
PS	100

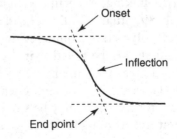

Figure 9.6   Determination of glass transition temperature from DSC curves

Polymers also exhibit a crystalline melting temperature (T_m), the temperature range over which crystalline polymers melt. Such polymers exhibit melting over a range rather than at a sharp melting point due to the mixture of amorphous and crystalline phases that are present. As the crystalline melting temperature of a polymer corresponds to a change from a solid to a liquid state, it gives rise to an endothermic peak in the DSC curve. Such a peak enables the melting point and the heat of fusion to be determined using DSC. The width of the melting peak provides an indication of the range of crystal size and perfection. Above the crystalline melting temperature the polymer will degrade at the degradation temperature (or decomposition temperature) (T_d). This transition can be accompanied by either an exothermic or an endothermic peak. For crystallisable polymers, an additional transition is observed between the T_g and T_m. The crystallisation temperature (T_c) is the temperature at which ordering and the production of crystalline regions occur. The polymer chains have sufficient mobility at this particular temperature to crystallise and an exothermic peak is observed.

DSC has been used to examine the unknown degraded plastic line used in a sculpture by Naum Gabo with the view to identifying a replacement material [Stuart and Thomas, 2000]. In Figure 9.7 the DSC curve for a 5 mg sample of line removed from the Gabo sculpture is shown. The

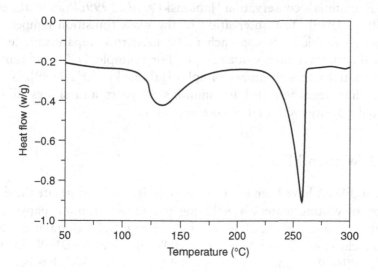

Figure 9.7 DSC curve for plastic taken from a Naum Gabo sculpture. Reprinted from Polymer Testing, Vol 19, Stuart and Thomas, 953–957, 2000 with permission from Elsevier

curve shows a broad negative peak at 135 °C due to free water and a melting transition in the range 250–260 °C, which is characteristic of nylon 6,6.

DSC can be used to study degradation of polymers, particularly oxidative degradation. Such experiments can provide an indication as to whether a polymer is vulnerable to or has suffered damage by reaction with oxygen. DSC data will show evidence of a loss of crystallinity in semi-crystalline polymers. The percentage crystallinity can be estimated using DSC by measuring the area of the sample melting peak and comparing it to that of a standard melting peak for that polymer.

9.3.2 Natural Materials

The reactions of organic natural materials have been widely studied using DSC and DTA. Protein-based materials, resins, glues, waxes, gums and oils have all been studied using these techniques [Prati et al., 2001]. For proteins, denaturation processes can be monitored using DSC/DTA. For example, the helical structure of collagen is destroyed on heating and shows a distinct endothermic transition at 40–45 °C [Haines, 1995]. More examples of the DSC/DTA of natural materials used in paintings, manuscripts and textiles are provided in the following sections.

DSC can be used to characterise the range of natural resins that are used in materials conservation [Jablonski et al., 1999; Prati et al., 2001; Schilling, 1989]. The observation of the glass transition temperatures for resins enables this approach to be used to compare samples and study the thermal history of a sample. For example, DSC has been used to estimate the age of amber samples [Jablonski et al., 1999]. Natural amber has been annealed for millions of years and shows different thermal transitions to that of younger resins.

9.3.3 Written Material

DSC and DTA have been used to characterise and investigate the degradation of writing materials including parchment, leather, papyrus and paper [Budrugeac et al., 2004; Chahine, 2000; Della Gatta et al., 2005; Franceschi et al., 2004; Marcolli and Weidemann, 2001; Weidemann et al., 1996; Weidemann and Baya, 1992]. DSC analysis has been used to differentiate historical papyri of different origin [Franceschi et al., 2004]. Important differences were observed in the DSC data obtained for ancient papyri sheets from the Greek-Roman age and the Egyptian

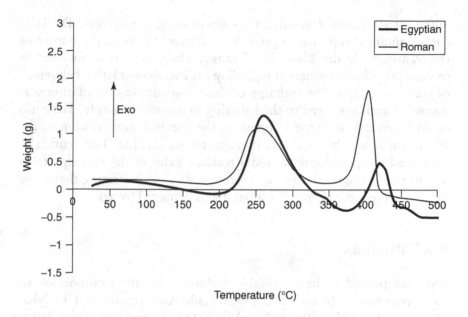

Figure 9.8 DSC curves for ancient Egyptian and Roman–Greek papyri sheets. Reprinted from Thermochimica Acta, Vol 418, Franseschi *et al.*, 39–45, 2004 with permission from Elsevier

Pharaonic age (Figure 9.8). Exothermic peak due to thermal cellulose decomposition are observed at 259 and 256 °C for the Egyptian and Greek–Roman samples, respectively. The shape of the Egyptian peak appears to indicate a contribution of three effects. The most notable differences between the curves in Figure 9.8 occur to the thermal lignin decomposition peaks at 421 and 405 °C for the Egyptian and Greek-Roman samples, respectively. The temperature at which this process occurs and the areas of the peaks due to lignin are significantly different. These differences indicate variation in the kind and quality of lignin in the papyri sheets from different eras and locations.

9.3.4 Textiles

Several studies have been used DSC to examine silk, woollen and linen textiles [Carr *et al.*, 2003; Odlyha *et al.*, 2005]. Modern, historic and artificially aged fibres have been examined. The study of silk involved the examination of 0.4–0.6 mg pieces of modern and historic silk threads from tapestries from Hampton Court Palace and the Royal Palace of Madrid [Odlyha *et al.*, 2005]. The samples were heated to

500 °C at 10 °C min^{-1} in either nitrogen or oxygen purge gas. The DSC curves obtained in nitrogen provided information about the nature of the transitions in the 230–250 °C range, while the curves obtained in oxygen provided information regarding the thermo-oxidative behaviour of the silk samples. The enthalpy of thermo-oxidative degradation was measured and compared to the enthalpy of a control sample. The ratio could be used to monitor a change in the chemical composition of the silk threads, and hence assess damage to the threads. The artificially light aged samples demonstrated a residual value of the enthalpy ratio and historic samples showed an even lower value, indicating that these samples were more chemically altered and damaged by age.

9.3.5 Paintings

DSC has proved to be a valuable technique for the examination of paint components [Burmester, 1992; Felder-Casagrande and Odlyha, 1997; Odlyha et al., 1989, 2000a, 2000b; Odlyha and Burmester, 1988; Odlyha and Scott, 1994; Odlyha, 1991, 1995, 1988, 2000; Prati et al., 2001]. The materials that are commonly present in paint media show characteristic transitions in DSC curves. Some common components of paint, such as resins, oils, waxes and proteins, and their associated DSC peaks are listed in Table 9.2. The appearance of DSC curves will, of course, be affected by the presence of other components and by parameters such as sample history, the rate of heating and the atmosphere.

DSC has been used to examine various paint samples taken from J.M.W. Turner's painting 'Opening of Valhalla' at the Tate Gallery, London [Odlyha and Scott, 1994]. In Figure 9.9 a resolved region of the DSC curve obtained for a white pigmented region of the Turner painting is illustrated. By comparing the DSC data with studies of known paint materials it is possible to assign the peaks shown in Figure 9.9 to the

Table 9.2 Characteristic DSC peaks for paint media

Component	Temperature range (°C)
Waxes	60, 220–250, 320, 350, 440
Proteins	110, 370
Resins	110, 450–500
Oils	220–250, 350, 370–400

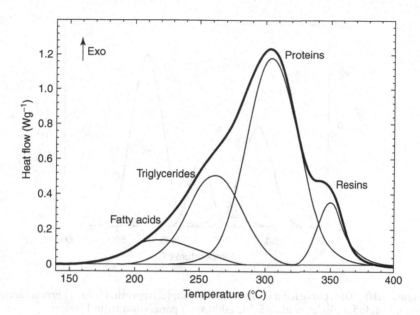

Figure 9.9 Resolved DSC curve for white paint from a Turner painting. Reprinted from Thermochimica Acta, Vol 234, Odlyha and Scott, 165–178, 1994 with permission from Elsevier

oxidative degradation temperatures of this component. In addition, the sample was shown to contain wax, protein, resin and drying oil.

DSC may be used to characterise binders used in paintings [Prati *et al.*, 2001; Odlyha, 1995; Odlyha and Scott, 1994; Odlyha and Burmester, 1988; Pagella and de Faveri, 1998]. Oils, glues, resins, gums, protein, waxes and polymers can also be characterised using DSC. In particular, studies of the oxidative degradation of oils, such as walnut and linseed, using DSC can be used to determine the type of oil used in the painting and thus provide information on the provenance of a painting. A comparison of the DSC traces of linseed and walnut oils reveals that linseed oil has a peak associated with its hydrocarbon chain at a higher temperature (496 °C) than that of walnut oil (475 °C). Another usual factor for differentiating oils is the ratio of the area of this peak with that observed near 305 °C. For instance, the ratio for linseed oil is 1.70 while that for walnut oil is 0.79.

DSC may be used to investigate the ageing of paintings. The technique can be used to identify the inorganic pigment, binders and protective layers, as well as to determine the degradation processes occurring in a paint sample [Cohen *et al.*, 2000a; Maines and de la Rie, 2005; Prati *et al.*, 2001; Odlyha, 1995, 2000; Odlyha *et al.*, 2000b, 2000a]. In

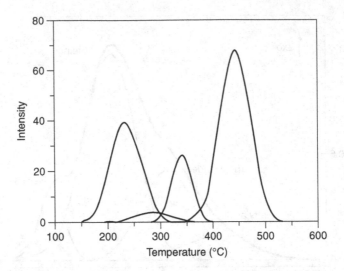

Figure 9.10 DSC curve for a smalt tempera sample. Reprinted from Thermochimica Acta, Vol 365, Odlyha *et al.*, 35-44, 2000 with permission from Elsevier

Figure 9.10 the DSC curve for a smalt tempera sample prepared with egg white binder is illustrated [Odlhya *et al.*, 2000a]. The exothermic peak observed is complex and occurs over a wide temperature range. The complex curve can be fitted with Gaussian bands to determine the components due to the thermo-oxidative degradation of the binding media. When the tempera is exposed to artificial ageing with light, temperature or pollutants, the DSC curve shows changes in the overall shape. By determining the peak ratios of the components, changes due to ageing can be quantified.

9.3.6 Ceramics

DTA can be used to characterise the firing temperatures and/or the presence of raw phases in ceramics [Moropoulou *et al.*, 1995b]. In clay minerals on endothermic peak near 100 °C is due to moisture, while peaks at 200–250 °C can be attributed to bound water. Gypsum shows an endothermic peak in the 120–160 °C range. An endothermic peak due to iron hydroxides losing water is observed at 300 °C, but may be made complex by overlap with an exothermic peak at 300–350 °C due to the recrystallisation of iron-oxy hydroxides. Any organic matter, such as binder, may show exothermic peaks in the 550–650 °C range.

9.3.7 Stone

The dehydration and degradation reactions of inorganic materials present in stone may be studied using DTA and DSC [Brown and Gallagher, 2003; Sorai, 2004]. The enthalpy values associated with transitions for inorganic materials may be used to determine mechanisms and to establish a fingerprint for the compounds under a given set of conditions.

Minerals have been widely studied using DSC/DTA [Brown and Gallagher, 2003; Sorai, 2004]. Simple minerals (e.g. quartz) show phase transitions, hydrated and hydroxyl minerals show dehydration peaks and carbonate minerals show peaks associated with the loss of carbon dioxide. The DSC of minerals is illustrated in Figure 9.11 which shows the trace of a kaolinite sample. The trace shows an endothermic peak near 100 °C due to the loss of water. A larger endotherm at 550–700 °C is associated with dehydroxylation. At 1000 °C the formation of mullite ($3Al_2O_3.2SiO_2$), a high temperature phase transformation of kaolinite, is observed.

Thermal analysis has proved successful for the characterisation of historic stone mortars [Bakolas et al., 1998; Biscontin et al., 2002]. DSC and TGA can be used to identify and quantify the amount of calcium carbonate, Calcium sulfate dihydrate, calcium hydroxide and magnesium hydroxide in mortar binder.

DTA has been demonstrated to be an effective means of quantifying salt efflorescence in historic building materials [Dei et al., 1998; Ramachandran and Polomark, 1978]. DTA has been used to characterise efflorescences such as potassium nitrate, sodium nitrate, calcium

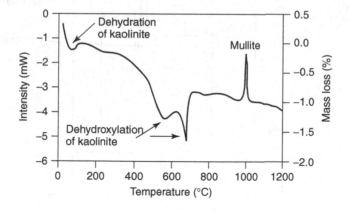

Figure 9.11   DSC curve for kaolinite

nitrate tetrahydrate, magnesium nitrate hexahydrate, calcium sulfate dihydrate and calcium oxalate monohydrate. For quantitative analysis, the ΔH for the particular transitions, such as dehydration processes, in calibration samples versus salt content may be applied. The advantage of DTA for these analyses is that there is no interference among salts and the method is sensitive, with samples of the order of 100 μg being examinable.

DSC is also a useful means of distinguishing precious stones. For example, it is possible to distinguish natural and cultural pearls using this technique because the temperatures at which the aragonite component of pearl transforms to calcite differ [Weidemann and Baya, 1992].

9.4   TENSILE TESTING

Tensile testing machines are designed to elongate materials at a constant rate [Charrier, 1991]. In Figure 9.12 the main components of a tensile testing apparatus are illustrated. The sample is mounted by its ends

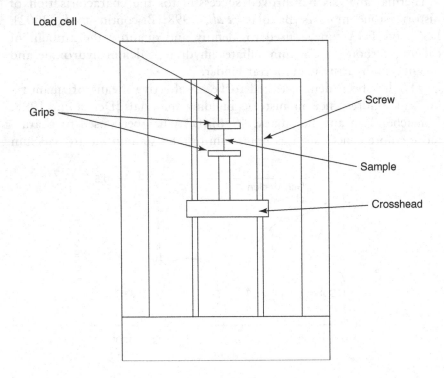

Figure 9.12   Schematic diagram of a tensile testing apparatus

into the grips of the testing apparatus and then elongated by a moving crosshead. The load cell measures the magnitude of the applied load on the sample and the extensometer measures the elongation of the sample. During testing, the deformation occurs to the narrowed central region of the sample with a uniform cross-sectional area along its length.

The degree to which a material strains depends on the magnitude of the imposed stress [Callister, 1997]. The stress (σ) is defined as the load (F) per unit area (A):

$$\sigma = \frac{F}{A} \tag{9.3}$$

Tensile stress is the resistance of a material to stretching forces. The strain (ε) is the amount of deformation per unit length of the material due to the applied load:

$$\varepsilon = \frac{(l_i - l_0)}{l_0} = \frac{\Delta l}{l_0} \tag{9.4}$$

where l_0 is the original length of the sample before any load is applied, l_i is the instantaneous length and Δl is the amount of elongation. The stress at break for fibres can be correlated with the fineness (weight per unit length) by measuring the tenacity. Tenacity is expressed in g den^{-1} (1den = 1 g per 9000 m).

Deformation where the stress and strain are proportional is called elastic deformation. In such a case, a plot of stress against strain produces a linear graph. The slope of such a plot provides the Young's modulus (also known as the modulus of elasticity or the tensile modulus) (E) of the material, a proportionality constant. Typical values of Young's modulus for some common materials are listed in Table 9.3. The modulus can be thought of as the stiffness of the material. Young's modulus can be evaluated from the slope of the linear elastic portion of a force-extension curve where:

$$E = \frac{slope \times gauge\ length}{cross-sectional\ area} \tag{9.5}$$

For certain materials, the initial elastic region is not linear and so it is not possible to determine Young's modulus from the slope. However, a tangent or secant modulus may be used. The secant modulus is often taken as the slope of the stress–strain curve at a specified strain (usually 0.2 % strain).

Plastic deformation may also be observed and is seen where the stress is no longer proportional to strain. The onset of plastic deformation and the stress at the maximum of this plot is known as the yield strength. The stress at which the fracture of the material occurs is known as the

Table 9.3 Young's modulus of some common materials

Material	Young's modulus (GPa)
Metals	
Al	69
Cast iron	90–172
Cu	110
Cu alloys	72–150
Au	78
Fe	208
Mg and alloys	45
Mn	191
Ni and alloys	207
Pt	172
Ag	74
Stainless steels	193–200
Steels	207
Sn	44
Ti and alloys	104–116
Zn and alloys	63–97
Ceramics	
Al_2O_3	393
Soda-lime glass	69
MgO	225
Polymers	
Epoxy	2.41
Nylon 6,6	1.58–2.79
Phenolic	2.76–4.83
PC	2.38
Polyester	2.07–4.41
HDPE	1.07–1.09
LDPE	0.17–0.28
PET	2.8–4.1
PMMA	2.24–3.24
PP	1.14–1.55
PS	2.28–3.28
PTFE	0.40–0.55
PVC	2.4–4.1
Silicone	6.2
Kevlar	1.31

ultimate tensile strength. This corresponds to the maximum stress that can be sustained by the material in tension; if the stress is maintained then fracture will result.

The degree of elongation of a material provides a measure of ductility. As there is often an amount of elastic recovery when some samples break, the percentage elongation at break is quoted. The percentage elongation is given by:

$$\% \text{ elongation} = \frac{\text{increase in gauge length} \times 100}{\text{original gauge length}} \quad (9.6)$$

Materials are often placed in environments where they are exposed to elevated temperatures and/or mechanical stresses. Deformation in such environments is known as creep and this phenomenon may affect the life of the material. Creep is the time-dependent and permanent deformation of materials when they are exposed to a constant load or stress. A creep test involves subjecting a sample to a constant load or stress while maintaining a constant temperature. The strain is measured and plotted as a function of time. Creep results are represented by the creep modulus given by the constant applied stress divided by the time-dependent strain.

9.4.1 Synthetic Polymers

Tensile testing is an established technique for characterising the mechanical properties of polymers [Callister, 1997; Stuart 2002]. Such tests have been used as part of a characterisation of polymers used as conservation coatings [Cocca et al., 2004; D'Orazio et al., 2001]. Tensile tests have also been used to study the degradation of nylon line taken from a sculpture by Naum Gabo [Stuart and Thomas, 2000]. Distortion of the sculpture was evident due physical changes to the nylon line. The nylon taken from the sculpture showed a considerable deterioration in mechanical properties, including a dramatic decrease of the tensile strength to 19 MPa. Potential replacement lines made of nylon 6,10 and poly(vinylidene fluoride) were also studied to compare mechanical properties.

As part of a project involving the study of the effects of temperature and relative humidity on photographic films, tensile tests were used to find the conditions that could extend the life of those materials of historic importance [Tumosa et al., 2001]. Studies over a range of temperature and humidity on cellulose nitrate and cellulose acetate film showed that the best storage environments for such materials involve a low relative humidity and/or low temperatures. The data indicated that these materials can tolerate storage at low temperatures to $-20\,°C$ and that the cycling of the film with the temperature range $+25$ to $-25\,°C$ has no adverse effect on the mechanical properties of the films.

9.4.2 Paintings

Tensile testing can be employed to examine painting supports. Such tests have been used to investigate naturally aged 19th century painting fragments and primed canvas samples from the Tate Gallery and the Courtauld Institute in London [Hedley, 1988]. The effect of relative humidity on the mechanical properties of the canvas samples was studied. The historic samples showed a pattern of high tension in dry conditions and a progressive tension loss with increasing humidity until the onset of canvas shrinkage. The stress changes were dependent on the varying contributions made by the paint, the ground, the fabric type and the painting size.

9.4.3 Written Material

Tensile testing is applicable to the study of ageing of document materials such as paper [Erhardt et al., 2001; Schaeffer et al., 1992]. The technique has been used to investigate the mechanical properties of naturally and artificially aged book paper [Erhardt et al., 2001]. Experiments were conducted on paper subjected to different temperatures (50–90 °C) and relative humidities (30–80 %) for different periods, as well as on naturally aged paper from the interior of a book printed in 1804. The tensile modulus, tensile strength and the elastic and plastic behaviour of the samples were compared. The stress-strain properties studied demonstrated that paper that is accelerated aged at moderate relative humidifies up to at least 90 °C can replicate natural ageing. Conclusions were made as result of the study that paper can be stored in a range of environments of moderate relative humidity and temperature. The optimum conditions were shown to be cooler temperatures and lower relative humidities.

9.4.4 Textiles

A measurement of elongation provides a useful means of characterising textile fibres [Timar-Balazsy and Eastop, 1998]. The elongation for some common dry fibres are estimated in Table 9.4. The elongation of a fibre will differ under wet and dry conditions.

Tensile testing has been carried out on a range of historic textiles fibres including linen, cotton, silk and wool to study the effects of ageing, dyeing and protective coatings [Hansen and Cunell, 1989; Kohara et al., 2001; Needles and Nowak, 1989; Odlyha et al., 2005; Yatagai et al.,

Table 9.4 Elongation of some common fibres

Fibre	Elongation %
Acrylic	15–45
Cotton	3–7
Nylon	18–25
Polyester	18–40
Silk	10–25
Viscose	9–30
Wool	25–35

2001]. One study that examined the degradation and colour fading of cotton fabrics dyed with natural dyes and mordants used tensile tests to evaluate the mechanical properties of the yarns [Kohara *et al.*, 2001]. The study demonstrated that the mordants and dyes did affect the tensile properties of the cotton fabrics. It also showed that the combination of mordant and dye, especially the type of mordant, significantly affects the strength of the fibres on light exposure.

9.5 FLEXURAL TESTING

Flexural or bending tests are used to measure the rigidity of materials [Callister, 1997; Charrier, 1991]. In such tests, samples to be evaluated are placed on supports and a load is applied to the sample at a specified rate. The three-point bending test is commonly used. The stress at fracture using the test is known as the flexural strength. For a rectangular cross-section, the flexural strength is given by:

$$\text{flexural strength} = \frac{3F_B L}{2bh^2} \tag{9.7}$$

where F_B is the load at break, L is the distance between the support points, b is the sample width and h is the sample height. When the sample cross-section is circular, then the flexural strength is calculated using:

$$\text{flexural strength} = \frac{F_B L}{\pi R^3} \tag{9.8}$$

where R is the radius of the sample. Flexural tests tend to be used for more brittle materials such as ceramics and glasses as it is often difficult to produce samples of such materials suitable for tensile testing.

9.6 THERMAL MECHANICAL ANALYSIS

Thermal mechanical analysis (TMA) is a technique that involves the deformation of a sample under a static load measured as a function of time or temperature [Haines, 1995; Odlyha, 2000]. A probe, such as quartz, is lowered onto the surface of a sample and the force exerted on the sample by the probe can be varied. Any change in the sample displacement with temperature is measured using a transducer attached to the sample probe. The set-up is enclosed in a programmable furnace. Liquid nitrogen can be used to cool samples. The stress may be compressive, tension, torsion or flexure.

Measurement of the glass transition temperature and softening temperatures of materials such as polymers or paint samples, can be made using TMA [Haines, 1995]. Below the glass transition temperature there is very little penetration by the probe, but above it, the material softens and the prober sinks into the sample. Complete penetration occurs when the crystalline melting temperature is reached. For softening temperatures, there are standard methods for materials such as polymers that involve the measurement of the temperature at which a particular penetration is obtained for a specified load on a particular sample.

For samples such as films and fibres, the material is generally held under slight tension and the movement of the probe indicates stretching or shrinkage of the sample. This is useful for indicating the glass transition temperature as the shrinkage of fibres or films often starts just below this temperature as the molecules become free to move.

9.6.1 Stone

TMA has potential as a technique for the study of stones, particularly the examination of dimensional changes associated with changes to the component inorganic materials. Precious stones may be studied using TMA. Thermal analysis provides a means of distinguishing jadeite and nephrite [Wiedemann and Baya, 1992]. The TMA curves of the two stones are shown in Figure 9.13. The TMA curve for nephrite indicates a strong increase in thermal expansion due to decomposition and to a change in the oxidation state of iron in the nephrite structure.

9.6.2 Ceramics

TMA has been used to determine the firing history of ceramics [Drebushchak *et al.*, 2006; Wiedemann and Baya, 1992]. This technique

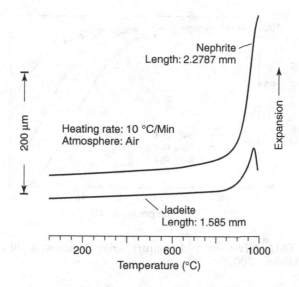

Figure 9.13 TMA curves for jadeite and nephrite. Reprinted from Thermochimica Acta, Vol 200, Weidemann and Baya, 215–255, 1992 with permission from Elsevier

has been used to show that a Chinese terracotta had not been fired [Wiedemann and Baya, 1992]. A comparison of TMA curves for the original and samples fired at 1000 °C demonstrates that the original sample shows a strong contraction while the fired sample does not show any contraction. The results indicate that the original terracotta had only been dried or slightly heated.

9.6.3 Paintings

TMA has been used to examine the effect of composition, solvents and relative humidity on the mechanical properties of prepared paint films [Hedley *et al.*, 1991; Odlyha, 2000;]. The technique has also been used to examine the effect of relative humidity on 19th century primed canvas sample [MacBeth *et al.*, 1993; Odlyha, 2000]. In Figure 9.14 the TMA curves for samples of the canvas at different values of relative humidity are illustrated. These curves clearly demonstrate a decrease in the softening temperature as the sample is exposed to higher value of relative humidity.

9.6.4 Textiles

TMA provides an excellent method for studying of the thermal history and the fibre-forming process of natural and synthetic fibres [Wiedemann

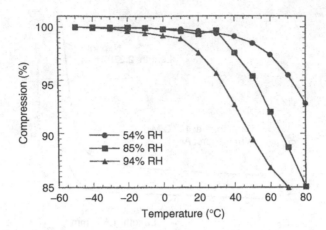

Figure 9.14 TMA curves for 19th century primed canvas at different relative humidities [Odlyha, 2000]

and Baya, 1992]. TMA is sensitive to the molecular chain orientation of the fibres, so changes due to heating can be detected. For example, a TMA study of PET fibres prepared by melting and quenching was used to show how this technique effectively demonstrates the thermal history of such fibres. In the TMA curve, the amorphous PET fibres are characterised by a strong thermal shrinkage from about 50–70 °C up to the melting region at 200–230 °C. If the fibres are preheated to about 150 °C and reheated a second time, the thermal shrinkage is almost eliminated due to the release of stresses which were introduced in the fibre-drawing process.

9.7   DYNAMIC MECHANICAL ANALYSIS

Dynamic mechanical analysis (DMA) is a method that determines the mechanical characteristics as a function of frequency and temperature [Haines, 1995; Odlyha, 2000; Sandler *et al.*, 1998]. DMA tests are carried out by vibrating the sample and varying the applied frequency. The changes can be related to the relaxation processes in a material. Generally, the stress is varied sinusoidally with time. As a result of time-dependent relaxation processes the strain lags behind the stress. DMA can be operated in various modes including flexure, tension, torsion, shear or compression.

For a frequency of oscillation $\omega/2\pi$ Hz the stress, σ, at any given time, t, is:

$$\sigma = \sigma_0 \sin \omega t \qquad (9.9)$$

where σ_0 is the maximum stress. The corresponding strain, ε, is given by:

$$\varepsilon = \varepsilon_0 \sin(\omega t - \delta) \tag{9.10}$$

where δ is the phase angle. The phase angle represents the amount that strain lags behind the stress.

The stress can be resolved into two parts; one in phase with the strain and one 90° out of phase with the strain. If these conditions apply to the torsion experiment, it is also possible to define two shear moduli, denoted E' and E''. The first shear modulus (E') represents the part of the stress in phase with the strain divided by the strain:

$$E' = \frac{\sigma_1}{\varepsilon_0} \tag{9.11}$$

E' is proportional to the recoverable energy and is called the storage modulus. The second shear modulus (E'') is the peak stress 90° out of phase with the strain divided by the peak strain:

$$E'' = \frac{\sigma_2}{\varepsilon_0} = \frac{\sigma_1 \tan \delta}{\varepsilon_0} \tag{9.12}$$

E'' is proportional to the energy dissipated as heat per cycle and is called the loss modulus. The ratio of the moduli is defined as $\tan \delta$:

$$\tan \delta = \frac{E''}{E'} \tag{9.13}$$

The two moduli can be combined to form the complex modulus (E^*):

$$E^* = E' + iE'' \tag{9.14}$$

where $i = \sqrt{-1}$. The moduli may be used to provide information about the properties of materials. For glassy materials, E' is high. Such materials have highly restricted structures and so exhibit good elasticity and, in these cases, no strain energy is lost as heat. For rubbery materials E' is low. For these materials there is a greater contribution from the viscous element and much strain is lost as heat.

A useful dynamic test used to study materials is dynamic mechanical thermal analysis (DMTA). During such an experiment a specimen is subjected to sinusoidal mechanical loading (stress) that induces a corresponding extension (strain) in material. The experiment is carried out over a range of temperature, which is varied typically $-100\,°C$ to $+200\,°C$. It is normal to define the dynamic mechanical behaviour of a material in terms of E' or $\tan \delta$.

DMA can be used to identify the relaxation processes in materials such as polymers. The test can be performed using a sinusoidal load over a range of temperature with controlled heating. When a material passes through its glass transition temperature, the storage modulus often decreases by two or three orders of magnitude and then passes through a maximum. DMA is much more sensitive than other thermal techniques for studying the glass transition. If DMA testing is performed using a static load under isothermal conditions, the Young's modulus and the creep behaviour of materials under controlled conditions can be determined.

Dielectric thermal analysis (DETA) is a complimentary method to DMA. In this technique, a sinusoidal oscillating electric field is applied to the sample. DETA measures the complex dielectric permittivity (ε^*) of a sample. The sample must contain or have an induced electrical dipole. The applications of DETA are similar to those of DMA, but the frequency range is extended.

9.7.1 Paintings

DMTA is a technique that can be employed when studying paintings, particularly the canvas [Foster et al., 1997; Hedley et al., 1991; Odlyha, 2000; Odlyha et al., 2000b, 1999, 2000, 1997a, 1997b]. This technique is able to reveal changes to cellulose structure, the main constituent of canvas or linen. Changes to the degree of hydrogen bonding in cellulose can be observed for naturally or artificially aged canvas or for treated canvas. These changes manifest themselves by a shift in the tan δ peak. In Figure 9.15 an example of the data obtained from a DMTA experiment in tensile mode of an oil-primed canvas subjected to different ageing conditions is provided [Foster et al., 1997]. The tan δ curves for an unaged canvas, as well as canvases heat and light aged for 17 and 34 days are shown. A peak observed near $0\,°C$ is more intense in the unaged sample, but reduces its magnitude after ageing. Another transition near $50\,°C$ appears as a shoulder for the unaged sample but becomes a distinct peak after ageing. The increase in stiffness of the sample on ageing can be quantified by calculating the log E' values at $30\,°C$: 8.28 Pa for unaged, 8.67 Pa for aged 17 days and 8.79 Pa for aged 34 days. DMTA has also been employed to test conservation treatments and canvas in different conditions of relative humidity [Foster et al., 1997; Odlyha, 2003].

9.7.2 Written Material

DMA can be used to characterise aged paper samples and changes in the DMA curves can be attributed to the onset of thermal degradation [Toth

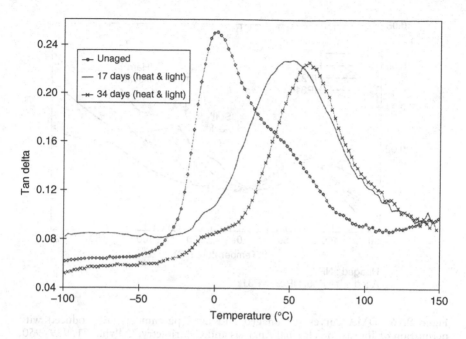

Figure 9.15 DMTA curves for aged canvas samples. Reprinted from Thermochimica Acta, Vol 294, Foster *et al.*, 81–89, 1997 with permission from Elsevier

et al., 1984, 1985]. DMA has also been used to examine historical parchment samples [Cohen *et al.*, 2000b; Odlyha *et al.*, 2003]. DMA curves for aged and unaged parchments are illustrated in Figure 9.16 [Odlyha *et al.*, 2003]. Each curve shows two major peaks and a shoulder peak, but the peaks occur at different temperatures. The peak below room temperature is attributable to collagen side chains. For unaged parchment, the peak at 25 °C is due to residual adsorbed water and the peak at 65 °C indicates the temperature at which motion of the main collagen backbone occurs. The transitions appear at different temperatures and show different intensities for the aged parchment samples. The shift of peaks to 90 and 140 °C for the aged samples reflect more rigid regions within the collagen structure.

9.7.3 Textiles

DMTA is a suitable technique for the examination of textile fibres. A study of wool and silk threads from model, artificially aged samples and historic tapestries used DMTA to characterise changes due to degradation and the effects of dyeing [Odlyha *et al.*, 2005]. DMTA tests

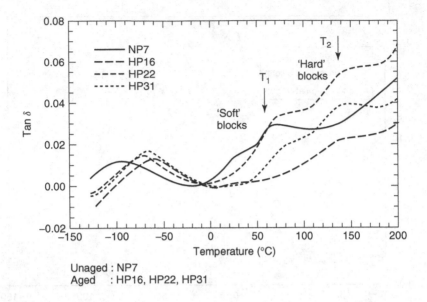

Figure 9.16 DMA curves for unaged and aged parchments. Reproduced with permission of Journal of Thermal Analysis and Calorimetry, Odlyha, 71, 939–950, 2003, Springer

were carried out using a sinusoidal load over a range of temperature with controlled heating to provide values for the glass transition temperature. Experiments were also carried out under static load at isothermal conditions to provide values of Young's modulus and the creep behaviour of the fibres under controlled environmental conditions. The DMTA results showed that dyeing reduces the glass transition temperature and so the amorphous structure of the thread is affected. The DMTA creep results on wool demonstrate different responses of dyed samples to wetting, shrinkage on heating at 90 °C and the relative humidity increase in the range 10–80 %.

9.7.4 Synthetic Polymers

DMA is often used to determine the glass transition temperature of polymers [Haines, 1995; Princi *et al.*, 2005; Stuart, 2002]. An example of the application of DMTA to conservation polymers is a study of the characterisation of a polyurethane used as coatings for artefacts [D'Orazio *et al.*, 2001]. DMTA was able to detect two glass transition temperatures values for the polyurethane attributable to the glass transitions of the polyether and urethanic segments, respectively. This technique provides more sensitivity than DSC in this case, as DSC could only detect one glass transition temperature for the sample.

9.8 HARDNESS

Hardness tests involve measuring of the resistance of a material to penetration by an indentor [Callister, 1997; Charrier, 1991]. The early qualitative hardness tests were based on the resistance of a material to scratching. The Mohs scale ranges from one for a soft material, such as talc, to ten for hard materials, such as diamond. This scale is still used in certain fields and the Mohs hardness values for a range of materials are listed in Table 9.5.

Table 9.5 Hardness values for some common materials

Material	Mohs hardness	Knoop hardness	Rockwell hardness
Glass	4.5–6.5	530	
Minerals and stone			
Agate	7		
Alabaster	1.7		
Alexandrite	8.5		
Amethyst	7		
Aquamarine	7.5–8		
Azurite	3.5		
Barite	3.3		
Beryl	7.8		
Calcite	3	135	
Coral	3.5		
Diamond	10	7000	
Emerald	7.5–8.0		
Feldspar	6	560	
Flint	7		
Garnet	6.5–8.5	1360	
Gypsum	1.6–2	32	
Haematite	5.5–6.5		
Jadeite	7		
Kaolinite	2.0–2.5		
Lapis lazuli	5.5		
Malachite	4		
Marble	3–4		
Nephrite	6.5		
Obsidian	5–5.5		
Opal	6		
Pearl	3		

Table 9.5 (*continued*)

Material	Mohs hardness	Knoop hardness	Rockwell hardness
Quartz	7	820	
Ruby	9		
Sapphire	9		
Topaz	8	1340	
Turquoise	5–6		
Metals			
Ag	2.5–4	60	
Al	2–2.9		
Au	2.5–3		
Brass	3–4		
Cd	2.0	37	
Cr	9	935	
Cu	2.5–3	163	
Fe	4–5		
Mg	2.0		
Ni		557	
Pb	1.5		
Pt	4.3		
Sb	3.0–3.3		
Sn	1.5–1.8		
Steel	5–8.5		
Zn	2.5	119	
Natural materials			
Amber	2.5		
Asphalt	1–2		
Ivory	2.5		
Jet	2.5		
Tortoiseshell	2.5		
Polymers			
Cellulose acetate			R49–R123
Epoxy resin			M75–M110
HDPE			R30–R50
Nylon			R108–R120
Phenolic resin			M93–M120
PC			M70–M180
Polyester resin			M80–M120
PMMA			M80–M105
PP			93
PS			M65–M85
PTFE			D50–D65
Silicone			M85–M95
UF resin			E94–E97

Quantitative hardness tests have been developed to study a range of material types. Indentors vary in geometry and the material from which they are produced depending on the sample to be studied. Hardness tests are carried out under a controlled load and rate of application. The size of the resulting indentation is measured and related to a hardness number. The larger and deeper the indentor, the lower the hardness value. Hardness tests have the advantages of being simple and inexpensive, as well as being visually non-destructive. The hardness of polymers is commonly measured against the Rockwell, Barcol and Shore scales, while the Knoop scale is used for ceramics and stone. Metals hardness is measured using the Rockwell, Brinell, Knoop or Vickers scales. The Knoop and Rockwell hardness values for some typical materials are also listed in Table 9.5 [Lide, 2006].

REFERENCES

A. Bakolas, G. Biscontin, A. Moropulou and E. Zendri, Characterisation of structural Byzantine mortars by thermogravimetric analysis, *Thermochimica Acta*, **321** (1998), 151–160.

G. Biscontin, M.P. Birelli and E. Zendri, Characterisation of binders employed in the manufacture of Venetian historical mortars, *Journal of Cultural Heritage*, 3 (2002), 31–37.

M. Brown and P. Gallagher (eds), *Handbook of Thermal Analysis and Calorimetry*, Vol. 2, Elsevier, Amsterdam (2003).

P. Budrugeac, L. Miu, V. Bocu, *et al.*, Thermal degradation of collagen-based materials that are supports of cultural and historical objects, *Journal of Thermal Analysis and Calorimetry*, **72** (2003), 1057–1064.

P. Budrugeac, L. Miu, C. Popescu and F.J. Wortmann, Identification of collagen-based materials that are supports of cultural and historical objects, *Journal of Thermal Analysis and Calorimetry*, 77 (2004), 975–985.

F. Burragato, S. Materazzi, R. Curini and G. Ricci, New forensic tool for the identification of elephant or mammoth ivory, *Forensic Science International*, **96** (1998), 189–196.

A. Burmester, Investigation of paint media by differential scanning calorimetry, *Studies in Conservation*, **37** (1992), 73–81.

W.D. Callister, *Materials Science and Engineering: An Introduction*, 4th edn, John Wiley & Sons, Inc., New York (1997).

D.J. Carr, M. Odlyha, N. Cohen, *et al.*, Thermal analysis of new, artificially aged and archival linen, *Journal of Thermal Analysis and Calorimetry*, 73 (2003), 97–104.

C. Chahine, Changes in hydrothermal stability of leather and parchment with deterioration: a DSC study, *Thermochimica Acta*, **365** (2000), 101–110.

J.M. Charrier, *Polymeric Materials and Processing: Plastics, Elastomers and Composites*, Hanser, Munich (1991).

E.L. Charsley and S.B. Warrington, *Thermal Analysis Techniques and Applications*, Royal Society of Chemistry, Cambridge (1992).

S.Z.D. Cheng, *Handbook of Thermal Analysis and Calorimetry*, Vol. 3, Elsevier, Amsterdam (2002).

M. Cocca, L. D'Arienzo, L. D'Orazio, et al., Polyacrylates for conservation: chemico-physical properties and durability of different commercial products, Polymer Testing, 23 (2004), 333–342.

N.S. Cohen, M. Odlyha, R. Campana and G.M. Foster, Dosimetry of paintings: determination of the degree of chemical change in museum exposed test paintings (lead white tempera) by thermal analysis and infrared spectroscopy, Thermochimica Acta, 365 (2000a), 45–52.

N.S. Cohen, M. Odlyha and G.M. Foster, Measurement of shrinkage behaviour in leather and parchment by dynamic mechanical thermal analysis, Thermochimica Acta, 365 (2000b), 111–117.

L. Dei, M. Mauro and G. Bitossi, Characterisation of salt efflorescence in cultural heritage conservation by thermal analysis, Thermochimica Acta, 317 (1998), 133–140.

G. Della Gatta, E. Badea, R. Ceccarelli, et al., Assessment of damage in old parchments by DSC and SEM, Journal of Thermal Analysis and Calorimetry (2006), 82 (2005), 637–649.

L. D'Orazio, G. Gentile, C. Mancarella, et al., Water-dispersed polymers for the conservation and restoration of cultural heritage: a molecular, thermal, structural and mechanical characterisation, Polymer Testing, 20 (2001) 227–240.

V.A. Drebushchak, L.N. Mylnikova, T.N. Drebushchak and V.V. Boldyrev, The investigation of ancient pottery: application of thermal analysis, Journal of Thermal Analysis and Calorimetry (2006), in press.

D. Erhardt, C.S. Tumusa and M.F. Mecklenburg, Chemical and physical changes in naturally and accelerated aged cellulose, in Historic Textiles, Papers and Polymers in Museums (eds J.M. Cardamone and M.T. Baker, American Chemical Society), Washington (2001), pp 23–37.

S. Felder-Casagrande and M. Odlyha, Development of standard paint films based on artists' materials, Journal of Thermal Analysis, 49 (1997), 1585–1591.

G. Foster. M. Odlyha and S. Hackney, Evaluation of the effects of environmental conditions and preventative conservation treatment of painting canvas, Thermochimica Acta, 294 (1997), 81–89.

E. Franceschi, G. Luciano, F. Carosi, et al., Thermal and microscope analysis as a tool in the characterisation of ancient papyri, Thermochimica Acta, 418 (2004), 39–45.

K.H. Friolo, B.H. Stuart and A.S. Ray, Characterisation of weathering of Sydney sandstones in heritage buildings, Journal of Cultural Heritage, 4 (2003), 211–220.

K.H. Friolo, A.S. Ray, B.H. Stuart and P.S. Thomas, Thermal analysis of heritage stones, Journal of Thermal Analysis and Calorimetry, 80 (2005), 559–563.

N. Ghedini, C. Sabbioni and M. Pantani, Thermal analysis in cultural heritage safeguard: an application, Thermochimica Acta, 406 (2003), 105–113.

P.J. Haines, Thermal Methods of Analysis: Principles, Applications and Problems, Chapman and Hall, London (1995).

E.F. Hansen and W.S. Cunell, The conservation of silk with Parylene-C, in Historic Textile and Paper Materials II (eds S.H. Zeronian and H.L. Needles), American Chemical Society, Washington (1989), pp 108–133.

G. Hedley, M. Odlyha, A. Burnstock, et al., A study of the mechanical and surface properties of oil paint films treated with organic solvents and water, Journal of Thermal Analysis, 37 (1991) 2067–2088.

G. Hedley, Relative humidity and the stress/strain response of canvas paintings: uniaxial measurements of naturally aged samples, Studies in Conservation, 33 (1998) 133–148.

J.W. Hellgeth, Thermal analysis–IR methods in *Handbook of Vibrational Spectroscopy*, Vol. 2, (eds J.M. Chalmers and P.R. Griffiths), John Wiley & Sons Ltd, Chichester (2002), pp 1699–1714.

P. Jablonski, A. Golloch and W. Borchard, DSC measurements of amber and resin samples, *Thermochimica Acta*, 333 (1999), 87–93.

N. Kohara, C. Sano, H. Ikuno, *et al.*, Degradation and colour fading of cotton fabrics dyed with natural dyes and mordants, in *Historic Textiles, Papers and Polymers in Museums* (eds J.M. Cardamone and M.T. Baker, American Chemical Society, Washington (2001), pp 74–85.

D.R. Lide (ed.), *CRC Handbook of Chemistry and Physics*, 87th edn, CRC Press, Boca Raton (2006).

R. MacBeth, M. Odlyha, A. Burnstock, *et al.*, in *ICOM Committee for Conservation Preprints*, Vol. 1 (ed J. Bridgland), James and James Science Publishers, London (1993), pp 150–156.

C.A. Maines and E.R. de la Rie, Size-exclusion chromatography and differential scanning calorimetry of low molecular weight resins used in varnishes for paintings, *Progress in Organic Coatings*, 52 (2005), 39–45.

C. Marcolli and H.G. Weidemann, Distinction of original and forged lithographs by means of thermogravimetry and Raman spectroscopy, *Journal of Thermal Analysis and Calorimetry*, 64 (2001), 987–1000.

A. Moropoulou, A. Bakalos and K. Bisbikou, Characterisation of ancient, Byzantine and later historic mortars by thermal and X-ray diffraction techniques, *Thermochimica Acta*, 269–270 (1995a), 779–795.

A. Moropoulou, A. Bakolas and K. Bisbikou, Thermal analysis as a method of characterising ancient ceramic technologies, *Thermochimica Acta*, 257 (1995b), 743–753.

H.L. Needles and K.C.J. Nowak, Heat-induced ageing of linen, in *Historic Textile and Paper Materials II* (eds S.H. Zeronian and H.L. Needles), American Chemical Society, Washington (1989), pp 159–167.

M. Odlyha, Characterisation of aged paint films by differential scanning calorimetry, *Thermochimica Acta*, 134 (1988), 85–90.

M. Odlyha and A. Burmester, Preliminary investigation of the binding media of paintings by differential thermal analysis, *Journal of Thermal Analysis*, 33 (1988), 1041–1052.

M. Odlyha, C.D. Flint and C.F. Simpson, The application of thermal analysis (DSC) to the study of paint media, *Analytical Proceedings*, 26 (1989), 52–56.

M. Odlyha, A novel approach to the problem of characterising the binding media in early Italian paintings (13th/16th century), *Journal of Thermal Analysis*, 37 (1991), 1431–1440.

M. Odlyha and R.P.W. Scott, The enthalpic value of paintings, *Thermochimica Acta*, 234 (1994), 165–178.

M. Odlyha, Investigation of the binding media of paintings by thermoanalytical and spectroscopic techniques, *Thermochimica Acta*, 269–270 (1995), 705–727.

M. Odlyha, J.J. Boon, O van den Brink and M. Bacci, Environmental research for art conservation (ERA), *Journal of Thermal Analysis*, 49 (1997a), 1371–1384.

M. Odlyha, G. Foster, S. Hackney and J. Townsend, Dynamic mechanical thermal analysis for the evaluation of deacidification treatment of painting canvases, *Journal of Thermal Analysis*, 50 (1997b), 191–202.

M. Odlyha, N.S. Cohen, R. Campana and G.M. Foster, Environmental research for art conservation and assessment for indoor conditions surrounding cultural objects, *Journal of Thermal Analysis and Calorimetry*, 56 (1999), 1219–1232.

M. Odlyha, Thermal Analysis, in *Modern Analytical Methods in Art and Archaeology* (eds E. Ciliberto and G. Spoto), John Wiley & Sons, Inc., New York (2000), pp 279–319.

M. Odlyha, N.S. Cohen and G.M. Foster, Dosimetry of paintings: determination of the degree of chemical change in museum exposed test paintings (smalt tempera) by thermal analysis, *Thermochimica Acta*, **365** (2000a), 35–44.

M. Odlyha, N.S. Cohen, G.M. Foster and R.H. West, Dosimetry of paintings: determination of the degree of chemical change in museum exposed test paintings (azurite tempera) by thermal and spectroscopic analysis, *Thermochimica Acta*, **364** (2000b), 53–63.

M. Odlhya, N.S. Cohen, G.M. Foster, *et al.*, Dynamic mechanical analysis (DMA), ^{13}C solid state NMR and micro-thermomechanical studies of historical parchment, *Journal of Thermal Analysis and Calorimetry*, **71** (2003), 939–950.

M. Odlyha, The application of thermoanalytical techniques to the preservation of art and archaeological objects in *Handbook of Thermal Analysis and Calorimetry*, Vol. 2 (eds M. Brown and P. Gallagher), Elsevier, Amsterdam (2003), pp 47–96.

M. Odlhya, Q. Wang, G.M. Foster, *et al.*, Thermal analysis of model and historic tapestries, *Journal of Thermal Analysis and Calorimetry*, **82** (2005), 627–636.

C. Pagella and D.M. de Faveri, DSC evaluation of binder content in latex paints, *Progress in Organic Coatings*, **33** (1998), 211–217.

S. Prati, G. Chiavari and D. Cam, DSC application in the conservation field, *Journal of Thermal Analysis and Calorimetry*, **66** (2001), 315–327.

E. Princi, S. Vicini, E. Pedemonte, *et al.*, New polymeric materials for paper and textile conservation I. Synthesis and characterisation of acrylic copolymers, *Journal of Applied Polymer Science*, **98** (2005), 1157–1164.

E. Ragazzi, G. Roghi, A. Giaretta and P. Gianolla, Classification of amber based on thermal analysis, *Thermochimica Acta*, **404** (2003), 43–54.

V.S. Ramachandran and G.M. Polomark, Application of DSC-DTA technique for estimating various constituents in white coat plasters, *Thermochimica Acta*, **25** (1978), 161–169.

M.P. Riccardi, P. Duminuco, C. Tomasi and P. Ferloni, Thermal, microscopic and X-ray diffraction studies on some ancient mortars, *Thermochimica Acta*, **321** (1998), 207–214.

C. Riotino, C. Sabbioni, N. Ghendini, *et al.*, Evaluation of atmospheric deposition on historic buildings by combined thermal analysis and combustion techniques, *Thermochimica Acta*, **321** (1998), 215–222.

K.A. Rodgers and S. Curne, A thermal analytical study of some modern and fossil resins from New Zealand, *Thermochimica Acta*, **326** (1999), 143–149.

S.R. Sandler, W. Karo, J. Bonesteel and E.M. Pearce, *Polymer Synthesis and Characterisation: A Laboratory Manual*, Academic Press, San Diego (1998).

T.T. Schaeffer, M.T. Baker, V. Blyth-Hill and D. Van der Reyden, Effects of aqueous light bleaching on the subsequent ageing of paper, *Journal of the American Institute of Conservation*, **31** (1992), 289–311.

M.R. Schilling, The glass transition of materials used in conservation, *Studies in Conservation*, **34** (1989), 110–116.

D.A. Silva, H.R. Wenk and P.J. Monteiro, Comparative investigation of mortars from Roman Colosseum and cistern, *Thermochimica Acta*, **428** (2005), 35–40.

M. Sorai, *Comprehensive Handbook of Calorimetry and Thermal Analysis*, John Wiley & Sons, Ltd, Chichester (2004).

B.H. Stuart, *Polymer Analysis*, John Wiley & Sons, Ltd, Chichester (2002).

B.H. Stuart and P.S. Thomas, The characterisation of plastic used in a Gabo sculpture, *Polymer Testing*, **19** (2000), 953–957.

A. Timar-Balazsy and D. Eastop, *Chemical Principles of Textile Conservation*, Butterworth-Heinemann, Oxford (1998).

F.H. Toth, G. Pokol, J. Gyore and S. Gal, A sample mounting technique for the dynamic mechanical analysis of cellulose and paper sheets, *Thermochimica Acta*, **80** (1984), 281–286.

F.H. Toth, G. Pokol, J. Gyore and S. Gal, Dynamic mechanical analysis of paper samples, *Thermochimica Acta*, **93** (1985), 405–408.

C.S. Tumosa, M.F. Mecklenburg and M.H. McCormick-Goodhart, The physical properties of photographic film polymers subjected to cold storage environments, in *Historic Textiles, Papers and Polymers in Museums* (eds J.M. Cardamone and M.T. Baker), American Chemical Society, Washington (2001), pp 126–144.

H.G. Weidemann and G. Baya, Approach to ancient Chinese artefacts by means of thermal analysis, *Thermochimica Acta*, **200** (1992), 215–255.

H.G. Weidemann, J.R. Gunter and H.R. Oswald, Investigation of ancient and new Japanese papers, *Thermochimica Acta*, **282–283** (1996), 453–459.

M. Yatagai, Y. Magoshi, M.A. Becker, *et al.*, 'Degradation and colour fading of silk fabrics dyed with natural dyes and mordants, in *Historic Textiles, Papers and Polymers in Museums* (eds J.M. Cardamone and M.T. Baker), American Chemical Society, Washington (2001), pp 86–97.

10

Nuclear Methods

10.1 INTRODUCTION

There are a number of nuclear processes that may be exploited in the analysis of materials. The majority of nuclei are unstable and exhibit radioactivity, that is, they spontaneously disintegrate by emitting radiation. Each type of unstable nucleus will exhibit a characteristic rate of radioactive decay. As the rate of decay can vary from a fraction of a second to billions of years, measurement of the amount of isotopes can provide a means of determining the age of an object. Carbon, lead and strontium dating methods can be used to determine the age of items of cultural significance. Another example of a nuclear process is neutron activation analysis. Neutrons are used to bombard a sample and convert a fraction of the atoms into radioisotopes. Such radioisotopes exhibit characteristic decay patterns that may be used to determine the elements present in the sample. Luminescence dating is another method which utilises radioactivity: the emission of light from crystalline materials is measured following absorption of radiation. In addition, a technique that exploits the generation of neutrons from a nuclear reactor is neutron diffraction. This technique parallels XRD, but the neutrons are scattered by the nuclei in a sample instead of by electrons. Although not readily available compared to XRD, access to neutron diffraction does enable the location of light atoms such as hydrogen to be determined and is a non-destructive approach for precious items.

Analytical Techniques in Materials Conservation   Barbara H. Stuart
© 2007 John Wiley & Sons, Ltd

10.2 RADIOISOTOPIC DATING

Radioisotopic dating uses radioisotopes to determine the age of an object. The technique exploits the half-life of the nuclides under study, which is a constant that can act as a nuclear clock. There are a number of isotopes that may be used to date objects, but the main isotopes of interest in materials conservation are ^{14}C, ^{238}U and ^{232}Th. Radiocarbon dating utilises the decay of ^{14}C, while lead isotope analyses involves the decay of ^{238}U and ^{232}Th.

Radioactive nuclei decay at a characteristic rate and follow a first-order kinetic process. The rate is proportional to the number of radioactive nuclei, N, in the sample:

$$\text{rate} = \frac{\Delta N}{\Delta t} = kN \tag{10.1}$$

where k is the decay constant. The decay rate is called its activity and is expressed in terms of disintegrations per second (a becquerel, Bq). A practical version of the rate law is:

$$\ln\left(\frac{N_t}{N_0}\right) = -kt \tag{10.2}$$

where t is the time interval of decay, N_0 is the initial number of nuclei and N_t is the number of nuclei remaining after the time interval. Decay rates are commonly expressed in terms of the half-life. The half-life, $t_{1/2}$, is the time it takes for half the nuclei to decay:

$$t_{1/2} = \frac{\ln 2}{k} \tag{10.3}$$

The half-lives of some common radioisotopes are listed in Table 10.1.

Radiocarbon dating is a technique that uses radioisotopes to determine the relative amounts of ^{14}C and ^{12}C in materials of biological origin [Hedges, 2000; Higham and Petchey, 2000]. The procedure is based on the formation of ^{14}C by neutron capture in the upper atmosphere:

$$^{14}_{7}N + ^{1}_{0}n \rightarrow ^{14}_{6}C + ^{1}_{1}p \tag{10.4}$$

This reaction provides a nearly constant amount of ^{14}C. It is generally assumed that the ratio of ^{14}C to ^{12}C in the atmosphere has been constant for at least 50 000 years. The ^{14}C is incorporated into carbon dioxide, which is then incorporated via photosynthesis into carbon-containing molecules in plants. When animals consume plants the ^{14}C becomes incorporated within them. As plants and animals have a constant intake

Table 10.1 Half-lives of some
common radioisotopes

Isotope	Half life (years)
$^{238}_{92}U$	4.5×10^9
$^{235}_{92}U$	7.0×10^8
$^{232}_{90}Th$	1.4×10^{10}
$^{14}_{6}C$	5.7×10^3
$^{87}_{38}Sr$	4.9×10^{10}

of carbon compounds, a constant ratio of ^{14}C to ^{12}C is maintained; it is the same as that of the atmosphere. However, once the plant or animal dies, carbon compounds are no longer ingested and the amount of ^{14}C decreases through radioactive decay:

$$^{14}_{6}C \rightarrow ^{14}_{7}N + ^{0}_{-1}\beta \qquad (10.5)$$

By measuring the ratio of ^{14}C to ^{12}C and comparing it to that of the atmosphere it is possible to estimate the age of an object.

Lead isotope analysis is based on the radioactive decay of ^{138}U and ^{232}Th [Brill, 1970; Gale and Stos-Gale, 2000]. This approach is useful for the dating of a range of materials known to contain lead. Lead is a common constituent of ore deposits and minerals. Natural lead consists of the four isotopes: ^{208}Pb, ^{207}Pb, ^{206}Pb and ^{204}Pb. The first three of these isotopes are partly derived from the decay of the long-lived naturally radioactive isotopes of uranium and thorium as follows:

$$^{238}_{92}U \rightarrow ^{206}_{82}Pb + 8^{4}_{2}\alpha + 6^{0}_{-1}\beta \qquad t_{1/2} = 4.5 \times 10^9 \text{ years} \quad (10.6)$$

$$^{235}_{92}U \rightarrow ^{207}_{82}Pb + 7^{4}_{2}\alpha + 4^{0}_{-1}\beta \qquad t_{1/2} = 7.0 \times 10^8 \text{ years} \quad (10.7)$$

$$^{232}_{90}Th \rightarrow ^{208}_{82}Pb + 6^{4}_{2}\alpha + 4^{0}_{-1}\beta \qquad t_{1/2} = 1.4 \times 10^{10} \text{ years} \quad (10.8)$$

The isotope of ^{204}Pb is nonradiogenic in origin. Lead isotope compositions are commonly expressed as $^{207}Pb/^{206}Pb$, $^{208}Pb/^{206}Pb$ or $^{206}Pb/^{204}Pb$ ratios, but other ratios may also be used. The ratio obtained for an artefact can be compared to ratios of possible ore deposits. The lead isotope compositions of rocks in the external parts of the earth have evolved over time. It is also possible to estimate the age of materials from the mass ratio of $^{206}_{82}Pb$ to $^{238}_{92}U$.

Another method of radioisotopic dating, strontium analysis, exploits the decay of ^{87}Rb to ^{87}Sr [Corfield, 2002]. The radiogenic ^{87}Sr isotope is produced as a result of β decay:

$$^{87}_{37}Rb \rightarrow {}^{87}_{38}Sr + {}^{0}_{-1}\beta \qquad (10.9)$$

In this method the ratio of ^{87}Sr to the stable ^{86}Sr is measured. The technique is suitable for the analysis of minerals obtained from igneous rock, such as those of volcanic origin. As the rock cooled, minerals precipitated out in a particular order and concentrated specific elements in the process. Strontium is concentrated in many of the first precipitates, while rubidium is gradually concentrated in the last phase. However, because any reheating of the minerals tends to change the rubidium/strontium ratio, this technique is limited to rocks which have not gone through significant heat cycles.

There are several means of analysing radioactive isotopes. Decay counting methods measure the concentration of isotopes and work by counting the number of decays that occur in a given period [Hedges, 2000]. A common method is liquid scintillation counting, which involves dissolving the samples in a solvent that fluoresces when struck by a charged particle (a phosphor). The counters are based on the measurement of the tiny flashes of light produced. Liquid scintillation counting is used to detect α and β emitters.

Mass spectrometry techniques may also be used to determine isotopic ratios. Traditional mass spectrometers, such as thermal ionisation and gas source mass spectrometers, may be used. In addition, accelerator mass spectrometry (AMS) has emerged as a suitably sensitive technique for measuring the concentration of carbon isotopes. In this technique, negatively charged carbon ions from the samples are injected into a nuclear particle accelerator. The negative ions are accelerated towards the positive potential and are separated from other atoms based on the nuclear charge and mass. AMS is considerably more sensitive than decay counting methods and also requires much smaller quantities of sample for successful analysis.

A plot of the natural logarithm of the specific activity versus time provides a useful graphical means of radioisotopic dating. The specific activity is the number of disintegrations per second for 1 g of sample. Such a plot is linear with a slope of $-k$. In Figure 10.1 a radiocarbon plot that may be used to estimate the age of objects is illustrated.

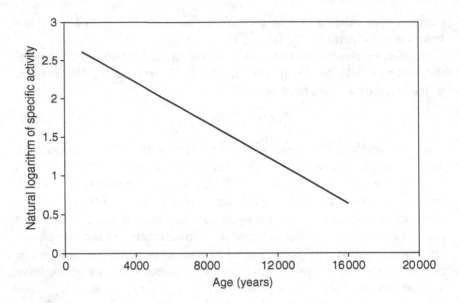

Figure 10.1 Radiocarbon plot of the natural logarithm of specific activity versus time

10.2.1 Textiles

Cellulose and protein-based textiles such as linen, cotton, silk and wool can be studied using radiocarbon dating [Donahue *et al.*, 1990; Hedges, 2000; Jull *et al.*, 1995; Tuniz *et al.*, 2000; Turnbull *et al.*, 2000]. A sample of at least 5–10 g is required for decay counting, while a sample of 5–10 mg is sufficient when using AMS ^{14}C dating. Probably the most well known case of radiocarbon dating is the analysis of the Shroud of Turin housed in Turin Cathedral. The shroud is a piece of linen that many people believe was the burial cloth used to wrap the body of Jesus Christ. Since the cloth first publicly emerged in 1345 there has been much debate about the authenticity of the shroud. Testing of the cloth by radiocarbon dating was carried out and three laboratories independently measured the $^{14}C/^{12}C$ ratio of a 50 mg piece of the linen. The analyses determined that the flax from which the cloth was made was grown between 1260 and 1390, indicating that the age of the shroud is consistent with the proposal of a medieval forgery [Corfield, 2002; Donahue *et al.*, 1990; Gove, 1996; Tuniz *et al.*, 2000].

10.2.2 Written Material

Writing materials including paper, parchment and papyrus lend themselves to examination by radiocarbon dating [Bonani et al., 1992; Higham and Petchey, 2000; Jull et al., 1995; Oda et al., 2000; Tuniz et al., 2000]. Radiocarbon dating has been used to date the Dead Sea Scrolls, the collection of 1200 parchment and papyrus manuscripts found in 1947 in caves close to the Dead Sea [Bonani et al., 1992; Jull et al., 1995; Tuniz et al., 2000]. Several manuscripts were dated using the AMS technique and the measured ages were is good agreement with palaeographic estimates. The dating confirmed that the scrolls predate Christianity.

10.2.3 Paintings

Prehistoric rock art can be dated using ^{14}C dating [Hedges, 2000; Nelson et al., 1995; Tuniz et al., 2000; Valladas et al., 1992, 2001; Valladas, 2003]. Thanks to the sensitivity of AMS, samples weighing as little as 1 mg may be used to date materials. Thus, dating of organic samples such as charcoal and beeswax can be carried out without visible damage to the painting. AMS ^{14}C dating may also be applied to the study of canvas paintings. This approach has been applied to linen fibres from 16th century Belgian paintings [Van Strydonck et al., 1998].

10.2.4 Metals

Lead isotope analysis is applicable to provenance studies of metal objects [Farquhar et al., 1995; Gale and Stos-Gale, 2000]. Pb isotope analysis has been carried out on different ore deposits that could have potentially supplied a metal item of interest. The ^{207}Pb/^{206}Pb, ^{208}Pb/^{206}Pb and ^{206}Pb/^{204}Pb ratios are determined. Plots of ^{206}Pb/^{204}Pb versus ^{207}Pb/^{206}Pb and ^{208}Pb/^{206}Pb versus ^{207}Pb/^{206}Pb are produced from the ore deposits. If all the lead isotope ratios of the item of interest fall with the field for an ore source in both diagrams, then that ore source can be the ore responsible for providing the metal for the item. Of course, the reliability of such an approach depends on comprehensive data obtained for all possible ore deposits. Although it is widely believed that the isotope composition of lead in an ancient metal object will be the same as that of the ore from which it is derived, this theory has been questioned [Budd et al., 1995; Pollard and Heron, 1996]. It has been suggested that there may be changes to the lead isotope ratio during the smelting process. Lead isotope analysis can be carried out on objects where lead is a major component, such as lead metal and bronzes, or where it is a minor component such as in copper, iron, silver and zinc artefacts.

10.2.5 Stone

Strontium isotope analysis has been used to examine marbles from a number of well known quarry areas in the Mediterranean, including Carrara [Brilli *et al.*, 2005]. The ratios fall in a narrow range which precludes them from being used to distinguish marbles from different quarry locations. However, some quarry areas were found to show unusual distributions in their isotopic values which may be useful for determining the provenance of a marble artefact.

10.2.6 Ceramics

Stable lead isotope analysis measured using ICP–MS can be used to identify the origin of lead ores used in the production of ceramic paints and glazes [Habicht-Mauche *et al.*, 2000; Wolf *et al.*, 2003]. Lead ores from different geological deposits can be fingerprinted by their stable lead isotope concentration because there is no significant fractionation of lead during glaze production or weathering. In a study of glaze-painted pottery from the Rio Grande, lead isotope values were compared to those of lead ore samples from historic mines in New Mexico to identify the source [Habicht-Mauche *et al.*, 2000].

10.2.7 Glass

Isotopic dating techniques that can be used for the characterisation of historic glasses are lead, strontium and oxygen isotope analyses [Degryse *et al.*, 2005a, 2005b; Freestone *et al.*, 2003; Henderson *et al.*, 2005; Wedepohl *et al.*, 1995]. Isotopes can be used to distinguish glasses based on the different sources of glass constituents such as fluxes and stabilisers, and thus can be used to characterise the production of the glass.

It is possible to obtain information regarding the source of silica and lime used in glass production using strontium analysis [Henderson *et al.*, 2005]. Plant ash glasses may be characterised using their $^{87}Sr/^{86}Sr$ ratio because plants take up strontium and this ratio reflects the geological conditions where the plant is grown. The strontium isotope ratio of silica depends upon age. Lime sources can be identified, as strontium generally substitutes for calcium and lime from old marine limestones can be readily distinguished from modern sea shell sources. For example, in a study of ancient glasses from Egypt and Israel was able to distinguish glasses based on strontium isotope analysis [Freestone et al., 2003]. The calcium source of the Egyptian glass was identified as limestone, while that of the Israeli glass was believed to be modern marine shells.

Lead isotope analysis has been used to identify the sources of historic glasses [Degryse et al., 2005a, 2005b; Henderson et al., 2005; Wedepohl et al., 1995]. A plot of $^{208}Pb/^{206}Pb$ versus $^{207}Pb/^{206}Pb$ can be used to distinguish the sources of lead in different glass objects.

Oxygen isotopes may be used to identify the source of silica in glass as there is a range of isotopes in silica depending on the geological conditions responsible for the quartz formation [Brill, 1970; Henderson et al., 2005].

10.3 NEUTRON ACTIVATION ANALYSIS

Neutron activation analysis (NAA) is a technique that involves bombarding a nonradioactive sample with neutrons [Neff, 2000; Pollard and Heron, 1996; Skoog and Leary, 1992]. A small fraction of the atoms are converted to radioisotopes and the characteristic decay patterns are recorded to identify the elements present. Until the advent of ICP and PIXE, NAA was a standard analytical method for elemental analysis at the ppm level. The technique was developed in the 1950s and has been used widely to study materials such as ceramics and coins. While NAA offers high sensitivity and minimal sample preparation, this approach does require access to a reactor.

The elements to be analysed using NAA must be able to undergo a nuclear reaction which results in a radioactively unstable product. The resulting product should have a half-life in the range from a few days to a few months and emit a particle with a characteristic energy. Neutron activation methods use a neutron capture reaction. In such a reaction a neutron is captured by the analyte nucleus to produce an isotope with a mass number increased by one. The isotope is in an excited state as it has acquired 8 MeV of energy in binding with the neutron. This energy is released almost instantaneously by prompt γ ray emission. For example, when ^{23}Na captures a neutron the following reaction occurs:

$$^{23}_{11}\text{Na} + ^{1}_{0}\text{n} \rightarrow ^{24}_{11}\text{Na} + \gamma \tag{10.10}$$

The ^{24}Na isotope is unstable and decays by β emission:

$$^{24}_{11}\text{Na} \rightarrow {}^{24}_{12}\text{Mg} + {}^{0}_{1}\beta + \gamma \qquad (10.11)$$

The gamma particle shows a characteristic energy of 1369 keV and the decay, in this case, of ^{24}Na may be monitored by measuring the β and γ particles. The half-life of ^{24}Na is 0.623 days. This information is summarised using the following notation:

$$^{23}\text{Na}(n, \gamma)\ {}^{24}\text{Na};\ 0.623\text{d},\ 1369\ \text{keV}$$

Another type of reaction used in this method is a transmutation reaction. When the nucleus captures a neutron, internal rearrangement occurs and a proton is immediately ejected from the nucleus. For example, titanium undergoes a transmutation reaction as follows:

$$^{47}_{22}\text{Ti} + {}^{1}_{0}\text{n} \rightarrow {}^{47}_{21}\text{Sc} + {}^{1}_{1}\text{p} \qquad (10.12)$$

$$^{47}_{21}\text{Sc} \rightarrow {}^{47}_{22}\text{Ti} + {}^{0}_{1}\beta + \gamma \qquad (10.13)$$

The notation for the process is: $^{47}\text{Ti}(n,p)^{47}\text{Sc}$; 3.43d, 159 keV. As titanium does not produce an isotope with a suitable half-life for measurement, the transmutation reaction is more useful for analysis.

Access to a nuclear reactor is required for NAA. Nuclear reactors are a source of high fluxes of neutrons. Research reactors usually have a neutron flux of 10^{11}–10^{14} neutrons cm^{-2}s^{-1} and the high densities produce detection limits in the order of 10^{-3}–10 μg. Most irradiations in NAA are carried out using thermal neutrons, relatively low energy neutrons with kinetic energies < 0.04eV. Other studies may employ higher energy neutrons using fast neutron activation analysis.

The gamma detectors used to monitor γ particles are similar to those used in EDXRF. The detector is cooled with liquid nitrogen. The spectra produced are very similar to those produced in EDXRF, but with a higher energy range.

Solids, liquids or gases may be examined using NAA. During an experiment the sample and a standard are irradiated simultaneously with neutrons. The samples and standard are contained in small PE vials or glass tubes. The irradiation time varies, but often it is 3–4 times the half-life of the element being examined. This may vary from several minutes to several hours. After irradiation the samples and standards are allowed to decay for a period varying from minutes to several hours. This procedure allows short-lived potential interferences to decay so that they will not interfere with the analysis.

Table 10.2   Some isotopes determined in NAA

Determined element	Activated product	Irradiation time	Half-life	Recommended γ ray / keV
^{27}Al	^{28}Al	25 min	2.24 min	1778.99
^{51}V	^{52}V	25 min	3.75 min	1434.08
^{50}Ti	^{51}Ti	25 min	5.76 min	320.08
^{48}Ca	^{49}Ca	25 min	8.72 min	3084.54
^{138}Ba	^{139}Ba	25 min	84.63 min	165.85
^{55}Mn	^{56}Mn	25 min	2.58 h	1810.72
^{41}K	^{42}K	25 min	12.36 h	1524.58
^{23}Na	^{24}Na	25 min	14.96 h	1368.60
^{75}As	^{76}As	7–8 days	26.32 h	559.1
^{86}Rb	^{85}Rb	21–28 days	18.66 days	1076.6
^{50}Cr	^{51}Cr	21–28 days	27.7 days	320.08
^{58}Fe	^{59}Fe	21–28 days	44.5 days	1099.25
^{123}Sb	^{124}Sb	21–28 days	60.2 days	1690.98
^{84}Sr	^{85}Sr	21–28 days	64.84 days	514.00
^{58}Co	^{58}Ni	21–28 days	70.82 days	810.77
^{64}Zn	^{65}Zn	21–28 days	243.9 days	1115.55
^{133}Cs	^{134}Cs	21–28 days	2.06 years	795.85
^{59}Co	^{60}Co	21–28 days	5.27 years	1173.24

The weight of the analyte can be determined using:

$$w_x = \frac{R_x}{R_s} w_s \qquad (10.14)$$

where w_x and w_s are the weights of the samples and standard, respectively, and R_x and R_s are the decay rate of the samples and standard, respectively. NAA is applicable to the determination of 69 elements [Skoog and Leary, 1992]. A number of the isotopes determined using NAA are listed in Table 10.2 [Neff, 2000].

10.3.1 Ceramics

NAA has been widely employed to determine the provenance of ceramic objects [Cogswell et al., 1996; Garcia-Heras et al., 1997; Glascock and Neff, 2003; Grimanis et al., 1997; Hancock et al., 1986; Hein et al., 1999; Meloni et al., 2000; Mommsen et al., 1987; Munito et al., 2000; Neff, 2000; Tenorio et al., 2000, 2005; Yellin, 1995]. As the source materials used to produce ceramics often come from widespread locations it is impractical to sample and characterise all possible sources. However, using statistical analysis it is possible to use a group of

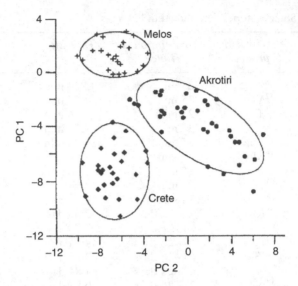

Figure 10.2 PCA plot for Greek ceramics. Reproduced with permission of Journal of Radioanalytical and Nuclear Chemistry, A P Grimanis *et al.*, 219, 177–185, 1997, Springer

unknown samples to create reference groups. The use of NAA for provenance studies of ceramics is illustrated by Figure 10.2 [Grimanis *et al.*, 1997], which shows a PCA plot from a study of ancient ceramics from locations in Greece, Melos, Crete and Akrotiri. The analysis is based on the principle components of the concentrations of samarium, lutetium, ytterbium, chromium and hafnium.

10.3.2 Glass

NAA has proved a useful technique for characterising obsidian [Almazin-Torres *et al.*, 2004; Glascock and Neff, 2003; Grimanis *et al.*, 1997; Jimenez-Reyes *et al.*, 2001; Neff, 2000; Oddone *et al.*, 2000, 1997]. It has been found that the source of obsidian may be determined based on the concentration of particular elements. Sodium and manganese have been shown to be the most useful elements for discriminating sources, with barium also sometimes providing additional discrimination [Neff, 2000].

10.3.3 Stone

NAA has been employed to study limestone and marble used in sculpture [Grimanis *et al.*, 1997]. This approach was used to examine a limestone sculptured head held by the Cleveland Museum of Art in the United States that was purported to be dated 5th century BC Greece. NAA was also used to study samples from two authenticated Acropolis limestone sculptures. The results showed that the concentrations of lanthium, lutetium, samarium, ytterbium, were significantly higher and the concentrations of arsenic, caesium, chromium and uranium lower in the stone under study compared to the genuine stone samples. Thus, any relationship between the Cleveland piece and the Acropolis pieces was ruled out.

10.3.4 Paintings

NAA can be used to investigate pigments found in paintings [Cotter, 1981; Ortega-Aviles *et al.*, 2005]. This technique formed part of an investigation of a canvas painting 'Virgin of Sorrows' purported to be painted in Mexico during the Colonial period (1535–1810) [Ortega-Aviles *et al.*, 2005]. NAA was used to determine the presence of vermillion in paint samples as PIXE could not be used to exclude the presence of natural or synthetic mercury sulfide. There was no evidence of a mercury isotope, so vermillion was discarded as a component of red paint taken from the painting. Together with information obtained using a range of analytical techniques, this find allowed the painting to be dated from the mid–18th century to the early part of the 19th century.

10.4 LUMINESCENCE

Luminescence dating is based on the principle that crystal inclusions in a material are able to accumulate electrons in metastable states and store for a long time the dose received from natural radioelements [Feathers, 2003; Troja and Roberts, 2000]. The process of luminescence dating is illustrated in Figure 10.3 [Feathers, 2003]. At A all the metastable energy levels are filled and between points A and B a zeroing event occurs. Such an event, such as heating, empties the metastable states and during the process luminescence is emitted. Via exposure to natural radiation, the states are gradually filled over time. This process is illustrated by the B to C line in Figure 10.3 and although the process is shown as linear, it may not be such at all times, particularly as saturation is neared. The metastable states are emptied in the process shown from C to D and this is where luminescence is measured in the laboratory. The measured light is proportional to the amount of trapped charge. If the amount of

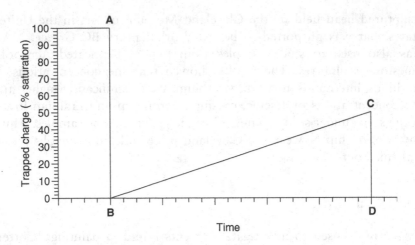

Figure 10.3 Luminescence dating processes. Reprinted from Measurement Science and Technology, Vol 14, J K Feathers, 1493–1509, 2003 with permission from IOP Publishing

radiation required to arrive at point C, the equivalent dose divided by the average dose rate, the time from B to D can be determined.

Stimulation of the system can be by heat in the case of thermoluminescence (TL) or by light in the case of photostimulated luminescence (PSL) or optically stimulated luminescence (OSL). TL is usually measured by heating a sample at a steady rate and detecting the light emission with a sensitive photomultiplier. As the temperature increases, the probability of escape for the metastable states increases and the number of electrons remaining in such states decreases. This produces a peak for a particular state and a composite of peaks is produced, called a glow curve. Another approach, isothermal TL, involves raising the temperature rapidly to a particular point, where it is held to produce a decay curve. OSL is usually measured as a decay curve when a sample is illuminated at a constant rate.

Luminescence dating usually employs either fine grains or coarse silt-sized or sand-sized grains of either quartz or feldspars [Troja and Roberts, 2000; Feathers, 2003]. Coarse grains are used to eliminate the influence of α radiation, since neither quartz nor feldspar have significant sources of α radiation. Course grain dating has the advantage of involving only a single mineral with well understood luminescence properties. Fine grain dating does require measurement of α radiation, but does have the advantage that there is less reliance on the more uncertain external dose rate.

10.4.1 Ceramics

Luminescence methods may be employed to date objects with timescales ranging from hundreds of years to several hundred millennia [Becker and Goediske, 1978; Galli *et al.*, 2004a, 2004b; Stoneham, 1991; Troja and Roberts, 2000; Yang *et al.*, 2005]. This approach can be used to authenticate a manufacturing date of ceramics and so can differentiate a modern copy from a genuine artefact. However, the detection of fakes may not be possible if the time gap is too small, such as when examining 19th century reproductions of 18th century ceramics.

10.5 NEUTRON DIFFRACTION

Although neutron diffraction is not considered a first choice for examining of museum objects because of the need for a source of neutrons from a reactor, it does provide supplementary information to XRD experiments [Kockelman *et al.*, 2000; Kockelman and Kirfel, 2004]. While XRD cannot provide information regarding light elements such as hydrogen, it is possible to study such elements using neutron diffraction. Neutron diffraction also has the advantage is that it is non-destructive.

The thermal neutrons described for NAA are used in neutron diffraction. Neutrons generated from a reactor possess high kinetic energy, but these are slowed through collisions with a moderator (a substance such as water). A velocity selector device is used to obtain a monochromatic beam of neutrons suitable for a diffraction study of crystals.

Neutron diffraction has the advantage over XRD in sampling when a bulky object is to be examined in a non-destructive manner. The sample does not need to be moved and large objects can be examined. The neutron TOF technique is used for such samples. The neutrons easily penetrate a painted, glazed or corroded surface. The relationship between the crystallographic d-spacing and the TOF is given by:

$$d = \frac{t}{505.56L \sin \theta} \qquad (10.15)$$

where t is the flight time in μs, L is the flight path from the source in m and d is the spacing measured in Å. Thus, for a given path at a scattering angle 2θ, the TOF measurements produce a diffraction pattern with intensities versus d-spacing that may be transformed into an intensity versus 2θ representation.

10.5.1 Ceramics

Neutron diffraction provides a quantitative method for determining the phase contents in ceramics [Kockelman et al., 2000; Kockelman and Kirfel, 2004]. The characterisation of the phase abundance can provide a means of understanding the provenance or determining the authenticity of ceramic objects. Additionally, information about the firing conditions is provided by the presence or absence of firing materials. For example, diffraction patterns may exhibit minerals such as quartz and mullite [Kockelman et al., 2000; Kockelman and Kirfel, 2004]. Quartz is present as a source material, while mullite is a product of the firing processes. Thus, the mullite-to-quartz ratio determined from the diffraction pattern is expected to increase with a higher firing temperature.

10.5.2 Metals

Quantitative phase identification may be carried out on metals, such as coins, using neutron diffraction techniques [Creagh et al., 2004; Creagh, 2005; Kockelman and Kirfel, 2004; Pantos et al., 2005]. The technique may also be used for texture analyses, which is helpful for determining the authenticity of items. For example, 16th century silver coins from the Kunsthistorische Museum in Vienna have been studied using neutron diffraction [Kockelman and Kirfel, 2004]. The minting process used for the genuine coins involved cold-rolling the two coin faces onto an silver/copper metal sheet of 90/10 wt-% composition. Using neutron diffraction a number of the coins in the collection were found to show irregular grain distributions and to be rich in copper, pointing to the proposition that these coins were fakes.

REFERENCES

M.G. Almazin-Torres, M. Jimenez-Reyes, F. Monroy-Guzman, et al., Determination of the provenance of obsidian samples collected in the archaeological site of San Miguel Ixtapan, Mexico State, Mexico by means of neutron activation analysis, Journal of Radioanalytical and Nuclear Chemistry, 260 (2004), 533–542.

K. Becker and C. Goedicke, A quick method for authentication of ceramic art objects, Nuclear Instruments and Methods, 151 (1978), 313–316.

G. Bonani, S. Ivy, W. Wolfli, et al., Radiocarbon dating of fourteen Dead Sea Scrolls, Radiocarbon, 34 (1992), 843–849.

R.H. Brill, Lead and oxygen isotopes in ancient objects, Philosophical Transactions, The Royal Society, London, 269 (1970) 143–164.

M. Brilli, G. Cavazzini and B. Turi, New data of $^{87}Sr/^{86}Sr$ ratio in classical marble: an initial database for marble provenance determination, *Journal of Archaeological Science*, 32 (2005), 1543–1551.

P. Budd, A.M. Pollard, B. Scaife and R.G. Thomas, The possible fractionation of lead isotopes in ancient metallurgical processes, *Archaeometry*, 37 (1995), 143–150.

J.W. Cogwell, H. Neff and M.D. Glascock, The effect of firing temperature on the elemental characterisation of pottery, *Journal of Archaeological Science*, 23 (1996), 283–287.

R. Corfield, Searching for real time, *Chemistry in Britain*, 38 (2002), 22–26.

M.J. Cotter, Neutron activation analysis of paintings, *American Scientist*, 69 (1981), 17–27.

D.C. Creagh, G. Thorogood, M. James and D.L. Hallam, Diffraction and fluorescence studies of bushranger armour, *Radiation Physics and Chemistry*, 71 (2004), 839–840.

D.C. Creagh, The characterisation of artefacts of cultural heritage significance using physical techniques, *Radiation Physics and Chemistry*, 74 (2005), 426–442.

P. Degryse, J. Schneider, U. Haack, *et al.*, Evidence for glass recycling using Pb and Sr isotopic ratios and Sr-mixing lines: the case of early Byzantine Sagalassos, *Journal of Archaeological Sciences*, 33 (2005a), 494–501.

P. Degryse, J. Schneider, J. Poblome, *et al.*, A geochemical study of Roman to early Byzantine glass from Sagalassos, south-west Turkey, *Journal of Archaeological Science*, 32 (2005b), 287–299.

D.J. Donahue, A.J.T. Jull and T.W. Linick, Some archaeologic applications of accelerator radiocarbon analysis, *Nuclear Instruments and Methods in Physics Research B*, 45 (1990), 561–564.

R.M. Farqhuar, J.A. Walthall and R.G.V. Hancock, 18th century lead smelting in central north America: evidence from lead isotope and INAA measurements, *Journal of Archaeological Science*, 22 (1995), 639–648.

J.K. Feathers, Use of luminescence dating in archaeology, *Measurement Science and Technology*, 14 (2003), 1493–1509.

I.C. Freestone, K.A. Leslie, M. Thirlwall and Y. Gorin-Rosen, Strontium isotopes in the investigation of early glass production: Byzantine and early Islamic glass from the near east, *Archaeometry*, 45 (2003), 19–32.

N.H. Gale and Z. Stos-Gale, Lead isotope analyses applied to provenance studies, in *Modern Analytical Methods in Art and Archaeology* (eds E. Ciliberto and G. Spoto), John Wiley & Sons, Inc., New York (2000), pp 503–583.

A. Galli, M. Martini, C. Montanari and E. Sibilia, Thermally and optically stimulated luminescence of glass mosaic tesserae, *Applied Physics A*, 79 (2004a), 253–256.

A. Galli, M. Martini, E. Sibilia, *et al.*, Luminescence properties of lustre decorated majolica, *Applied Physics A*, 79 (2004b), 293–297.

M. Garcia-Heras, R. Fernandez-Ruiz and J.D. Tornero, Analysis of archaeological ceramics by TXRF and contracted with NAA, *Journal of Archaeological Science*, 24 (1997), 1003–1014.

M.D. Glascock and H. Neff, Neutron activation analysis and provenance research in archaeology, *Measurement Science and Technology*, 14 (2003), 1516–1526.

H.E. Gove, *Relic, icon or hoax: carbon dating the Turin Shroud*, Institute of Physics Publishing, London (1996).

A.P. Grimanis, N. Kalogeropoulos, V. Kilikoglou and M. Vassilaki-Grimani, Use of NAA in marine environment and in archaeology in Greece, *Journal of Radioanalytical and Nuclear Chemistry*, 219 (1997), 177–185.

J.A. Habicht-Mauche, S.T. Glenn, H. Milford and A.R. Flegal, Isotopic tracing of prehistoric Rio Grande glaze-paint production and trade, *Journal of Archaeological Science*, **27** (2000), 709–713.

R.G.V. Hancock, N.B. Millet and A.J. Mills, A rapid INAA method to characterise Egyptian ceramics, *Journal of Archaeological Science*, **13** (1986), 107–117.

R.E.M. Hedges, Radiocarbon dating, in *Modern Analytical Methods in Art and Archaeology* (eds E. Ciliberto and G. Spoto), John Wiley & Sons, Inc., New York (2000), pp 465–501.

A. Hein, H. Mommsen and J. Maran, Element concentration distributions and most discriminating elements for provenancing by neutron activation analyses of ceramics from Bronze Age sites in Greece, *Journal of Archaeological Science*, **26** (1999), 1053–1058.

J. Henderson, J.A. Evans, H.J. Sloane, *et al.*, The use of oxygen, strontium and lead isotopes to provenance ancient glasses in the Middle East, *Journal of Archaeological Science*, **32** (2005), 665–673.

T. Higham and F. Petchey, Radiocarbon dating in archaeology: methods and applications, in *Radiation in Art and Archaeometry* (eds D.C. Creagh and D.A. Bradley), Elsevier, Amsterdam (2000), pp 255–284.

M. Jimenez-Reyes, D. Tenorio, J.R. Esparza-Lopez, *et al.*, Neutron activation analysis of obsidians from quarries of the Central Quaternary Trans-Mexican Volcanic Axis, *Journal of Radioanalytical and Nuclear Chemistry*, **250** (2001), 465–471.

A.J.T. Jull, D.J. Donahue, M. Broshi and E. Tov, Radiocarbon dating of scrolls and linen fragments from the Judean Desert, *Radiocarbon*, **37** (1995), 11–20.

W. Kockelman, E. Pantos and A. Kirfel, Neutron and synchrotron radiation studies of archaeological objects, in *Radiation in Art and Archaeometry* (eds D.C. Creagh and D.A. Bradley), Elsevier, Amsterdam (2000), pp 347–377.

W. Kockelman and A. Kirfel, Neutron diffraction studies of archaeological objects on ROTAX, *Physica B*, **350** (2004), 581–585.

S. Meloni, M. Oddone, N. Genova and A. Cairo, The production of ceramic materials in Roman Pavia: an archaeometric NAA investigation of clay sources and archaeological artefacts, *Journal of Radioanalytical and Nuclear Chemistry*, **244** (2000), 553–558.

H. Mommsen, A. Kreuser, J. Weber and H. Busch, Neutron activation analysis of ceramics in the X-ray energy region, *Nuclear Instruments and Methods in Physics Research A*, **257** (1987), 451–461.

C.S. Munito, R.P. Paiva, M.A. Alves, *et al.*, Chemical characterisation by INAA of Brazilian ceramics and cultural implications, *Journal of Radioanalytical and Nuclear Chemistry*, **244** (2000), 575–578.

H. Neff, Neutron activation analysis for provenance determination in archaeology, in *Modern Analytical Methods in Art and Archaeology* (eds E. Ciliberto and G. Spoto), John Wiley & Sons, Inc., New York (2000), pp 81–134.

D.E. Nelson, G. Chaloupka, C. Chippendale, *et al.*, Radiocarbon dating for beeswax figures in the prehistoric rock art of Northern Australia, *Archaeometry*, **37** (1995), 151–156.

H. Oda, Y. Yoshizawa, T. Nakamura and K. Fujita, AMS radiocarbon dating of ancient Japanese sutras, *Nuclear Instruments and Methods in Physics Research B*, **172** (2000), 736–740.

M. Oddone, Z. Yegingil, G. Bigazzi, *et al.*, Chemical characterisations of Anatolian obsidians by instrumental and epithermal neutron activation analysis, *Journal of Radioanalytical and Nuclear Chemistry*, **224** (1997), 27–38.

M. Oddone, G. Bigazzi, V. Keheyan and S. Meloni, Characterisation of Armenian obsidians: implications for raw material supply for prehistoric artefacts, *Journal of Radioanalytical and Nuclear Chemistry*, **243** (2000), 673–682.

M. Ortega-Aviles, P. Vandenabeele, D. Tenorio, *et al.*, Spectroscopic investigation of a 'Virgin of Sorrows' canvas painting: a multi-method approach, *Analytica Chimica Acta*, **550** (2005), 164–172.

E. Pantos, W. Kockelmann, L.C. Chapon, *et al.*, Neutron and X-ray characterisation of the metallurgical properties of a 7th century BC Corinthian-type bronze helmet, *Nuclear Instruments and Methods in Physics Research B*, **239** (2005), 16–26.

A.M. Pollard and C. Heron, *Archaeological Chemistry*, Royal Society of Chemistry, Cambridge (1996).

D.A. Skoog and J.J. Leary, *Principles of Instrumental Analysis*, Harcourt Brace College Publishers, Fort Worth (1992).

D. Stoneham, in *Scientific Dating Methods* (eds H.Y. Goksu, M. Oberhofer and D. Regulla), Kluwer Academic, Dordrecht (1991), pp 175–192.

D. Tenorio, M. Jimenez-Reyes, A. Cabral-Prieto, *et al.*, Archaeometry of pre-Hispanic pottery from San Luis Potosi, Mexico, *Hyperfine Interactions*, **128** (2000), 381–396.

D. Tenorio, M.G. Almazan-Torres, F. Monroy-Guzman, *et al.*, Characterisation of ceramics from the archaeological site of San Miguel Ixtapan, Mexico State, Mexico, using NAA, SEM, XRD and PIXE techniques, *Journal of Radioanalytical and Nuclear Chemistry*, **266** (2005), 471–480.

S.O. Troja and R.G. Roberts, Luminescence dating, in *Modern Analytical Methods in Art and Archaeology* (eds E. Ciliberto and G. Spoto), John Wiley & Sons, Inc., New York (2000), pp 585–640.

J. Turnbull, R. Sparks and C. Prior, Testing the effectiveness of AMS radiocarbon pretreatment and preparation on archaeological textiles, *Nuclear Instruments and Methods in Physics Research B*, **172** (2000), 469–472.

C. Tuniz, U. Zoppi and M. Banbetti, AMS dating in archaeology, history and art, in *Radiation in Art and Archaeometry* (eds D.C. Creagh and D.A. Bradley), Elsevier, Amsterdam (2000), pp 444–471.

H. Valladas, H. Cachier, P. Maurice, *et al.*, Direct radiocarbon dates for prehistoric paintings at the Altamira, El Castillo and Niaux caves, *Nature*, **357** (1992), 68–70.

H. Valladas, N. Tisnerat-Laborde, H. Cachier, *et al.*, Radiocarbon AMS dates for Paleolithic cave paintings, *Radiocarbon*, **43** (2001), 977–986.

H. Valladas, Direct radiocarbon dating of prehistoric cave paintings by accelerator mass spectrometer, *Measurement Science and Technology*, **14** (2003), 1487–1492.

M.J.Y. Van Strydonck, L. Masschelein-Kleiner, C. Alderliesten and A.F.M. de long, Radiocarbon dating of canvas paintings: two case studies, *Studies in Conservation*, **43** (1998), 209–214.

K.H. Wedepohl, I. Krueger and G. Hartmann, Medieval lead glass from north western Europe, *Journal of Glass Studies*, **37** (1995), 65–82.

S. Wolf, S. Stos, R. Mason and M.S. Tite, Lead isotope analysis of Islamic pottery glazes from Fustat, Egypt, *Archaeometry*, **45** (2003), 405–420.

X.Y. Yang, A. Kadereit, G.A. Wagner, *et al.*, TL and IRSL dating of Jiahu relics and sediments: clue of 7th millennium BC civilisation in central China, *Journal of Archaeological Science*, **32** (2005), 1045–1051.

J. Yellin, Neutron activation analysis: impact on the archaeology of the Holy Land, *Trends in Analytical Chemistry*, **14** (1995), 37–44.

Appendix Infrared Spectra of Polymers

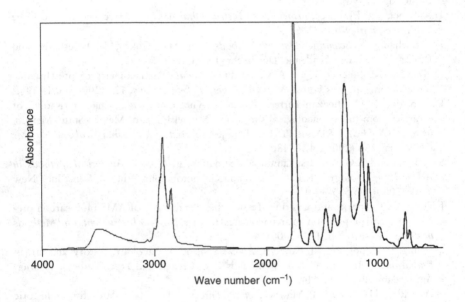

Figure A1   Alkyd resin

Analytical Techniques in Materials Conservation   Barbara H. Stuart
© 2007 John Wiley & Sons, Ltd

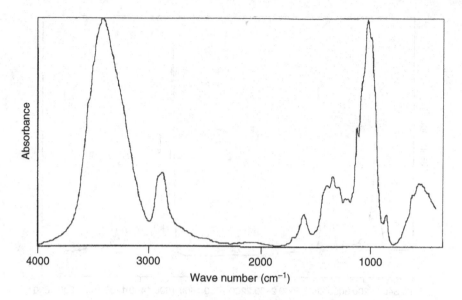

Figure A2   Cellophane

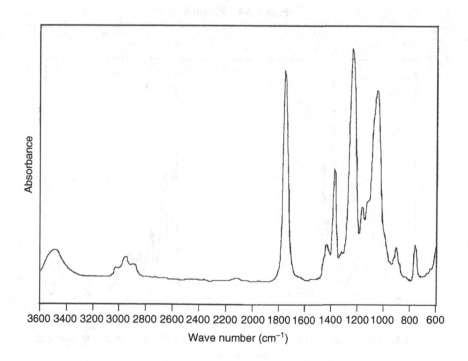

Figure A3   Cellulose acetate

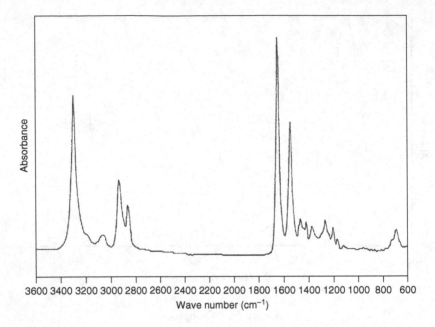

Figure A4   Nylon 6

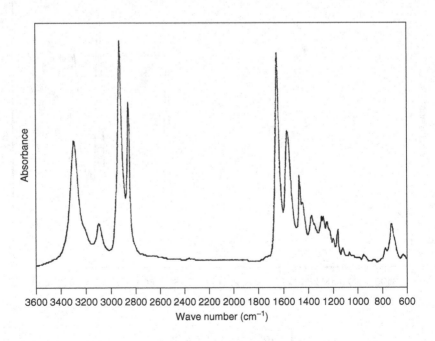

Figure A5   Nylon 12

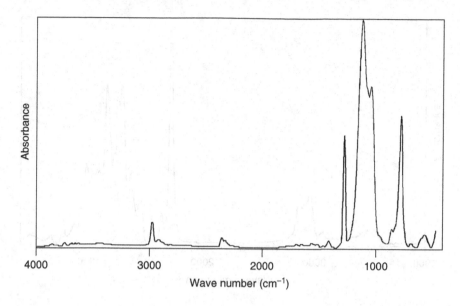

Figure A6   Poly(dimethyl siloxane)

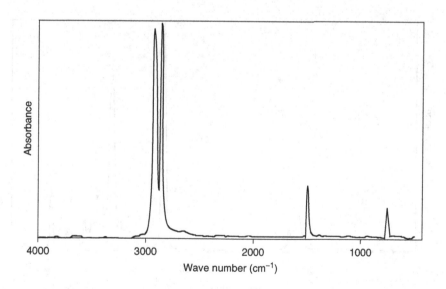

Figure A7   Polyethylene

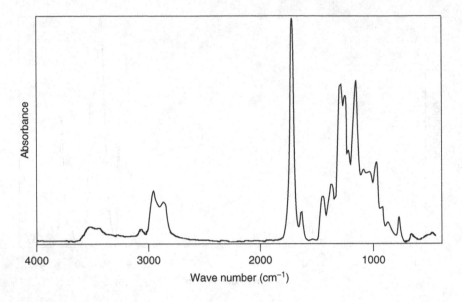

Figure A8   Polyester resin (unsaturated)

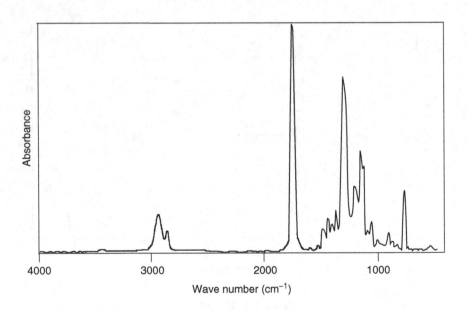

Figure A9   Poly(ethylene terephthalate)

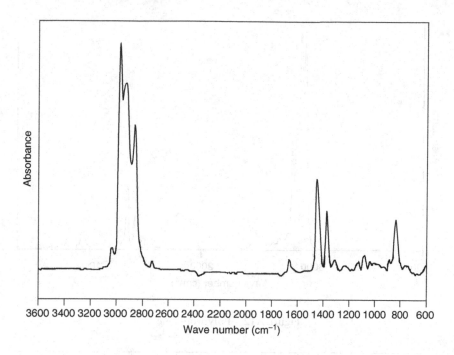

Figure A10 Poly(*cis*-isoprene)

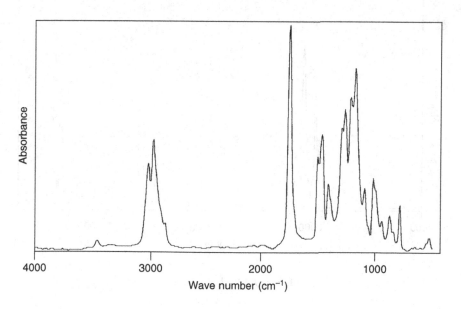

Figure A11 Poly(methyl methacrylate)

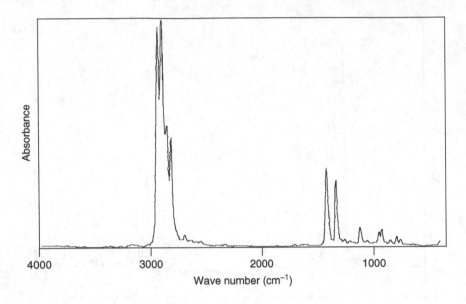

Figure A12   Polypropylene

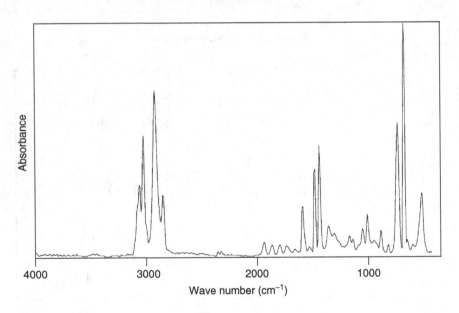

Figure A13   Polystyrene

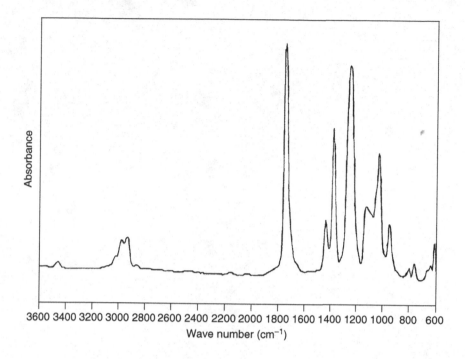

Figure A14   Poly(vinyl acetate)

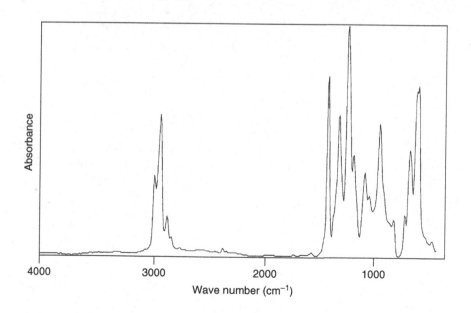

Figure A15   Poly(vinyl chloride)

Index